OPTIMAL CONTROL THEORY

OPTIMAL CONTROL THEORY

A Course in Automatic Control Theory

R. PALLU DE LA BARRIÈRE

Professor of Mathematics,
University of Caen, France

Translated by
SCRIPTA TECHNICA

Edited by
BERNARD R. GELBAUM

DOVER PUBLICATIONS, INC.
NEW YORK

Published in Canada by General Publishing Company, Ltd., 30 Lesmill Road, Don Mills, Toronto, Ontario.
Published in the United Kingdom by Constable and Company, Ltd., 10 Orange Street, London WC2H 7EG.

This Dover edition, first published in 1980, is an unabridged and unaltered republication of the English translation first published in 1967 by W. B. Saunders Company.
The work was originally published in 1966 by Dunod, S. A., Paris, under the title *Cours D'Automatique Théorique.*

International Standard Book Number: 0-486-63925-8
Library of Congress Catalog Card Number: 79-055910

Manufactured in the United States of America
Dover Publications, Inc.
180 Varick Street
New York, N.Y. 10014

Foreword

This book may be divided roughly into two parts. The first half is devoted to a rapid and extremely clear presentation of the mathematical facts required in the modern theory of optimal control. The latter half discusses this theory in great detail and with numerous illustrations.

The first part (Chapters 1 through 7) dealing with the mathematical background of the theory is extremely modern and conducts the exposition in terms of the most up-to-date vocabulary, the newest techniques (e.g., the theory of distributions) and the most powerful theorems.

The treatment of automatic control theory (Chapters 8 through 18) is quite general and includes Pontryagin's principle.

Students of engineering and of applied mathematics will find a wealth of interesting material in this book. Prerequisites for its reading are found in most junior and senior level courses in analysis.

Bernard R. Gelbaum

Preface

The present text represents in book form the course in automatic control theory that I taught in the Department of Sciences at Caen in 1962-63 and 1963-64. Thus, its contents do not constitute a definition of automatic control theory, which is a discipline in full evolution both as regards its scope and its connections with other branches of mathematics.

I have combined in a single volume on the one hand an exposition of the general theories such as Schwartz' distributions and harmonic analysis, which have become valuable tools for traditional automatic control, and, on the other, new developments from control theory, notably, the contributions of Pontryagin's school.

An important part of the book is devoted to optimization problems, which have led me to include linear and nonlinear programming, an indispensable basis for subsequent study of optimization of dynamic systems.

With regard to stochastic problems, I thought it would be useful to include a chapter giving the fundamentals of probability theory, since the Department of Sciences has no compulsory instruction on that subject. The latter chapters of the book deal with the theory of random processes and associated optimization problems.

I wish to thank my graduate students in the Department of Automatic Control for their assistance and valuable comments regarding the first drafts of this work.

Contents

OPTIMAL CONTROL THEORY

1
Preliminaries

1. SET THEORY

The concepts and notations that we shall use are those commonly employed. Let f denote a mapping of a set E into a set F. We say that the *domain of definition* of f is E and we write $\operatorname{def}(f) = E$. The set of all $y = f(x)$ for $x \in E$ is called the *range of values* (or simply the *range*) of f and is denoted by $\operatorname{val}(f)$. A mapping f of a set E into a set F is said to be *injective* if the equation $f(x) = f(y)$ implies $x = y$. It is said to be *surjective* if $\operatorname{val}(f) = F$. It is said to be *bijective* if it is both injective and surjective. An injective (resp. surjective, bijective) mapping is also called an *injection* (resp. *surjection, bijection*).

We denote by

- **R** the field of real numbers,
- **C** the field of complex numbers,
- **Z** the ring of integers,
- **T** the one-dimensional torus (the quotient space of **R** under the equivalence relation "$x \sim y$ if $x - y$ is an integral multiple of 2π").

The identity mapping on a set E is denoted by $\mathbf{1}_E$ or simply by $\mathbf{1}$ if there is no danger of confusion.

The characteristic function of a subset A of a set E is denoted by ε_A.

Ordering relations. A relation $\leqslant$ is called an *ordering relation* on a set E if it has the following properties:

(1) $x \leqslant x$
(2) $x \leqslant y$ and $y \leqslant z \Rightarrow x \leqslant z$
(3) $x \leqslant y$ and $y \leqslant x \Rightarrow x = y$.

The set E is then said to be *ordered*. We write $x < y$ if $x \leqslant y$ and $x \neq y$.

If, in addition,

(4) $\forall x, \quad y \in E$, either $x \leqslant y$ or $y \leqslant x$,

we say that the relation is a *total ordering* and that E is a *totally ordered set*. A relation satisfying only properties (1) and (2) is called *quasiordering*. The relation "$x \leqslant y$ and $y \leqslant x$" is thus an equivalence relation, and preordering induces an ordering relation on the quotient space of E by this equivalence relation. In an ordered set E, an element x is said to be *maximal* if $x \leqslant y$ implies $y = x$. Every finite ordered set has at least one maximal element.

On $\mathbf{R}^n$, let us consider the ordering relation defined as follows: If x^i and y^i are respectively the components of x and y, we set

$$x \leqslant y \quad \text{if} \quad x^i \leqslant y^i \quad (\forall i = 1, \ldots, n).$$

We shall write $x \ll y$ if $x^i < y^i$ ($\forall i = 1, \ldots, n$). We denote by $\mathbf{R}^n_+$ the set of all $x \in \mathbf{R}^n$ such that $x \geqslant 0$.

2. TOPOLOGICAL VECTOR SPACES

A real (resp. complex) *topological vector space* is defined as a vector space E equipped with a topology such that the mappings $x, y \to x + y$ (for $x, y \in E$) and $\lambda, x \to \lambda x$ (for $\lambda \in \mathbf{R}$ (resp. $\mathbf{C}$), $x \in E$) are continuous.

A topological vector space structure is completely defined if we know a base of neighborhoods of 0. We then obtain a base of neighborhoods of any point x by a translation to x.

A *seminorm* on a vector spece E is defined as a mapping $x \to p(x)$ of E into $\mathbf{R}$ and satisfying the following properties:

(1) $p(x) \geqslant 0$
(2) $p(\lambda x) = |\lambda| \, p(x)$
(3) $p(x + y) \leqslant p(x) + p(y)$.

If, in addition,

(4) $p(x) = 0 \Rightarrow x = 0$

then p is called a *norm*.

A seminorm p_1 is said to be *less fine* than a seminorm p_2 if there exists a constant k such that $p_1(x) \leqslant kp_2(x)$. If, in addition, p_2 is less fine than p_1, the seminorms p_1 and p_2 are said to be *equivalent*. For two given seminorms p_1 and p_2, the seminorms p_3 and p_4 defined by

$$p_3(x) = p_1(x) + p_2(x)$$
$$p_4(x) = \max\big(p_1(x), p_2(x)\big)$$

are equivalent to each other and are finer than p_1 and p_2. Every seminorm that is finer than p_1 and p_2 is finer than p_3 and p_4.

Every seminorm on a vector space E defines a topological vector space structure for which the sets $p(x) \leqslant \rho$ (for $\rho > 0$) constitute a base of neighborhoods of 0. A necessary and sufficient condition for the topology thus defined to be separated (T_2) is that p be a norm.

Two equivalent seminorms define the same topology.

If a normed space (that is, one equipped with a norm) is complete (for the topology defined by that norm), it is called a *Banach space*.

An arbitrary family $\{p_\alpha\}$ of seminorms on a vector space E defines a topological vector space structure for which the sets $V(\alpha_1, ..., \alpha_k ; \rho)$ defined by

$$x \in V(\alpha_1, ..., \alpha_k ; \rho) \quad \Leftrightarrow \quad p_{\alpha_i}(x) \leqslant \rho \; (i = 1, ..., k)$$

constitute a base of neighborhoods of 0, where ρ is an arbitrary positive number and $\alpha_1, ..., \alpha_k$ is any k-tuple of values of α. The topology thus obtained is separable if, in addition,

$$p_\alpha(x) = 0 \; (\forall \alpha) \quad \Rightarrow \quad x = 0 \, .$$

In particular, if the family $\{p_\alpha\}$ of seminorms is denumerable, we can index it with integral indices and substitute the notation p_n (for $n = 1, 2, ...$) for the notation p_α. Thus, we can take for the base of neighborhoods of 0 the sets $V(N ; \rho)$ (for $\rho > 0$ and $N = 1, 2, ...$) defined by

$$x \in V(N ; \rho) \quad \Leftrightarrow \quad p_n(x) \leqslant \rho \; (\forall n \leqslant N) \, .$$

In particular, if p_n is finer than p_{n-1} (in which case, we say that the seminorms p_n are ***directed by increasing fineness***), we can take as the base of neighborhoods of 0 the sets $W(n, \rho)$ (for $\rho > 0$ and $n = 1, 2, ...$) defined by

$$x \in W(n, \rho) \quad \Leftrightarrow \quad p_n(x) \leqslant \rho \, .$$

Finally, if the family of seminorms p_α is finite, the topology that it defines can be defined by a unique seminorm, for example,

$$\sum_\alpha p_\alpha \quad \text{or} \quad \sup_\alpha p_\alpha \, .$$

Two families $\{p_\alpha\}$ and $\{q_\beta\}$ of seminorms are said to be equivalent if they define the same topology. For this, it is necessary and sufficient that

(*a*) for every α, there exist a k-tuple $\beta_1, ..., \beta_k$ of values of β and a constant K such that

$$p_\alpha(x) \leqslant K \sup_i q_{\beta_i}(x) \, ;$$

(b) for every β, there exist a k-tuple $\alpha_1, ..., \alpha_k$ of values of α and a constant K such that

$$q_\beta(x) \leqslant K \sup_i p_{\alpha_i}(x) .$$

Every sequence $\{p_n\}$ of seminorms p_n is equivalent to a sequence $\{q_n\}$ of seminorms q_n directed according to increasing fineness. We may take, for example,

$$q_n = \sup_{k \leqslant n} p_k .$$

Every separable topological vector space whose topology is defined by a denumerable family of seminorms is metrizable. Such a space is said to be ***metrizable*** and ***locally convex.***

A complete, metrizable, locally convex space is called a ***Fréchet space.***

Let E and F denote two spaces with topologies defined respectively by families $\{p_\alpha\}$ and $\{q_{\beta\cdot}\}$ of seminorms. For a mapping f of E into F to be continuous, it is necessary and sufficient that, for every seminorm q_β, there exist a k-tuple $\alpha_1, ..., \alpha_k$ of values of α and a number $\rho > 0$ such that

$$\left\{ \begin{array}{c} p_{\alpha_1}(x) \leqslant \rho \\ \cdots\cdots\cdots \\ p_{\alpha_k}(x) \leqslant \rho \end{array} \right\} \quad \Rightarrow \quad q_\beta(x) \leqslant 1 .$$

Fundamental theorems on Banach and Fréchet spaces. Let E and F denote two topological vector spaces. We denote by $L(E, F)$ the space of continuous linear mappings of E into F. If E is a real (resp. complex) Fréchet space, its dual $L(E, \mathbf{R})$ (resp. $L(E, \mathbf{C})$) will be denoted by E'. The topology of simple convergence on E is called the ***weak topology*** on E'. If E and F are Banach spaces, then $L(E, F)$ is a Banach space for the norm

$$A \to \| A \| = \sup_{\|x\| \leqslant 1} \| Ax \| .$$

In particular, the dual of a Banach space is a Banach space.

In this connection, we have the following fundamental theorems:

Theorem. *Let E and F denote two Banach spaces. On every ball in L(E, F), the topology of simple convergence on E coincides with the topology of simple convergence on every dense subset of E.*

Theorem (Banach). *Every closed ball in the dual of a Banach space E is weakly compact.*

Theorem (Hahn-Banach). *Let E denote a Banach space, let F denote a closed vector subspace of E, and let u denote a continuous* *linear form on F. Then u can be extended as a continuous linear form defined on E.*

Suppose that F is of finite codimension; that is, suppose that it is the intersection of a finite number k of closed hyperplanes defined by equations $v_i(x) = 0$ (for $i = 1, \ldots, k$), where each v_i is a continuous linear form.

In this case, the statement of the Hahn-Banach theorem is elementary since we can easily construct a continuous projection P on F and we may set

$$u(x) = u(Px), \quad \text{for } x \in E .$$

We shall also use the following:

Theorem (closed-graph). *Let E and F denote two Fréchet spaces and let f denote a linear mapping of E into F. If the graph of f (that is, the subset of $E \times F$ consisting of all ordered pairs of the form $(x, f(x))$ as x ranges over E) is closed, the mapping f is continuous.*

Since the spaces E and F are metrizable, the hypothesis can be put in the following form:

$$\text{if } \lim_{n\to+\infty} x_n = x \quad \text{and} \quad \lim_{n\to+\infty} f(x_n) = y \quad \text{then} \quad y = f(x) .$$

We then need only verify it for $x = 0$.

From the closed-graph theorem, we get the following

Theorem. *If f is a continuous bijection of E into F, where E and F are Fréchet spaces, then f^{-1} is continuous.*

Pre-Hilbert and Hilbert spaces. A mapping $x, y \to A(x, y)$ of the Cartesian product $E \times F$ of two complex vector spaces into a complex vector space G is said to be ***sesquilinear*** if it is linear with respect to y and conjugate linear with respect to x.

A real (resp. complex) vector space E on which a bilinear (resp. sesquilinear) form $x, y \to \langle x, y \rangle$ is called a ***separable real*** (resp. ***complex***) ***pre-Hilbert space*** if

(1) $\langle x, y \rangle = \langle y, x \rangle$ (resp. $\langle x, y \rangle = \langle \overline{y, x} \rangle$).
(2) $\langle x, x \rangle \geqslant 0$
(3) $\langle x, x \rangle = 0 \Rightarrow x = 0$.

Throughout the present book, we shall write "pre-Hilbert" for "separable pre-Hilbert."

Every pre-Hilbert space becomes a normed space when we set

$$\| x \| = \sqrt{\langle x, x \rangle}.$$

A complete pre-Hilbert space is called a ***Hilbert space***. Every finite-dimensional pre-Hilbert space is a Hilbert space. A real (resp. complex) finite-dimensional Hilbert space is called a *Euclidean* (resp. *Hermitian*) space. We assume the classical properties of pre-Hilbert and Hilbert spaces are known.

Matrix calculus. The elements x of $\mathbf{C}^n$ (called complex n-tuples) are represented in the following form:

$$x = \begin{bmatrix} x^1 \\ \vdots \\ x^n \end{bmatrix}.$$

We denote by ε_i the n-tuple of which the ith component is equal to 1 and all the other components of which are equal to 0.

We shall call a mapping A of a space $\mathbf{C}^n$ into a space $\mathbf{C}^p$ a *matrix.* A matrix A may be represented by any of the following forms:

$$A = \begin{bmatrix} A_1^1 \dots A_n^1 \\ \vdots \qquad \vdots \\ A_1^p \qquad A_n^p \end{bmatrix}, \quad A = [A_1 \mid \dots \mid A_n], \quad A = \begin{bmatrix} A^1 \\ \hline \vdots \\ \hline A^p \end{bmatrix},$$

where

$$A_i = \begin{bmatrix} A_i^1 \\ \vdots \\ A_i^p \end{bmatrix} \quad \text{and} \quad A^j = [A_1^j \dots A_n^j].$$

If $y = Ax$, we have

$$y^j = A^j x = \sum_{i=1}^{n} A_i^j x^i .$$

The columns A_i are equal to $A\varepsilon_i$.

Every unit matrix of whatever dimensions is denoted by **1.** We denoted by $\overline{A}$ the matrix defined by

$$(\overline{A})_i^j = \overline{A_j^i} .$$

If A is real (resp. complex), $\overline{A}$ is called the ***transpose*** (resp. the ***Hermitian transpose***) of A. If $A = \overline{A}$, then A is said to be ***symmetric*** in the real case and ***Hermitian*** in the complex case.

Let E denote a vector space of finite dimension n and let $S = [e_1, \dots, e_n]$ denote a basis for E. Every element $x \in E$ has a unique representation in the form

$$x = \sum_{i=1}^{n} x_S^i e_i .$$

We set

$$x_S = \begin{bmatrix} x_S^1 \\ \vdots \\ x_S^n \end{bmatrix}.$$

If S' is another basis for E, we set

$$x_S = K x_{S'},$$

where K is an $n \times n$ matrix called a ***similarity matrix.***

If E_1 and E_2 are two finite-dimensional vector spaces with bases S_1 and S_2, respectively, and A is a linear mapping of E_1 into E_2, there exists a matrix $A_{S_1S_2}$ such that

$$y_{S_2} = A_{S_1S_2}\, x_{S_1}\,.$$

If S_1' and S_2' are two other bases of E_1 and E_2 respectively, we have

$$A_{S_1'S_2'} = \overset{-1}{K_2}\, A_{S_1S_2}\, K_1\,,$$

where K_1 and K_2 are similarity matrices in E_1 and E_2.

Suppose that E is an n-dimensional Euclidean or Hermitian space with basis

$$S = [e_1, \ldots, e_n].$$

To every $x \in E$ we can assign the n-tuple

$$x_{\hat{S}} = \begin{bmatrix} \langle e_1, x \rangle \\ \vdots \\ \langle e_n, x \rangle \end{bmatrix},$$

and $x_{\hat{S}}$ can be regarded as the components of x relative to a basis $\hat{S}$ called the ***supplementary basis*** of S. We may set

$$x_{\hat{S}} = G(S)\, x_S\,.$$

The matrix $G(S)$ is called the ***Gram matrix*** of the basis S. We have

$$G(S) = \begin{bmatrix} \langle e_1, e_1 \rangle \ldots \langle e_1, e_n \rangle \\ \cdots\cdots\cdots\cdots\cdots \\ \langle e_n, e_1 \rangle \ldots \langle e_n, e_n \rangle \end{bmatrix}$$

and

$$\langle x, y \rangle = \bar{x}_S\, G(S)\, y_S\,.$$

If S' is another basis of S, we have

$$G(S') = \overline{K}G(S)\,K$$

where K is a similarity matrix.

More generally, if $e_1, \ldots, e_n$ are n independent vectors in a pre-Hilbert space H, they generate an n-dimensional Euclidean or Hermitian subspace E for which they constitute a basis S. Let $G(S) = G(e_1, \ldots, e_n)$ denote the corresponding Gram matrix.

Let x denote an arbitrary vector in H and let Px denote its projection onto E. The relations

$$x - Px \perp e_i \ (\forall i = 1, \ldots, n)$$

imply

$$(Px)_{\hat{S}} = \begin{bmatrix} \langle e_1, Px \rangle \\ \cdots\cdots \\ \langle e_n, Px \rangle \end{bmatrix} = \begin{bmatrix} \langle e_1, x \rangle \\ \cdots\cdots \\ \langle e_n, x \rangle \end{bmatrix}.$$

From this, we derive the components of Px with respects to $S = [e_1, \ldots, e_n]$:

$$(Px)_S = \overset{-1}{G}(S) \begin{bmatrix} \langle e_1, x \rangle \\ \cdots\cdots \\ \langle e_n, x \rangle \end{bmatrix}.$$

Remark. If $e_1, \ldots, e_n$ are n vectors (independent or otherwise) of a pre-Hilbert space, we can still call the matrix with coefficients $\langle e_i, e_j \rangle$ the Gram matrix of these vectors.

A necessary and sufficient condition for $e_1,\ldots,e_n$ to be independent is that their Gram matrix be nonsingular.

Quadratic forms. A matrix $A \in L(\mathbf{R}^n, \mathbf{R}^n)$ is said to be ***positive-semidefinite*** (resp. ***positive-definite***) if $\overline{x}Ax \geqslant 0$ for every $x \in \mathbf{R}^n$ (resp. $\overline{x}Ax > 0$ for every nonzero $x \in \mathbf{R}^n$). Let E denote a real finite-dimensional vector space. A mapping $x \to \varphi(x, x)$, where φ is a bilinear form, is called a ***quadratic form*** on E. We can always assume φ to be symmetric.

For every basis S, there exists a matrix $\mathcal{A}(S)$ such that

$$\varphi(x, x) = \overline{x}_S\, \mathcal{A}(S)\, x_S\,.$$

If S' is another basis, we have

$$\varphi(x, x) = \overline{x}_{S'}\, \mathcal{A}(S')\, x_{S'}\,,$$

with

$$\mathcal{A}(S') = \overline{K}\mathcal{A}(S)\,K\,,$$

where K is a similarity matrix.

A quadratic form is said to be positive-semidefinite if $\varphi(x, x) \geqslant 0$ for all $x \in E$, that is, if $\mathcal{A}(S)$ is positive-semidefinite. It is said to be positive-definite if, in addition, $\varphi(x, x) = 0$ implies $x = 0$, that is, if $\mathcal{A}(S)$ is positive-definite.

If φ is bilinear and if $x \to \varphi(x, x)$ is positive-definite, the mapping $x, y \to \varphi(x, y)$ defines a Euclidean structure of which $\mathcal{A}(S)$ is the Gram matrix.

If E is a Euclidean space, then, for every quadratic form $x \to \varphi(x, x)$, there exists an $A \in L(E, E)$ such that

$$\varphi(x, x) = \langle x, Ax \rangle$$

and we can take A to be self-adjoint.

Spectral decomposition. Let E denote a vector space of finite dimension n. An operator $H \in L(E, E)$ is said to be ***nilpotent*** if there exists an integer k such that $H^k = 0$. We then have $H^n = 0$. Let A denote an element of $L(E, E)$. Let $P(s) = \det(s - A)$ denote the characteristic polynomial of A. We recall the

Theorem (Cayley-Hamilton). $P(A) = 0$.

Let $s_1, \ldots, s_p$ denote the eigenvalues of A and let $k_1, \ldots, k_p$ denote their multiplicities. We shall give two forms of the spectral-decomposition theorem:

Spectral-decomposition theorem (geometric form). *There exists a decomposition of E as a direct sum of subspaces $E_1, \ldots, E_p$ satisfying the following properties:*

(1) $E_1, \ldots, E_p$ *are stable under* A,
(2) $\dim(E_i) = k_i$,
(3) *the restriction* A_{E_i} *of* A *to* E_i *has the unique eigenvalue* s_i *and is hence of the form* $s_i + h_i$, *where* h_i *is nilpotent* ($h_i^{k_i} = 0$),
(4) $E_i = \ker((s_i - A)^{k_i})$.

Spectral decomposition theorem (analytic form). *There exist projections* π_i *and nilpotent operators* H_i *such that*

(1) $\sum_i \pi_i = \mathbf{1}$, $\quad \pi_i \pi_j = 0$ for $i \neq j$,
(2) $\mathrm{val}(\pi_i)$ *is stable under* A,
(3) $H_i \pi_i = H_i$,
(4) $A = \sum_i (s_i + H_i) \pi_i$.

From the results of the two theorems, we have the following relations: $E_i = \mathrm{val}(\pi_i)$, and h_i is the restriction of H_i to E_i. The subspace E_i is called the ***spectral subspace*** associated with the eigenvalue s_i. The projection π_i is called the ***spectral projection*** associated with the eigenvalue s_i.

Partitions. Suppose that $A \in L(\mathbf{C}^n, \mathbf{C}^m)$. Let α denote an injective mapping of the set $\{1, ..., m\}$ (where $0 < k \leqslant n$) into the set $\{1, ..., k\}$. We denote A_α by the matrix whose ith column is equal to $A_{\alpha(i)}$.

Similarly, if β is an injective mapping of the set $\{1, ..., h\}$ (where $0 < h \leqslant m$) into the set $\{1, ..., m\}$, we denote by A^β the matrix whose jth row is equal to $A^{\beta(i)}$.

We set $A_\alpha^\beta = (A_\alpha)^\beta = (A^\beta)_\alpha$. The matrix A_α^β is called a ***submatrix*** of the matrix A.

For convenience, we shall refer to α as an ordered subset of $\{1, .., n\}$ and we shall refer to β as an ordered subset of $\{1, ..., m\}$. Accordingly, we shall write $p \in \alpha$ (resp. $q \in \beta$) instead of $p \in \operatorname{val}(\alpha)$ (resp. $q \in \operatorname{val}(\beta)$) and we shall write $\alpha_2 = \complement\, \alpha_1$ instead of $\operatorname{val}(\alpha_2) = \complement \operatorname{val} \alpha_1$.

Suppose that $\alpha_1, ..., \alpha_\lambda$ constitute a partition of $\{1, ..., n\}$ and that $\beta_1, ..., \beta_\mu$ constitute a partition of $\{1, ..., m\}$. The set of matrices

$$A_{\alpha_r}^{\beta_s} \quad (r = 1, ..., \lambda, s = 1, ..., \mu)$$

constitutes a partition of the matrix A. Such a partition may be represented by the following scheme:

$$A = \left[\begin{array}{c|c|c} A_{\alpha_1}^{\alpha_1} & \cdots & A_{\alpha_\lambda}^{\beta_1} \\ \hline \cdots & \cdots & \cdots \\ \hline A_{\alpha_1}^{\beta_\mu} & \cdots & A_{\alpha_\lambda}^{\beta_\mu} \end{array}\right].$$

If $\alpha_1, ..., \alpha_\lambda$ constitutes a partition of $\{1, ..., n\}$ and if $B \in L(\mathbf{C}^q, \mathbf{C}^n)$, we have the following partitions for A and B:

$$A = [A_{\alpha_1} | \cdots | A_{\alpha_\lambda}], \qquad B = \left[\begin{array}{c} B^{\alpha_1} \\ \hline \vdots \\ \hline B^{\alpha_\lambda} \end{array}\right],$$

and the rule for multiplication of matrices implies the following formula:

$$AB = \sum_r A_{\alpha_r} B^{\alpha_r}.$$

3. MEASURES

Consider a space $\mathbf{R}^n$. We shall denote by $\mathfrak{D}^0(\mathbf{R}^n)$ or, more simply, $\mathfrak{D}^0$, the vector space of continuous complex functions with compact support. This space, equipped with the norm of uniform convergence

$$f \to \| f \|_u = \sup_x | f(x) |.$$

is a dense subspace of the space $\mathbf{L}_\infty$ of continuous complex functions that approach 0 as their argument becomes infinite.

We shall denote by $\mathfrak{D}_K^0$ the space of continuous complex functions with compact support. Equipped with the norm of uniform convergence, $\mathfrak{D}_K^0$ is a Banach space.

We apply the term ***measure*** to any linear form μ on $\mathfrak{D}^0$ satisfying the following continuity condition:

(CC_0) For every compact set K, the restriction of μ to $\mathfrak{D}_K^0$ is a continuous linear form for the topology of $\mathfrak{D}_K^0$.

We denote by $\int f \cdot \mu$ the value of the form μ for the element $f \in \mathfrak{D}^0$. It will also be denoted by $\int f(x) \cdot \mu^x$ when it is necessary to show the "variable of integration". The condition (CC_0) can also be written:

For every compact set K, there exists a nonnegative constant M_K such that

$$f \in \mathfrak{D}_K^0 \quad \Rightarrow \quad \left| \int f \cdot \mu \right| \leqslant M_K \, \| f \|_u \, .$$

Let $\mathfrak{D}_+^0$ denote the set of nonnegative functions belonging to $\mathfrak{D}^0$. In particular, if $\int f . \mu \geqslant 0$ for every $f \in \mathfrak{D}_+^0$, condition (CC_0) is necessarily satisfied. In this case, we say that μ is a positive measure.

We define an ordering relation among measures by writing

$$\mu \leqslant \nu \quad \Leftrightarrow \quad \int f \cdot \mu \leqslant \int f \cdot \nu \ (\forall f \in \mathfrak{D}_+^0) \, .$$

For a measure μ to be positive, it is necessary and sufficient that $\mu \geqslant 0$. A measure μ is said to be real if $\int f \cdot \mu$ is real for every real f. Every measure μ can be written in the form $\mu = \mu_1 + i \, \mu_2$, where μ_1 and μ_2 are real measures known as the real part and the imaginary part of μ respectively.

Let μ be a real measure. There exists a positive measure μ^+ defined by

$$\int f \cdot \mu^+ = \sup_{\left\{ \substack{0 \leqslant \varphi \leqslant f \\ \varphi \in \mathfrak{D}^0} \right.} \int \varphi \cdot \mu \, , \qquad (\forall f \in \mathfrak{D}_+^0) \, .$$

We can then set $\mu = \mu^+ - \mu^-$; then, μ^- is also a positive measure. The measures μ^+ and μ^- are called the positive and negative parts of μ respectively. If ν is a positive measure such that $\mu \leqslant \nu$, then $\mu^+ \leqslant \nu$, hence, $\mu^+$is the "smallest of the positive measures exceeding μ".

Examples. I. The linear form

$$f \to \int_{\mathbf{R}^n} f(x)\, dx$$

is a positive measure known as the ***Lebesgue measure.*** Its value for the element f will simply be denoted by $\int f$.

II. The linear form $f \to f(a)$ is a positive measure denoted by δ_a. It is known as the ***Dirac measure*** at the point a. We set

$$f(a) = \int f \,.\, \delta_a \,.$$

We shall also write δ for δ_0.

Extension of a measure. Let μ denote a positive measure. Examples show that one can define $\int f \,.\, \mu$ for certain functions other than continuous with compact support. For every measure μ, we are led to replace the condition of continuity with a condition of measurability, in other words, to define a class of functions said to be ***measurable*** with respect to the measure μ. The manner in which the concept of measurability is introduced depends to a great extent on the method of exposition chosen. We shall not go into this here.

We shall confine ourselves to recalling the essential properties of measurable functions and their integration:

(1) Continuous functions are measurable.
(2) Measurable functions constitute a vector space.
(3) The greatest lower bound and the least upper bound of a finite family of measurable real functions is measurable.
(4) If f is measurable, $|f|$ is measurable.
(5) If a sequence $\{f_k(x)\}$ of measurable functions f_k tends to limit function $f(x)$ for all x, then $f(x)$ is measurable.

Let f denote a mapping of $\mathbf{R}^n$ into $\mathbf{R}^p$. For $y = f(x)$, let us set $y^j = f^j(x)$. The mapping f is said to be measurable if each of the functions f^j is measurable. Furthermore, if g is a continuous mapping of $\mathbf{R}^p$ into $\mathbf{R}^q$, then $g \circ f$ is measurable. If f is a positive measurable function, $\int f \,.\, \mu$ is defined but it may possibly be equal to $+\infty$. If $\int f \,.\, \mu \neq +\infty$, then f is said to be ***integrable.*** A complex function f is said to be integrable if it is measurable and if $|f|$ is integrable. We then have

$$\left|\int f \cdot \mu\right| \leqslant \int |f| \cdot \mu .$$

The set of integrable functions constitutes a vector space and the mapping $f \to \int f \cdot \mu$ is linear on that vector space.

A set E is said to be measurable if its characteristic function is measurable. We then write

$$\mu(E) = \int \varepsilon_E \cdot \mu .$$

If $\mu(E) \neq +\infty$ (that is, if ε_E is integrable), then E is said to be integrable.

Any countable union or intersection of measurable sets is measurable. The complement of a measurable set is measurable. A set is said to be μ-***negligible*** if it is of measure zero with respect to μ.

Two functions f and g are said to be ***equal almost everywhere*** if they are equal except on a set of measure zero. We then write $f \overset{\text{p.p}}{=} g$ or $f = g$ almost everywhere with respect to μ. If f is measurable, so is g.

A necessary and sufficient condition for two measurable functions to be equal almost everywhere is that

$$\int |f - g| \cdot \mu = 0.$$

The relation $f \overset{\text{p.p}}{=} g$ is an equivalence relation.

For $p \geqslant 1$, we call $\mathbf{L}^p(\mu)$ the ***quotient space*** under the equivalence relation $\overset{\text{p.p}}{=}$ of the set of measurable functions such that $|f^p|$ is integrable. For convenience, we shall write $f \in \mathbf{L}^p(\mu)$ if f is measurable and $|f^p|$ is integrable.

The set $\mathbf{L}^p(\mu)$ is a vector space and the mapping

$$f \to \|f\|_p = \left[\int |f|^p . \mu\right]^{1/p}$$

is a norm on $\mathbf{L}^p(\mu)$. With respect to this norm, $\mathbf{L}^p(\mu)$ is complete; thus, it is a Banach space.

A measurable function f is said to be ***essentially bounded*** if there exists a number m such that the set of all x verifying the inequality $|f(x)| > m$ is of measure zero.

We denote by $\mathbf{L}^\infty(\mu)$ the quotient space under the equivalence relation $\overset{\text{p.p}}{=}$ of the set of functions that are measurable and essentially bounded. We denote by $\|f\|_\infty$ the greatest lower bound of numbers m such that the set of all x verifying the inequality $|f(x)| > m$ is of measure zero. The mapping $f \to \|f\|_\infty$ is a norm on $\mathbf{L}^\infty(\mu)$, and $\mathbf{L}^\infty(\mu)$ is a Banach space for that norm.

For $p \in [1, +\infty]$, the space $\mathbf{L}^p(\mu)$ has the dual space $\mathbf{L}^q(\mu)$ with

$$\frac{1}{p}+\frac{1}{q}=1.$$

In particular, the dual of $\mathbf{L}^1(\mu)$ is $\mathbf{L}^\infty(\mu)$. For convenience, we shall write $f \in \mathbf{L}^\infty(\mu)$ if f is measurable and essentially-bounded.

If μ is a real measure, a function f will be called measurable (resp. integrable) with respect to μ if it is measurable (resp. integrable) with respect to μ^+ and μ^-. If f is integrable with respect to μ, we define

$$\int f\,.\,\mu=\int f\,.\,\mu^+ - \int f\,.\,\mu^-.$$

Similarly, if μ is a complex measure, a function f will be called measurable (resp. integrable) with respect to μ if it is measurable (resp. integrable) with respect to the real part $\mathfrak{R}(\mu)$ and the imaginary part $\mathfrak{I}(\mu)$ of μ.

If f is integrable with respect to μ, we define

$$\int f\,.\,\mu=\int f\,.\,(\mathfrak{R}\mu)+\mathrm{i}\int f\,.\,(\mathfrak{I}\mu).$$

Remark. If μ is Lebesgue measure on a space E, we shall replace the notation $\mathbf{L}^p(\mu)$ (for $1 \leqslant p \leqslant +\infty$) with the notation $\mathbf{L}^p(E)$ or $\mathbf{L}^p$.

Bounded measures. A positive measure μ is said to be bounded if $\int 1\,.\,\mu \neq +\infty$. We then write $\int 1\,.\,\mu = \int \mu$. This number is called the ***total mass*** of the measure. It can also be defined by the following formula:

$$\int \mu = \sup_{\substack{f\in\mathfrak{D}^0\\ 0\leqslant f\leqslant 1}} \int f\,.\,\mu.$$

A real measure is said to be bounded if μ^+ and μ^- are bounded. A complex measure is said to be bounded if its real and imaginary parts are both bounded.

If μ is a positive bounded measure, every measurable essentially bounded function is integrable. We then have

$$\left|\int f\,.\,\mu\right| \leqslant \|f\|_\infty\,.\int \mu.$$

In particular, if f is continuous and bounded, then f is integrable and

$$\left|\int f\,.\,\mu\right| \leqslant \|f\|_u\,.\int \mu.$$

If μ is a bounded positive measure, then $\mathbf{L}^p(\mu) \subset \mathbf{L}^q(\mu)$ whenever $q \leqslant p$.

We shall return to the study of bounded measures in section 1 of Chapter III.

The support of a measure. There exists a largest closed set F such that $\int f \cdot \mu = 0$ for every function $f \in \mathfrak{D}^0$ with support in the complement of F. This set F is called the ***support*** of μ.

Every measure of bounded (i.e., compact) support is a bounded measure.

Theorem (Lebesgue). *Let μ denote a measure on $\mathbf{R}^n$. Let $\{f_k\}$ denote a sequence of integrable functions that converge pointwise to a function f almost everywhere. If there exists an integrable function g such that $|f_k| \leqslant g$, then f is integrable and*

$$\lim_{k\to+\infty} \int f_k \cdot \mu = \int f \cdot \mu .$$

The product of a measure and a locally integrable function. Let μ denote a measure. A measurable function φ is said to be ***locally integrable*** if $\varepsilon_\Omega \varphi$ is integrable for every open integrable set Ω. In this case, we can define a measure $\varphi\mu$ such that

$$\int f \cdot \varphi\mu = \int f\varphi \cdot \mu , \qquad (\forall f \in \mathfrak{D}^0) .$$

It can be shown that a necessary and sufficient condition for a function f to be integrable with respect to the measure $\varphi\mu$ is that $f\varphi$ be integrable with respect to μ and that, in this case, the preceding formula remains valid.

A necessary and sufficient condition for $\varphi\mu$ to be a bounded measure is that $\varphi \in \mathbf{L}^1(\mu)$.

If the "variable of integration" x needs to be indicated, we shall write $\varphi(x)\mu^x$ instead of $\varphi\mu$.

If $\nu = \varphi\mu$, we shall say that φ is the ***density*** of ν with respect to μ.

The direct product of two measures. Let μ and ν denote measures defined on $\mathbf{R}^n$ and $\mathbf{R}^p$, respectively. There exists exactly one measure $\mu\nu$ defined on $\mathbf{R}^{n+p}$ such that

$$\int f \cdot \mu\nu = \int \left[\int f(x, y) \cdot \mu^x\right] \cdot \nu^y = \int \left[\int f(x, y) \cdot \nu^y\right] \cdot \mu^x , \qquad (\forall f \in \mathfrak{D}^0(\mathbf{R}^{n+p})) .$$

The rule for reversing the order of integration can be generalized as follows:

Theorem (Fubini). *Let μ and ν denote positive measures defined on $\mathbf{R}^n$ and $\mathbf{R}^p$ respectively. Let f denote a function in $\mathbf{L}^1(\mu\nu)$. Then*

(a) *except for values of* y *belonging to a* ν*-negligible set, the function* $x \to f(x, y)$ *belongs to* $\mathbf{L}^1(\mu)$;

(b) *the function* $y \to \int f(x, y) \cdot \mu^x$ *(defined almost everywhere) belongs to* $\mathbf{L}^1(\nu)$;

(c) $\int f \cdot \mu\nu = \int \left[\int f(x, y) \cdot \mu^x\right] \cdot \nu^y$.

Conversely, if f *is positive and if conditions (a) and (b) are satisfied, then* $f \in \mathbf{L}^1(\mu\nu)$ *and condition (c) is satisfied.*

If μ and ν are bounded measures, then $\mu\nu$ is bounded.
If μ and ν are positive measures, then $\mu\nu$ is positive.
If μ and ν are bounded and positive, then

$$\int \mu\nu = \int \mu \cdot \int \nu .$$

If φ and ψ are locally integrable with respect to μ and ν respectively, then the function $x, y \to \varphi(x)\,\psi(y)$ locally integrable with respect to $\mu\nu$ and

$$[\varphi(x)\,\psi(y)]\,\mu^x\,\nu^y = [\varphi(x)\,\mu^x]\,[\psi(y)\,\nu^y] .$$

From Fubini's theorem, we get the rules for differentiation of integrals depending on a parameter. We shall confine ourselves to the following statement, which will be sufficient for our purposes:

Theorem. *Let* μ *denote a measure on* $\mathbf{R}^n$ *and let* $x, y \to f(x, y)$ *denote a continuous function defined for* $x \in \mathbf{R}^n$ *and* $y \in]a, b[$. *Suppose that the following three conditions are satisfied:*

(1) *The function* $x \to f(x, y)$ *is integrable with respect to* μ *for all* $y \in]a, b[$;

(2) *the partial derivative* $f'_y(x, y)$ *exists and is continuous on* $\mathbf{R}^n \times]a, b[$;

(3) *there exists a positive function* g *that is integrable with respect to* μ *and satisfies the inequality*

$$|f'_y(x, y)| \leqslant g(x) .$$

Then, the function $y \to \int f(x, y) \cdot \mu^x$ *is continuously differentiable and its derivative is given by*

$$\frac{\mathrm{d}}{\mathrm{d}y} \int f(x, y) \cdot \mu^x = \int f'_y(x, y)\,\mu^x .$$

The image of a measure under a measurable mapping. Let μ denote a measure on $\mathbf{R}^n$ and let φ denote a measurable mapping of $\mathbf{R}^n$ into $\mathbf{R}^p$. Suppose that the function $f \circ \varphi$ is integrable with respect to μ for every $f \in \mathfrak{D}^0(\mathbf{R}^p)$. Then, we define a measure on $\mathbf{R}^p$, called the *image* of μ under φ, by

$$\int f \cdot \nu = \int (f \circ \varphi) \cdot \mu ,$$

that is,

$$\int f(y) \cdot \nu^y = \int f(\varphi(x)) \cdot \mu^x .$$

In the case in which μ is a bounded measure, its image under a measurable mapping is always defined and is a bounded measure.

The convolution product of two measures. Let μ and ν denote two measures on $\mathbf{R}^n$. Suppose that, for every $f \in \mathfrak{D}^0(\mathbf{R}^n)$, the function $x, y \to f(x + y)$ is integrable with respect to $\mu^x \nu^y$. Then we define the convolution product $\mu * \nu$ of the measures μ and ν (which is a measure on $\mathbf{R}^n$) by

$$\int f \cdot \mu * \nu = \int f(x + y) \cdot \mu^x \nu^y \qquad (\forall f \in \mathfrak{D}^0(\mathbf{R}^n)) .$$

In other words, $\mu * \nu$ is the image of $\mu^x \nu^y$ under the mapping $x, y \to x + y$.

The convolution product $\mu * \nu$ is defined, in particular, in the following cases:

(a) μ and ν are bounded measures,

(b) μ or ν has compact support,

(c) both μ and ν have support bounded below (or above).

The convolution product is commutative. The formula

$$(\mu * \nu) * \rho = \mu * (\nu * \rho)$$

holds in the following principal cases:

(a) μ, ν, and ρ are bounded,

(b) μ and ν have bounded support,

(c) μ, ν, and ρ have support bounded below.

If μ has a density f with respect to Lebesgue measure, we set $\mu * \nu = f * \nu$. Thus, $f * \nu$ itself has a density with respect to Lebesgue measure. If we identify $f * \nu$ with this density, we have

$$(f * \nu)(x) \overset{\text{p.p}}{=} \int f(x - y)\, \nu^y ,$$

so that the right-hand member is defined except for a set of values of x that is negligible with respect to Lebesgue measure.

If ν itself has a density g with respect to Lebesgue measure, we set $f * g = f * \nu$. We then have

$$(f * g)(x) \overset{\text{p.p}}{=} \int f(x - y)\, g(y)\, dy .$$

If the right-hand member is defined for all x, we define

$$(f * g)(x) = \int f(x - y) g(y) \, dy .$$

Measures on T. The same theory is applied to **T**. Furthermore, we can embed **T** in $\mathbf{R}^2$ and identify the measures on **T** with the measures on $\mathbf{R}^2$ with support in **T**. However, since **T** is compact, we have the following simplifications:

(1) Every measure is bounded;
(2) every measurable set is integrable;
(3) every essentially bounded measurable function is integrable;
(4) every locally integrable function is integrable.

4. DIFFERENTIABLE FUNCTIONS

Let E and F denote two Banach spaces. Let f denote a mapping of an open subset Ω of E into F. We shall say that f is *differentiable* at a point $x \in \Omega$ if there exists an $A \in L(E, F)$ such that

$$f(x + h) = f(x) + Ah + \| h \| \varepsilon(h),$$

where $\varepsilon(h)$ tends to 0 as h tends to 0. The element A is called the *derivative* of f at the point x.

We shall say that the mapping f is differentiable on Ω if it is differentiable at every point $x \in \Omega$. Its derivative at the point x is then denoted by $f'(x)$ and the mapping f' of Ω into $L(E, F)$ is called the derivative of the mapping f.

We then have

$$f(x + h) = f(x) + f'(x)(h) + \| h \| \varepsilon(h) .$$

If f' is a continuous mapping of Ω into the Banach space $L(E, F)$, we say that f is *continuously differentiable*.

If f' is also differentiable on Ω, its derivative is denoted by f''. We have

$$f''(x) \in L(E, L(E, F))$$

and consequently,

$$f''(x)(h) \in L(E, F)$$
$$f''(x)(h)(k) \in F .$$

Therefore, if f'' is continuous, the mapping $h, k \to f''(x)(h)(k)$ is symmetric.

We have (the Taylor-MacLaurin formula)

$$f(x + h) = f(x) + f'(x)(h) + f''(x)(h)(h) + \| h \|^2 \varepsilon(h),$$

where $\varepsilon(h)$ tends to 0 as h tends to 0.

We can define the derivatives of higher order $f^{(m)}$ in the same way. We shall not use them in the general case.

If $E = \mathbf{R}$ and $F = \mathbf{R}^p$, we set

$$y = \begin{bmatrix} y^1 \\ \vdots \\ y^p \end{bmatrix} = \begin{bmatrix} f^1(x) \\ \vdots \\ f^p(x) \end{bmatrix},$$

and we have

$$f'(x) = \begin{bmatrix} (f^1)'(x) \\ \vdots \\ (f^p)'(x) \end{bmatrix}.$$

If $E = \mathbf{R}^n$ and $F = \mathbf{R}$, we set $f_i'(x) = f'(x)(\varepsilon_i)$. This quantity is called the *partial derivative* of f with respect to x_i at the point x. It is also denoted by $f'_{x_i}(x)$. Thus, we have

$$f'(x) = [f_1'(x) \dots f_n'(x)].$$

Finally, if $E = \mathbf{R}^n$ and $F = \mathbf{R}^p$, we set $y^j = f^j(x)$, and we have

$$f'(x) = \begin{bmatrix} (f')_1^1(x) & \dots & (f')_n^1(x) \\ \vdots & & \vdots \\ (f')_1^p(x) & \dots & (f')_n^p(x) \end{bmatrix} = \begin{bmatrix} (f^1)'(x) \\ \hline \vdots \\ \hline (f^p)'(x) \end{bmatrix}$$

with

$$(f')_i^j(x) = (f^j)_i'(x).$$

For $E = \mathbf{R}^n$ and $F = \mathbf{R}$,

$$f''(x)(h)(k) = \sum_{i,j} f_{ij}''(x)\, h^i k^j$$

(where f'' is assumed continuous) with

$$f_{ij}''(x) = f''(x)(\varepsilon_i)(\varepsilon_j).$$

We can form a matrix with coefficients $H_i^j = f_{ij}''(x)$. We then have

$$f''(x)(h)(k) = \bar{h}Hk.$$

The matrix H is symmetric.

More generally, if $f^{(m)}$ is continuous,

$$f^{(m)}(x)(h_1)\dots(h_m) = \sum_{i_1,\dots,i_m} f_{i_1,\dots,i_m}^{(m)}(x)\, h_1^{i_1} \dots h_m^{i_m}$$

with

$$f_{i_1,\dots,i_m}^{(m)}(x) = f^{(m)}(x)(\varepsilon_{i_1})\dots(\varepsilon_{i_m}).$$

We can identify $f^{(m)}(x)$ with an m-linear symmetric form.

2

The Theory of Distributions

1. INTRODUCTION TO THE CONCEPT OF A DISTRIBUTION

We know that measures are linear forms defined on the space $\mathfrak{D}^0$ of continuous functions with compact support. It is possible to consider in a natural way linear forms that are defined on proper subspaces of $\mathfrak{D}^0$ but not on the entire space $\mathfrak{D}^0$.

In particular, in $\mathbf{R}$, instead of considering the Dirac measure δ_a at the point a and defined for all $f \in \mathfrak{D}^0$ by

$$f \xrightarrow{\delta_a} f(a) = \int f \,.\, \delta_a \,,$$

we can consider the linear form

$$f \to f'(a) \,,$$

which is not defined unless f is differentiable at the point a. We shall denote this linear form by $-\delta_a'$ (the reason for the minus sign will be clarified in what follows).

We shall keep a notation analogous to that used for measures. Thus, we shall define

$$f'(a) = -\int f \,.\, \delta_a' \,.$$

Similarly, we may consider the linear form

$$f \to f^{(p)}(a),$$

which will be denoted by $(-1)^p \delta_a^{(p)}$ (the reason for the factor $(-1)^p$ will be clarified in what follows). Thus, we shall define

$$f^{(p)}(a) = (-1)^p \int f \,.\, \delta_a^{(p)}.$$

The linear form $\delta_a^{(p)}$ is defined if f is p times differentiable at a.

The linear forms $\delta'_a, \ldots, \delta_a^{(p)}, \ldots$ are not measures since they are not defined on $\mathfrak{D}^0$. These are examples of distributions. The procedure for defining distributions will be as follows:

(1) First, one chooses a function space.

(2) Then, one defines a distribution as a linear form on the function space chosen so that a suitable continuity condition analogous to the one introduced in the definition of a measure is satisfied.

Thus, we shall give the following provisional and incomplete definitions, which are valid in $\mathbf{R}^n$.

Notations. We denote by $\mathfrak{D}$ the vector space of infinitely differentiable complex functions with compact support.

We denote by $\mathfrak{D}^m$ the vector space of m-times continuously differentiable complex functions with compact support.

If $p \geqslant m$, we have $\mathfrak{D}^p \subset \mathfrak{D}^m$.

Definitions (to be completed). A *distribution* is a linear form defined on $\mathfrak{D}$ and satisfying a continuity condition to be specified later.

A *distribution of order m* is a linear form defined on $\mathfrak{D}^m$ and satisfying a continuity condition to be specified below.

The vector space of distributions will be denoted by $\mathfrak{D}'$.

The vector space of distributions of order m will be denoted by $(\mathfrak{D}^m)'$.

To avoid any contradiction in terminology, we shall choose the continuity conditions in such a way that (1) every distribution of order m restricted to $\mathfrak{D}^p$ will, for $p \geqslant m$, be a distribution of order p and (2) every distribution of order m restricted to $\mathfrak{D}$ will be a distribution.

Also, the continuity conditions imposed on distributions of order m will be chosen so that, for $m = 0$, it will reduce to continuity of measures, in other words, so that measures will be distributions of order 0.

For example, the linear forms $\delta_a, \delta'_a, \ldots, \delta_a^{(p)}$ are respectively distributions of order $0, 1, \ldots, p$.

2. STUDY OF CERTAIN FUNCTION SPACES

We denote by

$\mathfrak{D}_K$ the vector space of infinitely differentiable function with support in the compact set K.

$\mathfrak{D}_K^m$ the vector space of m-times continuously differentiable functions with support in the compact set K.

Figure 1 shows diagramatically the inclusion relations among $\mathfrak{D}$, $\mathfrak{D}_K$, $\mathfrak{D}^m$, $\mathfrak{D}_K^m$.

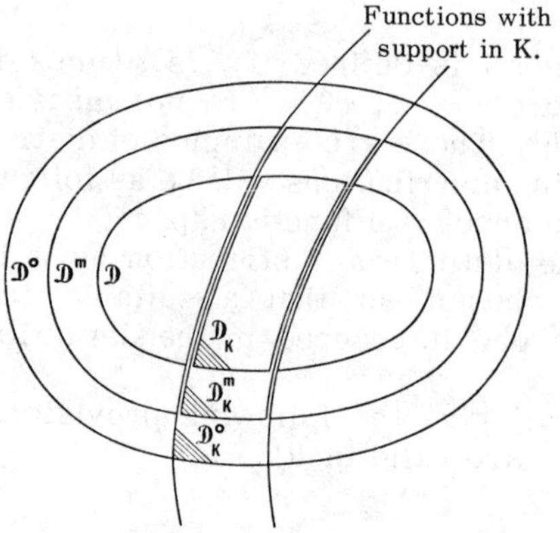

FIG. 1.

For any bounded continuous function φ, we denote by $\| \varphi \|_u$ the norm of uniform convergence defined by

$$\| \varphi \|_u = \sup_x | \varphi(x) | .$$

Study of the space $\mathfrak{D}_K^m$. We can equip $\mathfrak{D}_K^m$ with a normed vector space structure.

In $\mathbf{R}$, we define

$$\| f \| = \| f \|_u + \| f' \|_u + \cdots + \| f^{(m)} \|_u .$$

(By replacing the sum on the right with the least upper bound of the summands, we obtain an equivalent norm on $\mathfrak{D}_K^m$.)

Similarly, in $\mathbf{R}^n$, we define

$$\| f \| = \sum_{m_1 + \ldots + m_n \leqslant m} \| D_1^{m_1} \ldots D_n^{m_n} f \|_u$$

where D_i (for $i = 1, \ldots, n$) denotes differentiation with respect to the ith variable x^i. We state that $\mathfrak{D}_K^m$ equipped with this norm is a complete space, that is, a Banach space.

Remark. One can easily see that, in $\mathbf{R}$, the mapping $f \to \| f^{(m)} \|_u$ is a norm in $\mathfrak{D}_K^m$ and this norm by itself defines the topology we have defined on $\mathfrak{D}_K^m$ (in other words, that the addition to this norm of $\| f \|_u + \dots + \| f^{(m-1)} \|_u$ is superfluous). If the support of f is contained in $[a, b]$, we have

$$f(x) = \int_a^x f'$$

$$f'(x) = \int_a^x f''$$

$$\dots\dots\dots\dots\dots$$

$$f^{(m-1)}(x) = \int_a^x f^{(m)},$$

so that

$$\| f^{(m-1)} \|_u \leqslant | b - a | \, \| f^{(m)} \|_u$$
$$\dots\dots\dots\dots\dots\dots\dots\dots$$
$$\| f \|_u \leqslant | b - a |^m \, \| f^{(m)} \|_u .$$

Thus, the mappings

$$f \to \| f \|_u, \dots, f \to \| f^{(m-1)} \|_u$$

are less fine norms than the norm $f \to \| f^{(m)} \|_u$.

This remark can be extended to the case of functions with range in $\mathbf{R}^n$. We equip $\mathbf{R}^n$ with a norm (for example, the natural Euclidean norm). The space of p-linear forms on $\mathbf{R}^n$ will then be equipped with the norm defined as follows:

$$\| U \| = \sup_{\substack{\|h_1\| \leqslant 1 \\ \dots\dots \\ \|h_p\| \leqslant 1}} | U(h_1, \dots, h_p) | .$$

Then, for $f \in \mathfrak{D}_K^m$ and $p \leqslant m$, we can set

$$\| f^{(p)} \|_u = \sup_x \| f^{(p)}(x) \| .$$

Let d denote the diameter of a ball containing K. Then, by repeated use of the law of the mean, we obtain the inequality

$$\| f^{(p)} \|_u \leqslant d^{m-p} \| f^{(m)} \|_u, \qquad (p \leqslant m) .$$

The topology of $\mathfrak{D}_K^m$ can then be defined by the norm $f \to \| f^{(m)} \|_u$.

Study of the space $\mathfrak{D}_K$. On $\mathbf{R}$, the topology of the space $\mathfrak{D}_K$ can be defined by the sequence of norms

$$f \to \| f^{(p)} \|_u , \qquad p = 0, 1, 2, \dots$$

On $\mathbf{R}^n$, the topology of the space $\mathfrak{D}_K$ can be defined by the sequence of the seminorms

$$f \to \| D_1^{k_1} \dots D_n^{k_n} f \|_u \qquad (k_1, \dots, k_n = 0, 1, 2 \dots),$$

where D_i denotes differentiation with respect to the ith variable, or, more simply, by the sequence of seminorms

$$f \to \| f^{(p)} \|_u \qquad (p = 0, 1, 2 \dots).$$

Thus, $\mathfrak{D}_K$ is a locally convex metrizable space. Let us accept the fact that it is complete. Thus, it is a Fréchet space.

For $n = 1$, a necessary and sufficient condition for a sequence $\{f_i\}$ of elements of $\mathfrak{D}_K$ to approach 0 is that $\{f_i^{(p)}\}$ approach 0 uniformly for every $p \geqslant 0$ (that is, that $\{f_i\}$ and each sequence of its derivatives approach 0 uniformly) as $i \to \infty$.

For arbitrary n, a necessary and sufficient condition for a sequence $\{f_i\}$ of elements of $\mathfrak{D}_K$ to approach 0 is that, for arbitrary $p_1, \dots, p_n \geqslant 0$, the sequence $D_1^{p_1} D_2^{p_2} \dots D_n^{p_n} f_i$ approaches 0 uniformly (that is, that $\{f_i\}$ and each sequence of its partial derivatives converge uniformly to the zero function).

We shall not define a topology in $\mathfrak{D}$ or $\mathfrak{D}^m$; the topologies that one considers on these spaces are difficult to use (though fruitful for advanced studies).

We shall now study the relationships between the spaces $\mathfrak{D}_K^m$ and $\mathfrak{D}_K$. First, let us give an example of an infinitely differentiable function with compact support:

In $\mathbf{R}$ (see Fig. 2),

$$f(x) = \begin{cases} 0 & \text{if } |x| \geqslant 1, \\ \exp\left(-\dfrac{1}{1 - x^2}\right) & \text{if } |x| \leqslant 1; \end{cases}$$

in $\mathbf{R}^n$,

$$f(x) = \begin{cases} 0 & \text{if } \|x\| \geqslant 1, \\ \exp\left(-\dfrac{1}{1 - \|x\|^2}\right) & \text{if } \|x\| \leqslant 1, \\ \text{with } \|x\|^2 = (x^1)^2 + \cdots + (x^n)^2. & \end{cases}$$

By suitable expansion or contraction of the x- and y-axes, we can use the preceding example to construct a function ρ_k enjoying the following properties (see Fig. 3 for $n = 1$):

ρ_k is infinitely differentiable;
$\rho_k \geqslant 0$,
the support of ρ_k is the ball $\|x\| \leqslant 1/k$;
$\int \rho_k(x)\, dx = 1$.

Proposition 1. *Let f and g denote two continuous functions with compact support. If g is m times continuously differentiable, the convolution product $f * g$ is m times differentiable and*

$$(f * g)' = f * g', \ldots, (f * g)^{(m)} = f * g^{(m)}.$$

Proof. We have

$$(f * g)(x) = \int f(x - y)\, g(y)\, dy$$

$$= \int g(x - y)\, f(y)\, dy.$$

The conclusion follows from differentiating under the integral sign in the second integral.

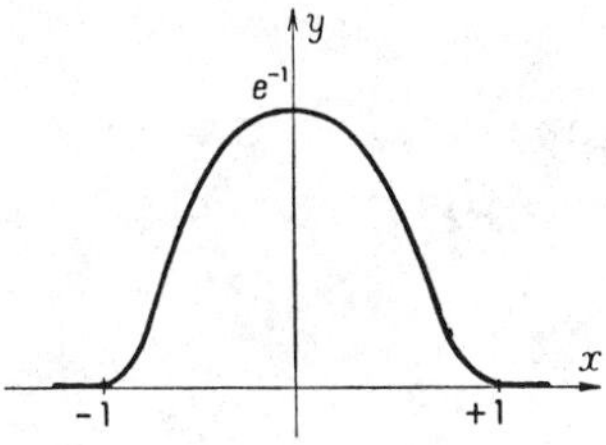

FIG. 2.

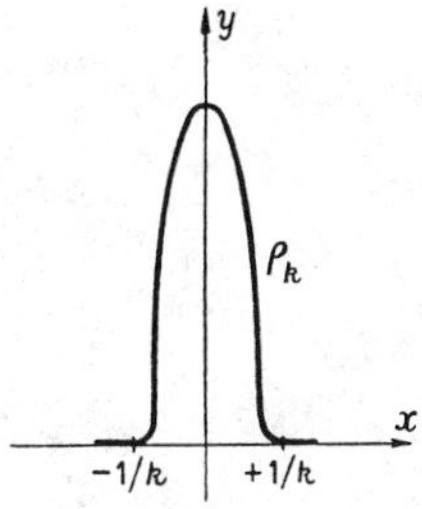

FIG. 3.

Remark. In $\mathbf{R}^n$, the preceding proposition means that

$$D_1^{m_1} \ldots D_n^{m_n}(f * g) = f * (D_1^{m_1} \ldots D_n^{m_n} g)$$

for $m_1 + \ldots + m_n \leqslant m$.

Proposition 2. *If $f \in \mathfrak{D}^0$, the sequence $\{f * \rho_k\}$ of convolution products $f * \rho_k$ tends uniformly to f as k approaches infinity.*

Proof. We have

$$(f * \rho_k)(x) = \int_{\|y\| \leqslant 1/k} f(x - y)\, \rho_k(y)\, dy = f(x_k) \int_{\|y\| \leqslant 1/k} \rho_k(y)\, dy = f(x_k),$$

where x_k is an element of the ball of radius $1/k$ with center at x. Consequently, since f is uniformly continuous, $\{f * \rho_k\}$ tends uniformly to f.

We might remark that $f * \rho_k \in \mathfrak{D}$ since $(f * \rho_k)^{(p)} = f * \rho_k^{(p)}$. Furthermore, if $f \in \mathfrak{D}^p$, then $(f * \rho_k)^{(p)} = f^{(p)} * \rho_k$ tends uniformly to $f^{(p)}$.

Furthermore, the proposition remains valid if the range of f is a finite-dimensional vector space (in this case, it is sufficient to study each component of f separately).

Theorem 1. *For every closed ball K, the set $\mathfrak{D}_K^m$ is dense in $\mathfrak{D}_K^p$ for $p \leqslant m$ and $\mathfrak{D}_K$ is dense in $\mathfrak{D}_K^p$.*

Proof. Since $\mathfrak{D}_K \subset \mathfrak{D}_K^m$, it suffices to prove the second part of the theorem. Suppose that $f \in \mathfrak{D}_K^m$. For $\lambda < 1$, the function $x \to f(x/\lambda)$ belongs to $\mathfrak{D}_K^m$, its support is contained in the ball λK, and it tends to f in $\mathfrak{D}_K^m$ as λ tends to 1. To prove the theorem, it suffices to show that for a function $f \in \mathfrak{D}_K^m$ with support in the ball λK (where $\lambda < 1$), there exists a sequence $\{g_k\}$ of functions $g_k \in \mathfrak{D}_K$ that converges to f in the sense of the topology of $\mathfrak{D}_K^m$.

Therefore, it suffices to take $g_k = f * \rho_k$. For k sufficiently large, the support of g_k is in K; thus, $g_k \in \mathfrak{D}_K$. Furthermore, g_k converges uniformly to f and, for $q \leqslant m$, the sequence $\{(g_k)^{(q)}\}$, where $(g_k)^{(q)} = f^{(q)} * \rho_k$, converges uniformly to $f^{(q)}$ as $k \to +\infty$. Consequently, the sequence $\{g_k\}$ converges to f in $\mathfrak{D}_K^m$.

We are now in a position to complete the definitions of distributions and distributions of order m.

Definition. A *distribution* is defined as a linear form $T : f \to \int f \cdot T$ defined on $\mathfrak{D}$ and enjoying the following continuity property:

(CC) *For every compact set K, the restriction of T to $\mathfrak{D}_K$ is a continuous linear form.*

The number $\int f \cdot T$ is called the *integral* of the function f with respect to the distribution T.

The continuity condition can be reworded as follows: For every sequence $\{f_i\}$ of functions $f_i \in \mathfrak{D}$ with support in a fixed compact set K, if the sequence $\{f_i\}$ and the sequence of each of the partial derivatives of the f_i converge uniformly to 0, the sequence $\left\{\int f_i \cdot T\right\}$ converges to 0.

The equivalence of the two statement is ensured by the fact that $\mathfrak{D}_K$ is a metrizable space.

A distribution of order m is defined as a linear form $T : f \to \int f \cdot T$ defined on $\mathfrak{D}^m$ and satisfying the following continuity condition:

(CC_m) *For every compact set K, the restriction of T to $\mathfrak{D}_K^m$ is a continuous linear form.*

The continuity condition can again be reworded as follows: For every sequence $\{f_i\}$ of functions $f_i \in \mathfrak{D}^m$ with support in a fixed compact set K, if the sequence $\{f_i\}$ and the sequence of each of its partial derivatives of order not exceeding m converge uniformly to 0, the sequence $\left\{\int f_i \cdot T\right\}$ converges to 0.

The equivalence of the two formulations is ensured by the fact that $\mathfrak{D}_K^m$ is a Banach and hence a metrizable space. We can replace the expression "each of its derivatives of order not exceeding m" with "the partial derivatives of order m" or even with "the derivative of order m".

Remark. To verify conditions (CC) or (CC_m), it suffices to take for K a family of balls of sufficiently great radius and with center at O.

Examples. Every measure is a distribution of order 0 and conversely. In particular, every function φ that is locally integrable will be identified with the measure of which it is the density with respect to Lebesgue measure and will thus be treated as a distribution. For $f \in \mathfrak{D}$, the value of this distribution is

$$\int f \cdot \varphi = \int_{-\infty}^{+\infty} f(x)\, \varphi(x)\, dx\, .$$

The distributions $\delta_a^{(p)}$ are the distributions of order p.

Let us now consider the distribution T defined in $\mathbf{R}$ by

$$\int f \cdot T = \sum_{i=0}^{+\infty} f^{(i)}(i)\, , \qquad \forall f \in \mathfrak{D}\, .$$

For every function $f \in \mathfrak{D}$, the series on the right is a finite sum. Thus, the quantity $\int f \cdot T$ is well-defined.

If $K = [-r, +r]$ with $r \in \mathbf{Z}_+$ and if $f_i \in \mathfrak{D}_K$, we have

$$\int f_i \cdot T = \sum_{i=0}^{r-1} f^{(i)}(i).$$

Consequently, if $\{f_i\}$ and the sequences of the derivatives of the f_i of order not exceeding $r-1$ converge uniformly to 0, then the sequence $\left\{\int f_i \cdot T\right\}$ converges to 0.

Consequently, T is a distribution. This distribution is not defined on any of the spaces $\mathfrak{D}^m$: It is a distribution "of infinite order".

It is important to know whether a distribution (defined on $\mathfrak{D}$) can be extended to be a distribution of order m (defined on $\mathfrak{D}^m$). The practical criterion is given by the following theorem:

Theorem 2. *A necessary and sufficient condition that a distribution T (defined on $\mathfrak{D}$), may be extended to a distribution of order m (defined on $\mathfrak{D}^m$) is that T satisfy the following condition: For every compact set K, the restriction of T to $\mathfrak{D}_K$ is continuous for the topology induced by $\mathfrak{D}_K^m$; in other words, for every sequence $\{f_i\}$ of functions $f_i \in \mathfrak{D}$ with support in a fixed compact set K, if $\{f_i\}$ and the* sequences of the *partial derivatives of the* f_i *of order not exceeding* m *converge uniformly to 0, the sequence* $\left\{\int f_i \cdot T\right\}$ *converges to 0.*

Proof. That the condition is necessary is obvious. Let us show that it is sufficient. For every ball K, let T_K denote the restriction of T to $\mathfrak{D}_K$. Since T_K is continuous for the topology induced by $\mathfrak{D}^m_K$, the restriction T_K can be extended as a linear form T^m_K that is continuous on $\mathfrak{D}^m_K$.

Now, let f denote a member of $\mathfrak{D}^m$ and let K denote a closed ball containing the support of f. Let us set

$$\int f \cdot T^m_K = \int f \cdot T^m .$$

It is easy to show that the definition of $\int f \cdot T^m$ is independent of K. Suppose that $K' \supset K$. Then,

$$\begin{array}{ccc} \mathfrak{D}_K & \subset & \mathfrak{D}_{K'} \\ \cap & & \cap \\ \mathfrak{D}^m_K & & \mathfrak{D}^m_{K'} \end{array}$$

and T^m_K induces on $\mathfrak{D}_{K'}$ the linear form $T_{K'}$ and on $\mathfrak{D}_K$ the linear form T_K (since T_K is the restriction of $T_{K'}$ to $\mathfrak{D}_K$).

Consequently, $T^m_{K'}$ induces on $\mathfrak{D}^m_K$ a linear form which, in turn, induces on $\mathfrak{D}_K$ the form T_K and which consequently cannot be the continuous extension of T_K to T^m_K. Thus, T^m_K is the restriction of $T^m_{K'}$ to $\mathfrak{D}^m_K$.

Consequently, if $f \in \mathfrak{D}^m_K$ and $K \subset K'$, we have

$$\int f \cdot T^m_K = \int f \cdot T^m_{K'} .$$

From this, we conclude that, if K and K' are two closed balls containing the support of f, we have

$$\int f \cdot T^m_K = \int f \cdot T^m_{K'} .$$

To see this, note that, if K'' is a closed ball containing K and K', we have

$$\int f \cdot T^m_{K''} = \begin{cases} \int f \cdot T^m_K \\ \int f \cdot T^m_{K'} . \end{cases}$$

Now that we have defined the mapping T^m, let us show that it is linear:

(i) $$\int (\lambda f) \cdot T^m = \lambda \int f \cdot T^m,$$

because, if K is a closed ball containing the support of f, we have

$$\int (\lambda f) \cdot T^m = \int (\lambda f) \cdot T^m_K = \lambda \int f \cdot T^m_K = \lambda \int f \cdot T^m .$$

(ii) $$\int (f+g) \cdot T^m = \int f \cdot T^m + \int g \cdot T^m .$$

This is true because, if K is a closed ball containing the supports of f and g, we have

$$\begin{aligned} \int (f+g) \cdot T^m &= \int (f+g) \cdot T^m_K \\ &= \int f \cdot T^m_K + \int g \cdot T^m_K \\ &= \int f \cdot T^m + \int g \cdot T^m . \end{aligned}$$

It remains to verify that T^m is continuous on $\mathfrak{D}^m$, which is obvious since the restriction T^m_K of T^m to $\mathfrak{D}^m_K$ is continuous in $\mathfrak{D}^m_K$.

Corollary. *Let T denote a distribution. Suppose that, for every compact set K, the restriction of T to $\mathfrak{D}_K$ is continuous for the topology of uniform convergence. Then, T can be extended (uniquely) to become a measure.*

For convenience, we shall say that T is a measure.

Remark. To show that a distribution T can be extended as a distribution of order m, it will be sufficient to verify the condition of Theorem 2 by taking, for example, for K a family of balls of sufficiently great radius with center at 0.

Definition. A distribution T is said to be positive if, for every positive function f belonging to $\mathfrak{D}$, we have $\int f \cdot T \geqslant 0$.

Theorem 3. *Every positive distribution is a positive measure.*

Proof. Let K denote a closed ball. Let ψ denote a function satisfying the following properties: $\psi \geqslant 0$, $\psi \in \mathfrak{D}$, and $\psi(x) = 1$ for $x \in K$.

Now let f denote a number of $\mathfrak{D}_K$. We have

$$-\|f\|_u \cdot \psi \leqslant f \leqslant \|f\|_u \cdot \psi,$$

so that

$$-\|f\|_u \int \psi \,.\, T \leqslant \int f \,.\, T \leqslant \|f\|_u \int \psi \,.\, T,$$

that is,

$$\left|\int f \cdot T\right| \leqslant \|f\|_u \cdot \int \psi \cdot T,$$

which, in accordance with the corollary to Theorem 2, ensures that T is a measure.

Let f denote a nonnegative member of $\mathfrak{D}^0$. Suppose that the support of f is contained in a ball A and let $K = \lambda A$, where $\lambda > 1$. Let us apply Proposition 2. We have $f * \rho_k \geqslant 0$ and $f * \rho_k \in \mathfrak{D}_K$ for sufficiently large k. Therefore,

$$\int f \cdot T = \lim_{k \to +\infty} \int f * \rho_k \cdot T \geqslant 0 .$$

Consequently, T is a positive measure. This completes the proof.

The support of a distribution. The extension of a distribution. Let A denote a closed set. We say that a distribution T has support in A if $\int f \cdot T = 0$ for every function $f \in \mathfrak{D}$ with support in $\complement A$. It is possible to show (but we shall not do it here) that there exists a smallest closed set A with this property. It is called the ***support*** of T. Actually, the concept "with support in" is sufficient for numerous theoretical results and is often the easier to handle than "support" itself. It frequently happens that we know that a distribution has its support in a closed set A without knowing just what that support is. We know also that, if T_1 and T_2 are two distributions with supports in A, then every linear combination of T_1 and T_2 has its support in A.

On the other hand, for a particular given distribution T, it is usually easy to determine its support. For example, we can easily verify that the support of each of the distributions $\delta_a, \delta'_a, \ldots, \delta_a^{(p)}$ is $\{a\}$.

Remark. For a measure μ, we have $\int f \cdot \mu = 0$ whenever f vanishes on the support of μ. This property does not carry over to distributions. For example, we can have

$$\int f \cdot \delta'_a (= -f'(a)) = 0$$

even if f vanishes on the support of δ'_a, that is, at a. On the other hand, if f vanishes on a neighborhood* of the support of T, it will follow that $\int f \cdot T = 0$. This property follows immediately from the definition. *A fortiori*, if T has support in A, we have $\int f \cdot T = 0$ whenever f vanishes in a neighborhood of A.

We shall use the preceding concepts to extend the domain of definition of a distribution to certain infinitely differentiable functions with noncompact support.

Definition. Let T denote a distribution with support in A and let f denote an infinitely differentiable function with support in B. If $A \cap B$ is bounded, we define

$$\int f \cdot T = \int f\psi \cdot T,$$

where ψ is a function in $\mathfrak{D}$, that is, equal to 1 in a neighborhood of $A \cap B$. To make this definition legitimate, we must naturally verify that it is independent of A, B, and ψ. Let us show, for example, that it is independent of ψ.

Let ψ_1 and ψ_2 denote two functions that are equal to 1 in an open set C containing $A \cap B$ (which we may suppose to be the same for ψ_1 and ψ_2). The function $f(\psi_1 - \psi_2)$ has support in $B \cap \complement C$ and hence in $\complement A$. Therefore,

$$\int f(\psi_1 - \psi_2) \cdot T = 0 .$$

We can easily verify that T can be extended as a linear form on the space of infinitely differentiable functions, the intersection of which with the support of T is a bounded set.

Examples. I. Let us denote by $\mathcal{E}$ the vector space of infinitely differentiable complex functions with arbitrary support and let us denote by $\mathcal{E}'$, the space of distributions with compact support. For T to belong to $\mathcal{E}'$, it is necessary and sufficient that there exist a compact set K such that the support of T is in K.

For $f \in \mathcal{E}$ and $T \in \mathcal{E}'$, we can define $\int f \cdot T$. In this case, the definition of an extension becomes

$$\int f \cdot T = \int f\psi \cdot T,$$

where ψ is a function of $\mathfrak{D}$ that is equal to 1 in a neighborhood of the support of T (or in a neighborhood of K if we know only that

*By a neighborhood of a set E is meant any set containing an open set containing E.

the support of T is in K). Every distribution with compact support can thus be extended as a linear form over $\mathcal{E}$.

Similarly, one can show that every distribution of order m with compact support can be extended as a linear form over the vector space $\mathcal{E}^m$ of m-differentiable complex functions with arbitrary support. We denote by $(\mathcal{E}^m)'$ the space of distributions of order m with compact support.

For example, $\delta_a^{(m)} \in (\mathcal{E}^m)'$, and we have

$$\int f \cdot \delta_a^{(m)} = (-1)^m f^{(m)}(a), \qquad \forall f \in \mathcal{E}^m.$$

II. A set A contained in $\mathbf{R}^n$ is said to be bounded above (resp. below) if the set of components of all $x \in A$ is bounded above (resp. below).

Suppose that $f \in \mathcal{E}$ and $T \in \mathcal{D}'$. We can then define $\int f \cdot T$ if the support of T is bounded below and the support of f is bounded above (or vice versa). Every distribution T with support bounded below can thus be extended as a linear form on the space of infinitely differentiable functions with support bounded above.

We denote by $\mathcal{D}'_+$ the vector space of distributions with support in $\mathbf{R}^n_+$ (we shall again say "with positive support"). If f is a member of $\mathcal{E}$ with support bounded above and if $T \in \mathcal{D}'_+$, then $\int f \cdot T$ is defined.

Topology in $\mathcal{D}'$. We shall equip $\mathcal{D}'$ with the topology of simple convergence on $\mathcal{D}$. According to the definition of this topology:

A necessary and sufficient condition for a sequence $\{T_i\}$ of distributions to converge to 0 in $\mathcal{D}'$ is that the sequence $\left\{\int f \cdot T_i\right\}$ converge to 0 for every $f \in \mathcal{D}$. Having defined a topology in $\mathcal{D}'$, we can consider series whose terms are distributions. We shall say that a series $\sum_i T_i$ *converges commutatively* to the sum S (which is a distribution) if, for every $f \in \mathcal{D}$, we have

$$\sum_i \int f \cdot T_i = \int f \cdot S,$$

where the series on the left is absolutely convergent.

For example, in $\mathbf{R}$, we have considered the distribution T defined by

$$\int f \cdot T = \sum_{i=0}^{+\infty} f^{(i)}(i).$$

We may write

$$T = \sum_{i=0}^{+\infty} (-1)^i \delta_i^{(i)}.$$

3. DIFFERENTIATION OF DISTRIBUTIONS

We shall now give a definition of distributions that will generalize the definition in the case of differentiable functions.

Let us consider the space **R**. Let φ denote a function having a continuous derivative. For every $f \in \mathfrak{D}$, we have

$$\int f \cdot \varphi' = [f\varphi]_{-\infty}^{+\infty} - \int f' \cdot \varphi = - \int f' \cdot \varphi .$$

This formula suggests the following definition:

Definition. The *derivative* of a distribution T is the distribution T' (alternately denoted by DT) defined by

$$\int f \cdot DT = \int f \cdot T' = - \int f' \cdot T \qquad (\forall f \in \mathfrak{D}) .$$

Thus, every distribution T on **R** has derivatives $D^p\, T = T^{(p)}$ of arbitrary order. We have

$$\int f \cdot D^p\, T = \int f \cdot T^{(p)} = (-1)^p \int f^{(p)} \cdot T.$$

Example. In **R**, the distribution δ'_a is indeed the derivative of δ_a since

$$\int f \cdot \delta'_a = -f'(a) = - \int f' \cdot \delta_a .$$

Similarly, the distribution that we have denoted by $\delta_a^{(p)}$ is the pth derivative of the distribution δ_a.

Similarly, in $\mathbf{R}^n$, we obtain a natural generalization of differentiation of functions by defining

$$\int f \cdot D_i\, T = - \int D_i\, f \cdot T ,$$

where D_i denotes differentiation with respect to the ith variable.

Every distribution T on $\mathbf{R}^n$ has derivatives $D_1^{p_1} \dots D_n^{p_n}\, T$ of all orders. We have

$$\int f \cdot D_1^{p_1} \dots D_n^{p_n}\, T = (-1)^{p_1 + \cdots + p_n} \int D_1^{p_1} \dots D_n^{p_n} f \cdot T .$$

We point out that one may reverse the order of the differentiations. For example, we have

$$\int f \cdot D_1^{p_1} \dots D_n^{p_n}\, \delta_a = (-1)^{p_1 + \cdots + p_n} (D_1^{p_1} \dots D_n^{p_n} f)(a) .$$

One can verify immediately that, if T is a distribution of order m, the above formulas enable us to define the pth derivatives of T as distributions of order $m + p$. In particular, on **R**, the first derivative of a measure is a distribution of order 1.

Finally, we note that if T has support in A, all its derivatives have support in A.

Let φ denote a function having only discontinuities of the first kind and having continuous derivative φ' in the usual sense everywhere except at these points of discontinuity. The theorem that follows shows that the derivative in the distribution sense is different from φ'. Therefore, we shall keep the notation $D\varphi$ for the derivative in the distribution sense.

Theorem 4. *If φ has discontinuities of the first kind at points a_i with salti σ_i and a continuous derivative φ' in the usual sense everywhere except at the points a_i, then its derivative in the distribution sense is*

$$D\varphi = \varphi' + \sum_i \sigma_i \delta_{a_i}.$$

Proof. For simplicity, we shall take the case of a single discontinuity at a point a, where there is a saltus σ. We have

$$\begin{aligned}\int f \,.\, D\varphi &= -\int f' \,.\, \varphi \\ &= -\int_{-\infty}^{a} f' \varphi - \int_{a}^{+\infty} f' \,.\, \varphi \\ &= -[f\varphi]_{-\infty}^{a} + \int_{-\infty}^{a} f \,.\, \varphi' - [f\varphi]_a^{+\infty} + \int_a^{+\infty} f \,.\, \varphi' \\ &= -f(a)\,\varphi(a-0) + f(a)\,\varphi(a+0) + \int_{-\infty}^{+\infty} f \,.\, \varphi' \\ &= f(a)\sigma + \int_{-\infty}^{+\infty} f \,.\, \varphi' \\ &= \int f \,.\, (\varphi' + \sigma\delta_a),\end{aligned}$$

which completes the proof.

Example. Heaviside's function $\mathcal{Y}$ defined by

$$\mathcal{Y}(x) = \begin{cases} 1 & \text{if} \quad x \geqslant 0 \\ 0 & \text{if} \quad x < 0 \end{cases}$$

has a derivative in the distribution sense $D\mathcal{Y} = \delta$.

An application. If $\varphi, \varphi', \varphi'', \ldots$ have discontinuities of the first kind at the origin with salti $\sigma_0, \sigma_1, \sigma_2, \ldots$, we have

$$\begin{aligned} D\varphi &= \varphi' + \sigma_0\,\delta\,, \\ D^2\varphi &= \varphi'' + \sigma_1\,\delta + \sigma_0\,\delta'\,, \\ D^3\varphi &= \varphi''' + \sigma_2\,\delta + \sigma_1\,\delta' + \sigma_0\,\delta'', \ldots \end{aligned}$$

Definition. A *cumulative* (or *primitive*) function of a measure μ on $\mathbf{R}$ is defined as any function F such that

$$a \leqslant b \Rightarrow \int_{[a,b[} \mu = F(b) - F(a).$$

From this definition, we get

$$F(x) = \begin{cases} F(a) + \int_{[a,x[} \mu, & \text{if } x \geqslant a, \\ F(a) - \int_{[x,a[} \mu, & \text{if } x \leqslant a. \end{cases}$$

A cumulative function of a measure is therefore determined up to a constant.

If μ is a bounded measure, we choose in general

$$F(x) = \int_{]-\infty,x[} \mu .$$

Theorem 5. *Every measure μ is the derivative of its cumulative function.*

Proof. Let us take $a = 0$ and $F(a) = 0$. It will be sufficient to show that, for every $f \in \mathfrak{D}$, we have $\int f \cdot \mu = -\int f' \cdot F$. Let Δ denote the set of all (x, y) such that either $0 \leqslant y < x$ or $x \leqslant y < 0$ (see Fig. 4). Let Δ_x and Δ_y denote the subsets of Δ adjacent to the origin and intercepted by the vertical segment of abscissa x and the horizontal ray of ordinate y respectively. We have

$$\begin{aligned} \int f' . F &= \int_{-\infty}^{+\infty} \left[f'(x) \operatorname{sgn}(x) \int_{\Delta_x} \mu^y \right] dx \\ &= \int \left[\int_{\Delta_y} \operatorname{sgn}(x) f'(x)\, dx \right] \mu^y . \end{aligned}$$

Now,

$$\int_{\Delta_y} \operatorname{sgn}(x) f'(x)\, dx = \left. \begin{cases} f(+\infty) - f(y) & \text{for } y \geqslant 0 \\ -[f(y) - f(-\infty)] & \text{for } y \leqslant 0 \end{cases} \right\} = -f(y),$$

so that

$$\int f' . F = -\int f(y) . \mu^y ,$$

which completes the proof.

Remark. According to Theorem 5, the integral $\int f \cdot \mu$ of a function of $\mathfrak{D}^0$ with respect to a measure μ can be written $\int f \cdot DF$, where F is a cumulative function of μ. In general, one finds in the literature the notation $\int f \cdot dF$ (Stieltjes integral). We point out that it can be shown that every function of bounded variation has a derivative that is a measure. This result enables us to interpret the theory of Stieltjes integrals in terms of the concepts of measure and distribution.

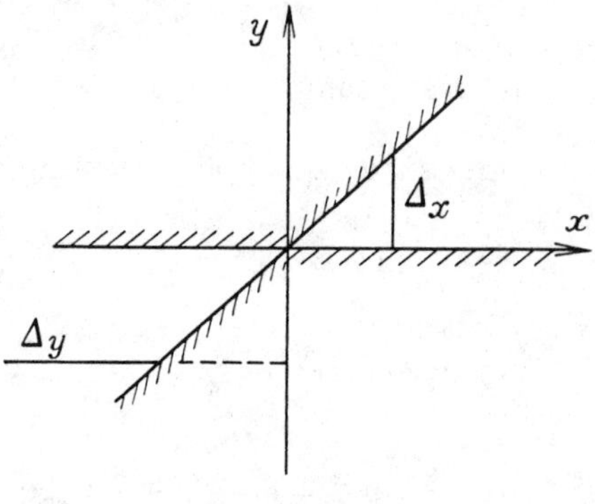

FIG. 4.

The differentiation of measures enables us to give the following theorem regarding the local structure of distributions:

Theorem 6. *Let T denote a distribution on* **R**. *For every compact set K, there exists a measure μ with support in K and a number m such that*

$$f \in \mathfrak{D}_K \Rightarrow \int f \,.\, T = \int f \,.\, D^m \mu$$

$$\left(= (-1)^m \int f^m \,.\, \mu \right).$$

We shall say that, on $\mathfrak{D}_K$, the distribution T can be represented by the derivative of some order of a measure.

In general, the number m depends on the compact set in question.

Proof. For any compact interval K, let us denote by T_K the restriction of T to $\mathfrak{D}_K$. From the definition of a distribution, T_K is a continuous linear form on $\mathfrak{D}_K$ the value $\int f \cdot T$ of which corresponding to a function $f \in \mathfrak{D}_K$ we denote by $T_K(f)$.

Since T_K is continuous, there exists a neighborhood V of 0 in $\mathfrak{D}_K$ such that

$$f \in V \quad \Rightarrow \quad |T_K(f)| \leqslant 1 .$$

Now, the norms $\{f \to \|f^{(p)}\|_u\}$ constitute a sequence arranged in increasing order of fineness. Consequently, the sets defined by $\{\|f^{(p)}\|_u \leqslant \eta\}$ constitute, for integral p and positive η, a basis of neighborhoods of $\mathfrak{D}_K$. Consequently, there exist an integer m and a positive number η such that

$$\|f^{(m)}\|_u \leqslant \eta \quad \Rightarrow \quad |T_K(f)| \leqslant 1 .$$

In other words, T_K is continuous for the topology induced on $\mathfrak{D}_K$ by $\mathfrak{D}_K^m$ and hence can be extended as a continuous linear form T_K^m onto $\mathfrak{D}_K^m$. For $f \in \mathfrak{D}_K$, we have

$$\int f \, . \, T = T_K^m(f) .$$

Now, let us consider the mapping $D^m\,(f \to f^{(m)})$ of $\mathfrak{D}_K^m$ onto $\mathfrak{D}_K^0$. Let $\mathfrak{M}$ denote the range of this mapping, that is, the set of $f^{(m)}$ for $f \in \mathfrak{D}_K^m$. Since the topology of $\mathfrak{D}_K^m$ is defined by the norm $f \to \|f^{(m)}\|_u$, D^m is a continuous one-to-one mapping of $\mathfrak{D}_K^m$ into $\mathfrak{D}_K^0$ and has a continuous one-to-one inverse (defined as its range $\mathfrak{M}$). (A necessary and sufficient condition for a sequence $\{f_i\}$ of functions in $\mathfrak{D}_K^m$ to converge to 0 is that the sequence $f_i^{(m)}$ converge to 0 in $\mathfrak{D}_K$.) Consequently, there exists a linear form L_1 on $\mathfrak{M}$ such that

$$T_K^m(f) = L_1(f^{(m)}) , \qquad \forall f \in \mathfrak{D}_K^m .$$

According to the Hahn-Banach theorem, L_1 can be extended as a continuous linear form L_2 onto $\mathfrak{D}_K^0$. We have again

$$T_K^m(f) = L_2(f^{(m)}) , \qquad \forall f \in \mathfrak{D}_K^m .$$

If we consider $\mathfrak{D}_K^0$ as a closed vector subspace of the space $\mathbf{L}_\infty$ of continuous functions that approach 0 as their argument becomes infinite and if we again apply the Hahn-Banach theorem, we can extend L_2 as a linear form on $\mathbf{L}_\infty$, that is, as a bounded measure μ. We then have

$$T_K^m(f) = \int f^{(m)} \, . \, \mu , \qquad \forall f \in \mathfrak{D}_K^m .$$

If we replace μ with $\varepsilon_K \mu$ (where ε_K is the characteristic function of K), we can choose μ with support in K. We have finally

$$\int f \, . \, T = T_K^m(f) = \int f^{(m)} \, . \, \mu = (-1)^m \int f \, . \, D^m \mu, \qquad \forall f \in \mathfrak{D} ,$$

which completes the proof.

Remark. The subspace $\mathfrak{M}$ of $\mathfrak{D}_K^0$ is defined thus: $g \in \mathfrak{D}_K^0$ if and only if

$$\int g = \int g' = \dots = \int g^{(m-1)} = 0 .$$

Thus, it is of finite codimension. Therefore, the extension of L_1 as a continuous linear form L_2 requires only the elementary formulation of the Hahn-Banach theorem. Furthermore if $K = [a, b]$, the form L_2 can be extended as a continuous linear form L_3 onto the subspace $\mathfrak{N}$ (of $\mathbf{L}_\infty$) consisting of functions φ that vanish at $x = 0$ and $x = b$ by setting

$$L_3(\varphi) = L_2(\varepsilon_K \varphi) .$$

The extension of L_3 as a continuous linear form on $\mathbf{L}_\infty$ (that is, as a measure) requires only the elementary formulation of the Hahn-Banach theorem since $\mathfrak{N}$ is of codimension 2.

Remarks. I. In $\mathbf{R}^n$, we can show that, for every compact set K, every distribution T coincides on $\mathfrak{D}_K$ with some sum of derivatives of measures.

II. We can, by starting with the preceding theorem, show that every distribution with compact support is of finite order.

4. THE DIRECT PRODUCT. THE MULTIPLICATIVE PRODUCT

We shall denote by $\mathfrak{D}(E)$ the space of infinitely differentiable functions with compact support in the vector space E and we shall denote by $\mathfrak{D}'(E)$ the space of distributions on E.

If f is a function of two variables, the notation $\int f(x, y) \cdot T^x$ will denote the value of the integral of the mapping $x \to f(x,y)$ with respect to the distribution T.

Proposition 3. *If $f \in \mathfrak{D}(E \times F)$ and $T \in \mathfrak{D}'(E)$, the function*

$$\varphi : \ y \to \int f(x, y) \cdot T^x$$

belongs to $\mathfrak{D}(F)$ and it has the derivative

$$y \to \int f_y'(x, y) \cdot T^x .$$

Proof. One can show immediately that if K and H are two compact subsets of E and F respectively such that f has compact support in $K \times H$, then φ has support in H. Furthermore,

$$\frac{\varphi(b+k)-\varphi(b)}{k} = \int \frac{f(x, b+k) - f(x, b)}{k} \cdot T^x .$$

The function

$$x \to \frac{f(x, b+k) - f(x, b)}{k}$$

tends in $\mathfrak{D}_K$ to the function

$$x \to f'_y(x, b)$$

as k approaches 0. Therefore,

$$\lim_{k\to 0} \frac{\varphi(b+k)-\varphi(b)}{k} = \int f'_y(x, b) \, . \, T .$$

Thus, we have

$$\frac{d}{dy}\int f(x, y) \, . \, T^x = \int f'_y(x, y) \, . \, T^x .$$

By successive application of this rule, we see that the function

$$y \to \int f(x, y) \, . \, T^x$$

is infinitely differentiable.

Theorem 7. *Suppose that*

$$f \in \mathfrak{D}(E \times F), \; S \in \mathfrak{D}'(E), \; T \in \mathfrak{D}(F).$$

Then,

$$(\alpha) \qquad \int \left[\int f(x, y) \, . \, S^x\right] . \, T^y = \int \left[\int f(x, y) \, . \, T^y\right] S^x .$$

Proof. We shall confine ourselves to the case $E = F = \mathbf{R}$. Let K and H denote two compact subsets of $\mathbf{R}$ such that f has support in $K \times H$. Let μ and ν denote two measures such that

$$\begin{cases} S = D^p \mu & \text{on} \quad \mathfrak{D}_K , \\ T = D^q \nu & \text{on} \quad \mathfrak{D}_H . \end{cases}$$

Then

$$\int \left[\int f(x, y) \, S^x \right] T^y = \int \left[\int f(x, y) \, . \, D^p \mu^x \right] . \, D^q \nu^y$$

$$= (-1)^p \int \left[\int D_x^p f \, . \, \mu^x \right] . \, D^q \nu^y$$

$$= (-1)^{p+q} \int [D_y^q D_x^p f \, . \, \mu^x] \, . \, \nu^y$$

$$= (-1)^{p+q} \int\int D_y^q D_x^p f \, . \, \mu^x \nu^y .$$

According to Fubini's theorem, we find the same expression for the second member, which completes the proof of the theorem.

Notation. We denote by $\int f(x, y) \cdot S^x T^y$ the common value of the two sides of (α). One can easily verify that $S^x T^y$ is a distribution (i.e., is a linear mapping satisfying the continuity condition (CC) of distributions). The mapping $S^x, T^y \to S^x T^y$ is thus a bilinear mapping of $\mathfrak{D}'(E) \times \mathfrak{D}'(F)$ into $\mathfrak{D}'(E \times F)$. The distribution $S^x T^y$ is called the *direct product* of S^x and T^y. One can also verify that, if S has support in K and if T has support in H, then $S^x T^y$ has support in $K \times H$. More precisely, one can show that the support of $S^x T^y$ is the product of the supports of S and T.

The multiplicative product. Definition. Suppose that φ belongs to $\mathcal{E}$ (that is, is infinitely differentiable) and that $T \in \mathfrak{D}'$. We define φT by

$$\int f \, . \, \varphi T = \int f\varphi \, . \, T, \qquad (\forall f \in \mathfrak{D}) .$$

To make this definition legitimate, we need to note that the right-hand member is defined for all $f \in \mathfrak{D}$ and that condition (CC) is satisfied.

One can generalize the definition to the case in which $\varphi \in \mathcal{E}^m$ (that is, is m times continuously differentiable) and $T \in (\mathfrak{D}^m)'$.

Theorem 8. *If T is a distribution on $\mathbf{R}$, then* $(\varphi T)' = \varphi' T + \varphi T'$.

Proof. We need to verify that, for every $f \in \mathfrak{D}$,

$$\int f \, . \, (\varphi T)' = \int f \, . \, \varphi' T + \int f \, . \, \varphi T',$$

that is, that

$$- \int f' \, . \, \varphi T = \int f\varphi' \, . \, T - \int (f\varphi)' \, . \, T,$$

or

$$\int (f\varphi)' \, . \, T = \int (f' \varphi + f\varphi') \, . \, T \, ,$$

which follows from the rule for differentiating the product of two functions.

Remark. For a distribution T on $\mathbf{R}^n$, we can show in the same way that

$$D_i(\varphi T) = (D_i \, \varphi) \, T + \varphi(D_i \, T) \, ,$$

where D_i denotes differentiation with respect to the ith variable.

5. INVOLUTION AND CONVOLUTION

Involutions. In the space of complex functions, there exist two common involutions $f \to \bar{f}$ and $f \to \tilde{f}$ defined by

$$\bar{f}(x) = \overline{f(x)}, \quad \tilde{f}(x) = \overline{f(-x)} \, .$$

Similarly, we shall define two involutions $T \to \bar{T}$ and $T \to \tilde{T}$ in $\mathfrak{D}'$ by

$$\int f \, . \, \bar{T} = \overline{\int \bar{f} \, . \, T}, \quad \int f \, . \, \tilde{T} = \overline{\int \tilde{f} \, . \, T}, \qquad (\forall f \in \mathfrak{D}) \, .$$

If T is a locally integrable function, these definitions coincide with those given for functions.

Also, we define

$$\check{f} = \bar{\tilde{f}} \; (\check{f}(x) = f(-x)) \quad \text{and} \quad \check{T} = \bar{\tilde{T}}.$$

We have

$$\int \check{f} \, . \, T = \int f \, . \, \check{T} \, .$$

It should be noted that the mapping $T \to \check{T}$ is not an involution since it is linear. We shall use it only to simplify certain notations.

If T has support in A, then $\bar{T}$ has support in A and $\tilde{T}$ and $\check{T}$ have support in the negative, $-A$ of A.

The convolution product. The convolution product of two distributions S and T will be defined for certain pairs S, T. In this chapter, we shall confine ourselves to modifying the definition of the convolution product of two measures in such a way as to get a definition for the convolution product of two distributions S and T with certain restrictions on the support of these two distributions.

In the following chapter, we shall define the convolution product under other conditions.

Definition. The convolution product $S * T$ of two distributions S and T on the same space $\mathbf{R}^n$ is defined by

$$\int f \,.\, S * T = \int f(x+y) \,.\, S^x T^y,$$

with the hypothesis that

(a) the right-hand member is defined for all $f \in \mathfrak{D}(\mathbf{R}^n)$,
(b) the continuity condition (CC) is satisfied.

Let A, B, and K denote three sets such that S, T, and f have supports respectively in A, B, and K. Let us define

$$\hat{f}(x, y) = f(x+y).$$

The function $\hat{f}$ is infinitely differentiable in $\mathbf{R}^{2n}$, and its support is in the strip $\hat{K}$ defined by $x + y \in K$. Also, $S^x T^y$ has support in $A \times B$. Consequently, if $(A \times B) \cap \hat{K}$ is bounded, then the integral

$$\int f(x+y) \,.\, S^x T^y = \int \hat{f}(x, y) \,.\, S^x T^y$$

will be defined. If $(A \times B) \cap \hat{K}$ is bounded for every compact set K, then condition (a) of the definition is satisfied. This will be the case, in particular, under the following conditions:

(1) Either A or B is compact (see Fig. 5);
(2) $A = B = \mathbf{R}^n_+$ (see Fig. 6) (or, more generally, A and B are both bounded below or both bounded above).

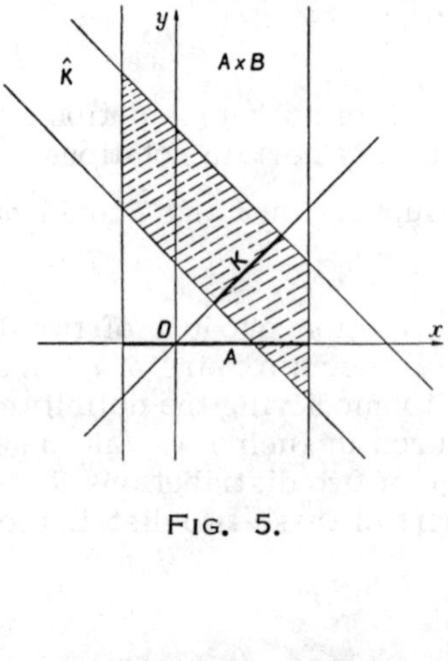

FIG. 5.

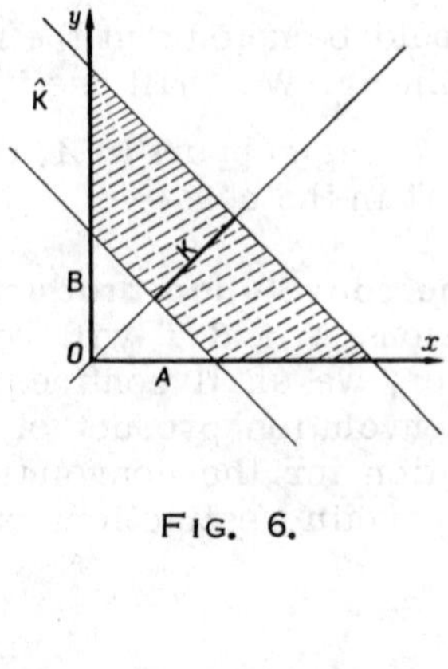

FIG. 6.

In Figs. 5 and 6, the units on the x- and y-axes and on the bisector of the first quadrant were chosen in such a way that the mapping $x, y \to x + y$ is represented by the projection onto the bisector of the first quadrant.

Theorem 9. *Suppose that S and T have supports respectively in A and B. Suppose that, for every compact set K, the intersection of the strip $\hat{K}$ defined by $x + y \in K$ and the set $A \times B$ is a bounded set. Then, the convolution product $S * T$ is defined. The hypothesis may also be worded as follows: For $x \in A$ and $y \in B$, if $x + y$ is bounded, so are x and y bounded.*

Proof. We have already seen that condition (a) is satisfied. Condition (b) is verified as follows: Let K denote a compact set. Let α denote a function in $\mathfrak{D}(\mathbf{R}^{2n})$ that assumes the value 1 in a neighborhood of $(A \times B) \cap \hat{K}$. Let f_i denote a sequence of functions f_i in $\mathfrak{D}_K(\mathbf{R}^n)$. Then, for each, f_i is infinitely differentiable with support in K. We have

$$\int f_i \cdot (S * T) = \int f_i(x + y) \cdot S^x T^y = \int \hat{f}_i(x, y) \cdot S^x T^y$$

$$= \int \alpha(x, y) \hat{f}_i(x, y) \cdot S^x T^y.$$

Now, $\alpha \hat{f}_i \in \mathfrak{D}_{(A \times B) \cap \hat{K}}(\mathbf{R}^{2n})$. If $\{f_i\}$ converges to 0 in $\mathfrak{D}_K(\mathbf{R}^n)$, then $\left\{\alpha \hat{f}_i\right\}$ converges to 0 in $\mathfrak{D}_{(A \times B) \cap \hat{K}}(\mathbf{R}^{2n})$ and hence $\left\{\int f_i(x + y) \cdot S^x T^y\right\}$ converges to 0.

Let us confine ourselves to the two principal cases in which $S * T$ is defined:

(1) Either S or T has compact support;
(2) S and T have positive support; that is, $S, T \in \mathfrak{D}'_+$.

From this definition we immediately get

Theorem 10. *The convolution product is commutative: if $S * T$ is defined, then $T * S$ is also defined and $T * S = S * T$.*

One can easily verify the following theorems:

Theorem 11. *The mapping $S * T$ is bilinear in the following two cases: (1) $S \in \mathcal{E}'$ and $T \in \mathfrak{D}'$ (or $S \in \mathfrak{D}'$ and $T \in \mathcal{E}'$); (2) $S \in \mathfrak{D}'_+$ and $T \in \mathfrak{D}'_+$.*

Theorem 12. *If S or T has compact support* or *if S and T both have supports bounded below, or both have supports bounded above, then*

$$\overline{S * T} = \overline{S} * \overline{T}$$

$$(S * T)^{\sim} = \tilde{S} * \tilde{T},$$

$$(S * T)^{\vee} = \check{S} * \check{T}.$$

To apply the definition of the convolution product, we often need to evaluate $\int f(x + y) \cdot S^x T^y$, where $f \in \mathfrak{D}$. Since the function $x, y \to f(x + y)$ does not have compact support, Theorem 7 is not directly applicable. The following lemma enables us to resolve this difficulty.

Lemma 1. *We have*

$$\int f(x + y) \,.\, S^x T^y = \int \left[\int f(x + y) \,.\, S^x\right] . \, T^y$$

in the following four cases:

(1) $f \in \mathfrak{D}$ *and* S *has compact support;*
(2) $f \in \mathfrak{D}$ *and* T *has compact support;*
(3) $f \in \mathcal{E}$ *and both* S *and* T *have compact support;*
(4) $f \in \mathcal{E}$ *and has support bounded avove (resp. below) and* S *and* T *have support bounded below (resp. above).*

Proof. Case (1). Let K denote the support of f and suppose that S has support in a compact set A. Let α denote a function in $\mathfrak{D}$ that is equal to 1 in a neighborhood of the second canonical projection C of $(A \times \mathbf{R}^n) \cap \hat{K}$. The function $x, y \to \alpha(y) f(x + y)$ belongs to $\mathfrak{D}(\mathbf{R}^{2n})$, and

$$\int f(x + y) \,.\, S^x T^y = \int \alpha(y) f(x + y) \,.\, S^x T^y$$

$$= \int \left[\int \alpha(y) f(x + y) \,.\, S^x\right] . \, T^y.$$

The function

$$y \to \int \alpha(y) f(x + y) \,.\, S^x = \alpha(y) \int f(x + y)\, S^x$$

belongs to $\mathfrak{D}$ (by Proposition 3). Since α can be chosen equal to 1 in a neighborhood of any arbitrary point, we conclude that the function

$$y \to \int f(x + y) \,.\, S^x$$

is infinitely differentiable. Furthermore, this function has support in C. Thus,

$$\int \left[\iint f(x+y) \,.\, S^x\right] . \, T^y = \int \left[\iint f(x+y)\, S^x\right] \alpha(y) \,.\, T^y$$

$$= \int \left[\iint f(x+y)\, \alpha(y) \,.\, S^x\right] . \, T^y .$$

Case (2). The proof is analogous. If T has support in a compact set B, we take $\alpha \in \mathfrak{D}$ equal to 1 in a neighborhood of B.

Case (3). Suppose that S (resp. T) has support in the compact set A (resp. B). Let α (resp. β) denote a function in $\mathfrak{D}$ that vanishes in a neighborhood of A (resp. B). We then have

$$\int f(x+y) \,.\, S^x T^y = \int \alpha(x)\, \beta(y) f(x+y) \,.\, S^x T^y$$

$$= \int \left[\iint \alpha(x)\, \beta(y) f(x+y) \,.\, S^x\right] . \, T^y$$

$$= \int \beta(y) \left[\iint \alpha(x) f(x+y)\, S^x\right] . \, T^y$$

$$= \int \left[\iint f(x+y) \,.\, S^x\right] . \, T^y .$$

Case (4). The proof is analogous to the preceding one. If S and T have support in A and B, where A and B are bounded below, we take an $\alpha \in \mathfrak{E}$ (resp. $\beta \in \mathfrak{E}$) that has support bounded below and that is equal to 1 in a neighborhood of A (resp. B).

Examples. I. Calculate $\delta * S$ for $S \in \mathfrak{D}'$. By definition, we have

$$\int f \,.\, \delta * S = \int f(x+y) \,.\, \delta^x S^y \qquad (\forall f \in \mathfrak{D})$$

$$= \int \left[\iint f(x+y) \,.\, \delta^x\right] . \, S^y$$

$$= \int f(y) \,.\, S^y = \int f \,.\, S ,$$

so that

$$\delta * S = S .$$

Consequently, δ is the neutral element of the convolution.

II. Calculate $\delta' * S$ for $S \in \mathfrak{D}'$. By definition,

$$\int f \,.\, \delta' * S = \int f(x+y) \,.\, (\delta')^x S^y$$

$$= \int \left[\iint f(x+y) \,.\, (\delta')^x\right] . \, S^y$$

$$= \int -f'(y) \,.\, S^y = \int f \,.\, S' ,$$

from which we get the fundamental formula

$$\delta' * S = S' .$$

In particular,

$$\delta' * \delta_a = \delta'_a .$$

III. We know that the function obtained from a function f by a translation of amplitude a is equal to $\delta_a * f$:

$$(\delta_a * f)(x) = f(x - a) .$$

Similarly, we shall call $\delta_a * S$ the translation of amplitude a of the distribution S. We have the following formula:

$$\int f \,.\, S = \int \delta_a * f \,.\, \delta_a * S , \qquad (\forall f \in \mathfrak{D}) ,$$

which expresses the invariance of the integral with respect to the same translation of the function and the distribution:

$$\begin{aligned}\int \delta_a * f \,.\, \delta_a * S &= \int f(x - a) \,.\, \delta_a * S \\ &= \iint f(x + y - a) \,.\, \delta_a^x \,.\, S^y \\ &= \int \left[\int f(x + y - a) \,.\, \delta_a^x \right] \,.\, S^y \\ &= \int f(y) \,.\, S^y = \int f \,.\, S .\end{aligned}$$

Theorem 13. *Suppose that* $S \in \mathfrak{D}'$ *and* $\varphi \in \mathcal{E}$. *Then,**

$$(S * \varphi)(u) = \int \varphi(u - x) \,.\, S^x$$

in the following three cases:

(1) *S has compact support;*
(2) *φ has compact support;*
(3) *S and φ have positive support.*

Proof. Let f denote a member of $\mathfrak{D}$. Then,

---

*More precisely, one should say that the function $u \to \int \varphi(u - x) \cdot S^x$ is a representative of $S * \varphi$.

$$\int f \,.\, (S * \varphi) = \int f(x + y) \,.\, S^x \, \varphi(y) \, dy$$

$$= \int \left[\int f(x + y) \, \varphi(y) \, dy\right] . \, S^x$$

$$= \int \left[\int f(u) \, \varphi(u - x) \, du\right] . \, S^x$$

$$= \int \left[\int \varphi(u + x) \,.\, f(u) \, du\right] . \, \check{S}^x$$

$$= \int \varphi(u + x) \,.\, \check{S}^x f(u) \, du$$

$$= \int \left[\int \varphi(u + x) \, \check{S}^x\right] f(u) \, du \, .$$

The third, fifth, and sixth equations follow from Lemma 1.* Thus,

$$(S * \varphi)(u) = \int \varphi(u + x) \,.\, \check{S}^x = \int \varphi(u - x) \,.\, S^x \, .$$

Theorem 14. *Let* S, T, *and* U *denote three distributions. If either*

(1) S *and* T *have compact support*

or

(2) S, T, *and* U *have positive support,*

then

$$(S * T) * U = S * (T * U).$$

Proof. Let f denote a member of $\mathfrak{D}$. According to Lemma 1,

$$\int f \,.\, (S * T) * U = \int f(u + z) \,.\, (S * T)^u \, U^z$$

$$= \int \left[\int f(u + z) \,.\, (S * T)^u\right] . \, U^z$$

$$= \int \left[\int f(x + y + z) \,.\, S^x \, T^y\right] . \, U^z$$

$$= \int \left[\int f(x + y + z) \,.\, S^x\right] . \, T^y \, U^z$$

$$= \int \left[\int f(x + v) \, S^x\right] . \, (T * U)^v$$

$$= \int f \,.\, S * (T * U) \, .$$

---

*The introduction of $\check{S}$ serves only to facilitate application of Lemma 1.

Example. If T has compact support or even if T and U have positive support, then

$$\delta' * (T * U) = (\delta' * T) * U ;$$

that is,

$$(T * U)' = T' * U ,$$

Similarly,

$$T * (\delta' * U) = (T * \delta') * U ;$$

that is,

$$T * U' = T' * U ,$$

so that we finally have

$$(T * U)' = T' * U = T * U' .$$

This result can be expressed by the following rule. To differentiate a convolution product, differentiate one or the other of the factors.

Notation. Let T denote a distribution and let f denote a function. We define

$$\langle f, T \rangle = \int \bar{f} \, . \, T$$

whenever the right-hand member is defined.

Theorem 15. *If either*

(1) *two of the three elements f, S, and T have compact support*

or

(2) *S and T have support bounded below and f has support bounded above,*

then

$$\langle f, S * T \rangle = \langle \tilde{S} * f, T \rangle .$$

Proof. We have

$$\langle f, S * T \rangle = \int \bar{f}(x+y) \,.\, S^x T^y = \int \left[\int \bar{f}(x+y) \,.\, S^x \right] . \, T^y$$

$$= \int \left[\int \bar{f}(y-x) \,.\, \check{S}^x \right] T^y$$

$$= \int (\bar{f} * \check{S}) \,.\, T$$

$$= \langle f * \tilde{S}, T \rangle .$$

Theorem 14 is a theorem on the associativity of the convolution product. In the following chapter, this associativity will be proven in other cases. In any case, since the convolution product is not always defined, when using the formula

$$(S * T) * U = S * (T * U)$$

one should check whether one or the other of the hypotheses under which it has been proven is verified. In practice, we shall need to consider "distribution spaces" (that is, vector subspaces of $\mathfrak{D}'$) on which the convolution product is always defined and induces an algebraic structure. We shall then say that such a space of distributions is a ***convolution algebra.*** We have two common examples: the space $\mathcal{E}'$ of distributions with compact support and the space $\mathfrak{D}'_+$ of distributions with positive support. Also, for a distribution space $\mathcal{H}$, we shall assign a distribution space C such that, for $S \in C$ and $U \in \mathcal{H}$, we have $S * U \in \mathcal{H}$. We shall then verify the following associativity property: If $S, T \in C$, then

$$S * T \in C \quad \text{and} \quad (S * T) * U = S * (T * U) .$$

On C the convolution product is always defined and C is a convolution algebra. The elements of C are called ***convolvers*** on $\mathcal{H}$, and C is called the algebra of convolvers on $\mathcal{H}$. For example, $\mathcal{E}'$ is the algebra of convolvers on $\mathfrak{D}'$. To apply the formula $(S * T) * U = S * (T * U)$, it suffices to verify that one or the other of the following two situations obtains:

(a) S, T, and U belong to a single convolution algebra;
(b) U belongs to a distribution space and S and T belong to the algebra of the convolvers of that space.

Let us consider a convolution algebra $\mathcal{A}$ containing δ. Since the distribution δ is the neutral element for the convolution, we can apply to $\mathcal{A}$ the general results regarding commutative algebras with a unit element. In particular,

Let A denote an element of $\mathcal{A}$. If there exists an A_{-1} such that $A * A_{-1} = \delta$, then A_{-1} is unique. The mapping A_{-1} is called the ***convolution inverse*** of A, and A is said to have an inverse in the convolution algebra $\mathcal{A}$.

If A has an inverse in $\mathcal{A}$, then the equation $A * X = B$ has, for $B \in \mathcal{A}$, a unique solution $X = A_{-1} * B$.

The convolution algebra $\mathfrak{D}'_+$ also has the following property, which we state without proof.

Theorem 16. *The convolution algebra $\mathfrak{D}'_+$ is an algebra without divisors of 0; that is, the equation $U * V = 0$ implies that either $U = 0$ or $V = 0$.*

From this we derive the following property:

In $\mathfrak{D}'_+$,

(1) *if the equation $A * X = B$ has a solution, this solution is unique (even if A does not have an inverse),*

(2) *every equation $U = V$ is equivalent to any equation obtained by convolving both members with a nonzero distribution.*

Example. We have $\mathcal{Y} \in \mathfrak{D}'_+$ and $\delta' * \mathcal{Y} = \delta$. Therefore, $\mathcal{Y}$ and δ' are convolution inverses of each other.

For $U, V \in \mathfrak{D}'_+$, the following relations are therefore equivalent:

$$U = DV = \delta' * V \quad \text{and} \quad V = \mathcal{Y} * U.$$

6. FURTHER REMARKS

Continuity of operations on distributions. 1. *Differentiation is a continuous mapping.* For example, let $\{T_i\}$ denote a sequence of distributions in $\mathbf{R}$. For every $f \in \mathfrak{D}$, we have

$$\int f \,.\, T_i' = - \int f' \,.\, T_i$$

Consequently, if $T_i \to 0$, we have $\int f' \cdot T_i \to 0$ and hence $T_i' \to 0$.

Since we have equipped $\mathfrak{D}$ with the topology of simple convergence on $\mathfrak{D}$, a topology that is not metrizable, this is not sufficient to prove the continuity of the mapping $T \to T'$. Therefore, to complete the preceding proof one goes back to the general definition of continuity. We leave this to the reader, who can get the idea for the procedure from the proof made in an analogous case in Chapter 3 (Theorem 12).

2. Multiplication $T \to \varphi T$ by a function $\varphi \in \mathcal{E}$ is a continuous mapping of $\mathfrak{D}'$ into $\mathfrak{D}'$. The proof, which is easy, is left to the reader.

3. We state without proof that, for a distribution U with compact support, the mapping $T \to U * T$ is a continuous mapping of $\mathfrak{D}'$ into $\mathfrak{D}'$.

Vector distributions. Let E denote a Banach space. Let T denote a linear mapping of $\mathfrak{D}$ into E such that, for every compact set K, the restriction of T to $\mathfrak{D}_K$ is a continuous linear mapping of $\mathfrak{D}_K$ into E. We shall refer to such a linear mapping T as a distribution on $\mathbf{R}^n$ with range in E.

We can continue to use the notation $\int f \cdot T$ to denote the value of the mapping T at the point $f \in \mathfrak{D}$. If $E = \mathbf{R}^q$, we set

$$\int f . T = \begin{bmatrix} \int f . T^1 \\ \cdots\cdots \\ \int f . T^q \end{bmatrix},$$

where the mappings T^i are scalar distributions. More generally, if E is a Banach space and if x' is an element of the dual E' of E, the linear form $f \to \langle \int f . T, x' \rangle$ is a scalar distribution.

We can define differentiation of a vector distribution in the same way as for scalar distribution; for example, in $\mathbf{R}$, we define it by the formula

$$\int \varphi . T' = - \int \varphi' . T, \qquad \forall \varphi \in \mathfrak{D}.$$

The multiplication product is generalized without difficulty. On the other hand, the convolution product is more difficult to defne and will be introduced only in one particular case (Chapter 7, section 5).

3

The Fourier Transform

The theory of the Fourier transform will be developed on **R**. We shall then indicate certain results that remain valid on $\mathbf{R}^n$ and that we shall need in the following chapters.

1. FURTHER REMARKS REGARDING MEASURES

We denote by **M** the set of bounded measures on **R**. The set of bounded positive measures will be denoted by $\mathbf{M}_+$. Every measure $\mu \in \mathbf{M}$ defines a continuous linear form on the space of continuous bounded functions f and, in particular, on the space $\mathbf{L}_\infty$ of continuous functions that approach 0 as their argument becomes infinite.

Thus, we may consider any $\mu \in \mathbf{M}$ as an element of the dual of $\mathbf{L}_\infty$. We define

$$\| \mu \| = \sup_{\substack{f \in \mathbf{L}_\infty \\ \|f\| \leq 1}} \left| \int f \cdot \mu \right| .$$

If $\mu \in \mathbf{M}_+$, we have $\| \mu \| = \int \mu$.

We can also consider any $\mu \in \mathbf{M}$ as an element of the dual of the space of bounded continuous functions. We again have

$$\left| \int f \, . \, \mu \right| \leqslant \| \mu \| \, . \, \| f \|_u , \qquad \forall \text{ continuous bounded } f,$$

where

$$\| f \|_u = \sup_{x \in \mathbf{R}} | f(x) | .$$

Theorem 1. *Let μ denote a measure. If there exists a constant k such that*

$$\left| \int f \, . \, \mu \right| \leqslant k \, \| f \|_u , \qquad \forall f \in \mathfrak{D}^0 ,$$

then μ is a bounded measure.

Proof. To simplify the proof, let us assume that μ is real. For every $f \in \mathfrak{D}^0_+$, we have

$$\int f \, . \, \mu^+ = \sup_{\left\{ \substack{\varphi \in \mathfrak{D}^0 \\ 0 \leqslant \varphi \leqslant f} \right.} \int \varphi \, . \, \mu .$$

Since

$$\left| \int \varphi \, . \, \mu \right| \leqslant k \, \| \varphi \|_u \leqslant k \, \| f \|_u ,$$

we have

$$\int f \, . \, \mu^+ \leqslant k \, \| f \|_u , \qquad \forall f \in \mathfrak{D}^0_+ ,$$

and similarly,

$$\int f \, . \, \mu^- \leqslant k \, \| f \|_u , \qquad \forall f \in \mathfrak{D}^0_+ .$$

Consequently,

$$\int \mu^+ = \sup_{\left\{ \substack{f \in \mathfrak{D}^0_+ \\ f \leqslant 1} \right.} \int f \, . \, \mu^+ \leqslant k ,$$

and similarly,

$$\int \mu^- \leqslant k .$$

Thus, μ^+ and μ^- are bounded measures. Consequently, μ is bounded.

Theorem 2. *$\mathbf{M}$ is the dual of the space $\mathbf{L}_\infty$ of continuous functions that approach 0 as their argument becomes infinite.*

Proof. Let m denote a continuous linear form on $\mathbf{L}_\infty$. The restriction of m to $\mathfrak{D}^0$ is a bounded measure μ in accordance with the preceding theorem. The extension of μ to $\mathbf{L}_\infty$ is a continuous linear form that coincides with m since $\mathfrak{D}^0$ is dense in $\mathbf{L}_\infty$.

Remark. $\mathbf{M}$ is a proper subspace of the dual of the space of bounded continuous functions. In other words, there exist continuous linear forms on the space of bounded continuous functions that are not measures. Proof of this fact would take us beyond the scope of the book.

Topology on measures. Definition. The *vague topology* on the set of measures is defined as the topology of pointwise convergence on $\mathfrak{D}^0$. A sequence $\{\mu_n\}$ of measures μ_n converges vaguely to a measure μ if, for every $f \in \mathfrak{D}^0$,

$$\lim_{n\to\infty} \int f \cdot \mu_n = \int f \cdot \mu .$$

In $\mathbf{M}$, we can consider two other common topologies.

Definition. The *weak topology* on $\mathbf{M}$ is defined as the weak topology in the usual sense on the dual of the space $\mathbf{L}_\infty$. A sequence $\{\mu_n\}$ of measures $\mu_n \in \mathbf{M}$ converges weakly to a measure $\mu \in \mathbf{M}$ if, for every $f \in \mathbf{L}_\infty$,

$$\lim_{n\to\infty} \int f \cdot \mu_n = \int f \cdot \mu .$$

Definition. The *narrow topology* on $\mathbf{M}$ is defined as the weak topology in the usual sense when we consider $\mathbf{M}$ as a subspace of the dual of the space of bounded continuous functions. A sequence $\{\mu_n\}$ of measures $\mu_n \in \mathbf{M}$ converges narrowly to a measure $\mu \in \mathbf{M}$ if, for every bounded continuous function f,

$$\lim_{n\to\infty} \int f \cdot \mu_n = \int f \cdot \mu .$$

On $\mathbf{M}$, the narrow topology is finer than the weak topology and the weak topology is finer than the vague topology. On every ball $\|\mu\| \leqslant A$, the vague and weak topologies coincide. This is true because $\mathfrak{D}^0$ is dense in $\mathbf{L}_\infty$ and we can apply the first fundamental theorem on Banach and Fréchet spaces (cf. section 2 of Chapter 1).

Examples. As $n \to +\infty$,

$$\begin{aligned} \delta_n &\to 0 \quad \text{weakly but not narrowly} \\ n\delta_n &\to 0 \quad \text{vaguely but not weakly.} \end{aligned}$$

Lemma 1. *Suppose that a sequence $\{\mu_n\}$ of bounded positive measures μ_n converges vaguely to a bounded measure μ_0 and that*

$$\lim_{n\to\infty}\int \mu_n = \int \mu_0.$$

Then, for every $\varepsilon > 0$, *there exists a compact set* K *and an integer* N *such that*

$$n \geqslant N \;\Rightarrow\; \mu_n(\complement K) \leqslant \varepsilon .$$

Furthermore,

$$\mu_0(\complement K) \leqslant \varepsilon .$$

Proof. We prove the first part of this lemma by contradiction. Suppose that there exists an $\varepsilon > 0$ such that, for every compact set K, there exists a sequence $\{n_k\}$ that approaches $+\infty$ and is such that

$$\mu_{n_k}(\complement K) > \varepsilon .$$

Now, let f denote a function in $\mathfrak{D}^0$ such that $0 \leqslant f \leqslant 1$. Let us apply the preceding property to the support of K of f. There exists a sequence $\{n_k\}$ that approaches $+\infty$ and that satisfies the inequality

$$\int f \cdot \mu_{n_k} \leqslant \int \mu_{n_k} - \varepsilon .$$

This implies

$$\int f \cdot \mu_0 \leqslant \int \mu_0 - \varepsilon .$$

Since this property holds for every $f \in \mathfrak{D}^0$ such that $0 \leqslant f \leqslant 1$, it contradicts the definition of $\int \mu_0$. The second part of the conclusion follows from the fact that the mapping $\mu \to \mu(\Omega)$ is, for every open set Ω, lower-semicontinuous for $\mu \in \mathbf{M}_+$ equipped with the vague topology. Specifically,

$$\mu(\Omega) = \sup_{\left\{\substack{f \in \mathfrak{D}^0 \\ f \leqslant \varepsilon_\Omega}\right.} \int f \cdot \mu$$

and the mappings $\mu \to \int f \cdot \mu$ are continuous for $\mu \in \mathbf{M}_+$ equipped with the vague topology.

Theorem 3. *Suppose that a sequence* $\{\mu_n\}$ *of bounded positive measures* μ_n *converges vaguely to a bounded measure* μ *and that*

$$\lim_{n\to\infty}\int \mu_n = \int \mu ,$$

Then, the sequence $\{\mu_n\}$ *tends narrowly to* μ.

Proof. Let f denote a continuous bounded function and M a constant such that $|f| \leqslant M$. According to the preceding lemma, for every $\varepsilon > 0$, there exists a compact set K such that

$$\mu(\complement K) \leqslant \varepsilon/M \quad \text{and} \quad \mu_n(\complement K) \leqslant \varepsilon/M \quad \text{for} \quad n \geqslant N.$$

Let φ denote a function in $\mathfrak{D}^0$ that satisfies the inequality $0 \leqslant \varphi \leqslant 1$ everywhere and that is equal to 1 on K. We then have

$$|f - f\varphi| \begin{cases} \leqslant M \text{ on } \complement K, \\ = 0 \text{ on } K, \end{cases}$$

so that

$$\left| \int f \cdot \mu_n - \int f\varphi \cdot \mu_n \right| \leqslant \varepsilon, \quad \text{for} \quad n \geqslant N;$$

and

$$\left| \int f \cdot \mu - \int f\varphi \cdot \mu \right| \leqslant \varepsilon.$$

Also, since $f\varphi$ has compact support, there exists an N_1 such that

$$n \geqslant N_1 \quad \Rightarrow \quad \left| \int f\varphi \cdot \mu_n - \int f\varphi \cdot \mu \right| \leqslant \varepsilon,$$

so that, for $n \geqslant \max(N_1, N)$,

$$\left| \int f \cdot \mu_n - \int f \cdot \mu \right| \leqslant 3\varepsilon,$$

which completes the proof.

Subsets of M. Consider the space $\mathbf{L}^1$ of integrable functions (defined modulo functions that vanish almost everywhere) with respect to Lebesgue measure.

To every f in $\mathbf{L}^1$, we can assign the measure $f(x)\,dx$ of density $f(x)$ with respect to Lebesgue measure. Thus, we can identify $\mathbf{L}^1$ with a subset of $\mathbf{M}$. The norm of the measure $f(x)\,dx$ is equal to the norm of f in $\mathbf{L}^1$:

$$\|f\|_1 = \int |f|.$$

Naturally, this identification enables us to treat every subset of $\mathbf{L}^1$ as a subset of **M** and, in particular, of

> the space $\mathfrak{D}^0$ of continuous functions with compact support and
>
> the space $\mathfrak{D}$ of infinitely differentiable functions with compact support.

2. THE FOURIER TRANSFORM OF BOUNDED MEASURES

Let μ denote a bounded measure. Its Fourier transform $\mathcal{F}\mu$ is a function of a variable ω and is defined by

$$(\mathcal{F}\mu)(\omega) = \int e^{i\omega x} \,.\, \mu^x .$$

The superscript x on the μ means that the integration is performed with ω treated as a parameter; in other words, x is the variable of integration.

Theorem 4. *$\mathcal{F}\mu$ is a continuous function and* $|(\mathcal{F}\mu)(\omega)| \leqslant \|\mu\|$.

Proof. This result follows from Lebesgue's theorem (or its corollary dealing with integrals depending on a parameter) since $|e^{i\omega x}| \leqslant 1$.

A particular case. If $\mu = f(x)\,dx$, let us set $\mathcal{F}\mu = \mathcal{F}f$. The Fourier transform of a function f in $\mathbf{L}^1$ is therefore defined by

$$(\mathcal{F}f)(\omega) = \int_{-\infty}^{+\infty} e^{i\omega x} f(x)\,dx .$$

Examples. I. Let δ_a denote the Dirac measure at the point a. We have

$$\mathcal{F}\delta_a(\omega) = \int e^{i\omega x} \,.\, \delta_a = e^{ia\omega} .$$

II. Let $f(x) = 1/(1 + x^2)$. Then,

$$\mathcal{F}f(\omega) = \int_{-\infty}^{+\infty} \frac{1}{1 + x^2} e^{i\omega x}\,dx = \pi\, e^{-|\omega|}$$

(as we can show, for example, by the method of residues).

3. DIFFERENTIABILITY OF THE FOURIER TRANSFORM

Theorem 5. *If μ is a bounded measure and if the identity function $x \to x$ is integrable with respect to μ, then $\mathcal{F}\mu$ is differentiable and**

$$(\mathcal{F}\mu)' = \mathrm{i}\,\mathcal{F}(x\mu)\,.$$

Proof. This result follows from the theorem on differentiation under the integral sign of integrals depending on a parameter.

Corollary. *If $f \in \mathbf{L}^1$ and $xf(x) \in \mathbf{L}^1$, then $\mathcal{F}f$ is differentiable and*

$$(\mathcal{F}f)' = \mathrm{i}\,\mathcal{F}\left(xf(x)\right).$$

Examples. I. Verify the above theorem for $\mu = \delta_a$. We have

$$(\mathcal{F}\delta_a)(\omega) = \mathrm{e}^{\mathrm{i}a\omega}$$

and

$$\mathrm{i}\,\mathcal{F}(x\delta_a) = \mathrm{i}\,\mathcal{F}(a\delta_a) = \mathrm{i}\,a\mathcal{F}\delta_a\,,$$

that is,

$$\mathrm{i}[\mathcal{F}(x\delta_a)](\omega) = \mathrm{i}\,a\,\mathrm{e}^{\mathrm{i}a\omega} = \frac{\mathrm{d}}{\mathrm{d}\omega}(\mathrm{e}^{\mathrm{i}a\omega})\,.$$

II. $f(x) = 1/(1+x^2) \in \mathbf{L}^1$, but $xf(x) = x/(1+x^2) \notin \mathbf{L}^1$. Therefore, we may not assert that $\mathcal{F}f$ is differentiable. In fact, $\mathcal{F}f(\omega) = \pi\,\mathrm{e}^{-|\omega|}$ is not differentiable.

4. THE FOURIER TRANSFORM OF THE DERIVATIVE OF A FUNCTION

Theorem 6. *If f is a continuously differentiable function such that f and f' belong to $\mathbf{L}^1$, then*

$$(\mathcal{F}f')(\omega) = -\,\mathrm{i}\,\omega(\mathcal{F}f)(\omega)\,.$$

Proof. We have

$$f(x) = f(0) + \int_0^x f'\,.$$

---

*This chapter contains many simplifications in notation. Here, $\mathcal{F}(x\mu)$ denotes the Fourier transform of the product measure of μ and the identity function $x \to x$. Later, we write $xf(x) \in \mathbf{L}^1$ to mean that the function $x \to xf(x)$ is integrable with respect to Lebesgue measure. We denote by $\mathcal{F}(xf(x))$ the Fourier transform of the function $x \to xf(x)$, etc.

Therefore, $f(x)$ has a limit both as $x \to +\infty$ and as $x \to -\infty$. Since $f \in \mathbf{L}^1$, these limits must be 0. Therefore, integration by parts yields

$$(\mathcal{F}f')(\omega) = \int_{-\infty}^{+\infty} e^{i\omega x} f'(x)\,dx$$

$$= \left[e^{i\omega x} f(x)\right]_{-\infty}^{+\infty} - \int_{-\infty}^{+\infty} i\,\omega\, e^{i\omega x} f(x)\,dx = -\,i\omega(\mathcal{F}f)(\omega)\,.$$

Consequences. I. If f is continuously differentiable and if f and f' belong to $\mathbf{L}^1$, we have $(\mathcal{F}f)(\omega) = O(1/\omega)$ as $\omega \to \infty$.

II. If f is infinitely differentiable and if all its derivatives belong to $\mathbf{L}^1$, in particular if $f \in \mathfrak{D}$, then $(\mathcal{F}f)(\omega) = O(1/\omega^k)$ for every k (and hence $f(\omega) = o(1/\omega^k)$ for every k). This property can be expressed by saying that f decreases rapidly at infinity.

These results can be improved with the aid of the following theorem:

Theorem 7 (Lebesgue). *For every* $f \in \mathbf{L}^1$,

$$\lim_{\omega \to \infty} (\mathcal{F}f)(\omega) = 0\,.$$

Proof. Consider the form u_ω defined on $\mathbf{L}^1$ by

$$u_\omega(f) = (\mathcal{F}f)(\omega)\,.$$

Since $|(\mathcal{F}f)(\omega)| \leqslant \|f\|_{\mathbf{L}^1}$, if follows that $\|u_\omega\| \leqslant 1$. (In fact, we have $\|u_\omega\| = 1$, where the norm of u_ω is the norm in $\mathbf{L}^\infty$ of the function $x \to e^{i\omega x}$.) For $f \in \mathfrak{D}$, we have $\lim\limits_{\omega \to \infty} u_\omega(f) = 0$. Since $\mathfrak{D}$ is dense in $\mathbf{L}^1$ and since u_ω belongs to the unit ball of $\mathbf{L}^1$, this implies that $\lim\limits_{\omega \to \infty} u_\omega(f) = 0$ for every f in $\mathbf{L}^1$.

5. PLANCHEREL'S THEOREM

Let us define a transformation $\mathcal{F}^\star$ by

$$(\mathcal{F}^\star \mu)(\omega) = \int e^{-i\omega x} \,.\, \mu^x \qquad \text{for} \quad \mu \in \mathbf{M}\,,$$

$$(\mathcal{F}^\star f)(\omega) = \int_{-\infty}^{+\infty} e^{-i\omega x} f(x)\,dx \qquad \text{for} \quad f \in \mathbf{L}^1\,.$$

We shall call this transformation the *conjugate Fourier transform.*

Under hypotheses analogous to those of the preceding theorems, we have

$$(\mathcal{F}^\star \mu)' = -\,i\,\mathcal{F}^\star(x\mu) \qquad (\mathcal{F}^\star f')(\omega) = i\,\omega(\mathcal{F}^\star f)(\omega)\,,$$
$$(\mathcal{F}^\star f)' = -\,i\,\mathcal{F}^\star(xf(x))\,.$$

Notation. Let μ denote a measure and let f denote a function that is integrable with respect to μ. We define

$$\langle f, \mu \rangle = \int \bar{f} \,.\, \mu\,,$$

and, if fg is integrable,

$$\langle f, g \rangle = \int \bar{f} g\,.$$

If f and g belong to the space $\mathbf{L}^2$ of measurable functions that are square-integrable with respect to Lebesgue measure, the expression $\langle f, g \rangle$ is the usual scalar product defining a Hilbert structure in $\mathbf{L}^2$.

In particular, the expression $\langle f, \mu \rangle$ is defined if μ is bounded and f is continuous and bounded.

Theorem 8. *Suppose that* $f \in \mathbf{L}^1$ *and* $\mu \in \mathbf{M}$. *Then,*

$$\langle \mathcal{F} f, \mu \rangle = \langle f, \mathcal{F}^\star \mu \rangle\,,$$

and

$$\langle \mathcal{F}^\star f, \mu \rangle = \langle f, \mathcal{F} \mu \rangle\,.$$

In particular, if f *and* g *belong to* $\mathbf{L}^1$, *then*

$$\langle \mathcal{F} f, g \rangle = \langle f, \mathcal{F}^\star g \rangle\,.$$

Proof. We have

$$\langle \mathcal{F} f, \mu \rangle = \int \left[\int \overline{f(x)}\, e^{-i\omega x}\, dx \right] .\, \mu^\omega$$
$$= \iint \overline{f(x)}\, e^{-i\omega x} \,.\, dx \,.\, \mu^\omega\,,$$

and

$$\langle f, \mathcal{F}^\star \mu \rangle = \int \left[\overline{f(x)} \int e^{-i\omega x} \,.\, \mu^\omega \right] dx$$
$$= \iint \overline{f(x)}\, e^{-i\omega x} \,.\, dx \,.\, \mu^\omega\,.$$

Remark. We shall also use the following formulas, which are equivalent to those in Theorem 8:

$$\int \mathcal{F} f.\mu = \int f.\mathcal{F}\mu\,, \qquad \int \mathcal{F}^\star f.\mu = \int f.\mathcal{F}^\star\mu \qquad (\forall f \in \mathbf{L}^1,\ \mu \in \mathbf{M}).$$

Theorem 9 (partial statement of Plancherel's theorem). *If* $f \in \mathfrak{D}$, *then*

$$2\pi \int |f|^2 = \int |\mathcal{F}f|^2 = \int |\mathcal{F}^\star f|^2 .$$

Proof. Suppose that $f \in \mathfrak{D}$. Let T denote a number such that $f(x) = 0$ outside the interval $[-T/2, +T/2]$. Consider the function f_T defined by

$$\begin{cases} f_T(x) = f(x) & \text{for} \quad x \in [-T/2, +T/2]\,, \\ f_T(x+T) = f_T(x)\,. \end{cases}$$

The function f_T has period T.

Its Fourier coefficients are given by the formula

$$a_p = \frac{1}{T}\int_{-T/2}^{+T/2} f(x) \exp\left(-\mathrm{i}\,p\frac{2\pi}{T}x\right) \mathrm{d}x = \frac{1}{T}\mathcal{F}^\star f\left(p\frac{2\pi}{T}\right).$$

Parseval's formula yields

$$\frac{1}{T}\int_{-T/2}^{+T/2} |f_T(x)|^2\,\mathrm{d}x = \sum_{-\infty}^{+\infty} |a_p|^2 = \frac{1}{T^2}\sum_{-\infty}^{+\infty}\left|\mathcal{F}^\star f\left(p\frac{2\pi}{T}\right)\right|^2,$$

that is,

$$\int_{-\infty}^{+\infty} |f(x)|^2\,\mathrm{d}x = \frac{1}{T}\sum_{-\infty}^{+\infty}\Phi\left(p\frac{2\pi}{T}\right), \quad \text{with } \Phi = |\mathcal{F}^\star f|^2 .$$

If we take $T = n\,.\,2\pi$, we have

$$2\pi\int_{-\infty}^{+\infty} |f(x)|^2\,\mathrm{d}x = \frac{1}{n}\sum_{-\infty}^{+\infty}\Phi\left(\frac{p}{n}\right).$$

The right-hand member is the total mass of the measure

$$\mu_n = \sum_{p=-\infty}^{+\infty}\frac{1}{n}\Phi\left(\frac{p}{n}\right)\delta_{p/n}\,.$$

Let E_k denote the interval $[k, k+1[$. Then $\mu_n(E_k)$ tends to

$$\int_k^{k+1}\Phi(\omega)\,\mathrm{d}\omega\,,$$

as $n \to +\infty$. Now, we have

$$\int \mu_n = \sum_{k=-\infty}^{+\infty} \mu_n(E_k).$$

Furthermore, since $\Phi(\omega)$ is of rapid decrease, we can majorize $\Phi(\omega)$ with an integrable function $\psi(\omega)$ that is constant on each interval E_k. Then

$$\sum_{k=-\infty}^{+\infty} \int_k^{k+1} \psi(\omega)\, d\omega = \int_{-\infty}^{+\infty} \psi(\omega)\, d\omega .$$

Therefore,

$$\mu_n(E_k) \leqslant \int_k^{k+1} \psi(\omega)\, d\omega .$$

Consequently, the series $\sum_k \mu_n(E_k)$ converges uniformly, and we have

$$\begin{aligned} \lim_{n\to\infty} \mu_n(\mathbf{R}) &= \lim_{n\to\infty} \sum_{k=-\infty}^{+\infty} \mu_n(E_k) \\ &= \sum_{k=-\infty}^{+\infty} \lim \mu_n(E_k) = \sum_{k=-\infty}^{+\infty} \int_k^{k+1} \Phi(\omega)\, d\omega = \int_{-\infty}^{+\infty} \Phi(\omega)\, d\omega ; \end{aligned}$$

that is,

$$2\pi \int_{-\infty}^{+\infty} |f(x)|^2\, dx = \int_{-\infty}^{+\infty} \Phi(\omega)\, d\omega ,$$

which completes the proof.

Corollary. *Let f and g denote two functions in $\mathfrak{D}$. Then,*

$$2\pi \int_{-\infty}^{+\infty} f\bar{g} = \int_{-\infty}^{+\infty} \mathcal{F}f \, . \, \overline{\mathcal{F}g} ;$$

that is,

$$2\pi \langle f, g \rangle = \langle \mathcal{F}f, \mathcal{F}g \rangle .$$

Similarly,

$$2\pi \langle f, g \rangle = \langle \mathcal{F}^\star f, \mathcal{F}^\star g \rangle .$$

6. EXTENSION OF THE FOURIER TRANSFORM TO L^2

The transforms $\mathcal{F}/\sqrt{2\pi}$ and $\mathcal{F}^\star/\sqrt{2\pi}$ can, by virtue of the preceding theorem, be extended as two isometric operators onto $\mathbf{L}^2$. We shall continue to denote these transforms by $\mathcal{F}/\sqrt{2\pi}$ and $\mathcal{F}^\star/\sqrt{2\pi}$.

The relation $\langle \mathcal{F}f, g \rangle = \langle f, \mathcal{F}^\star g \rangle$, which is valid for f and g in $\mathfrak{D}$, remains valid, because of the continuity, for f and g in $\mathbf{L}^2$. It shows that $\mathcal{F}$ and $\mathcal{F}^\star$ are adjoints of each other and hence that $\mathcal{F}/\sqrt{2\pi}$ and $\mathcal{F}^\star/\sqrt{2\pi}$ are unitary, and inverses of each other on $\mathbf{L}^2$:

$$\mathcal{F}^{-1} = \frac{1}{2\pi}\mathcal{F}^\star .$$

Furthermore, the relation

$$\langle \mathcal{F}f, g \rangle = \langle f, \mathcal{F}^\star g \rangle$$

which is valid for $f \in \mathbf{L}^1 \cap \mathbf{L}^2$ and $g \in \mathfrak{D}$, shows, whether we take for $\mathcal{F}$ the extension to $\mathbf{L}^2$ just defined or the original definition on $\mathbf{L}^1$, that the extension of the Fourier transform to $\mathbf{L}^2$ coincides on $\mathbf{L}^1 \cap \mathbf{L}^2$ with the original definition of this Fourier transform. This justifies the apparent abuse of notation. Figure 1 illustrates diagramatically the spaces on which the Fourier transform has been defined and the inclusion relationships among them.

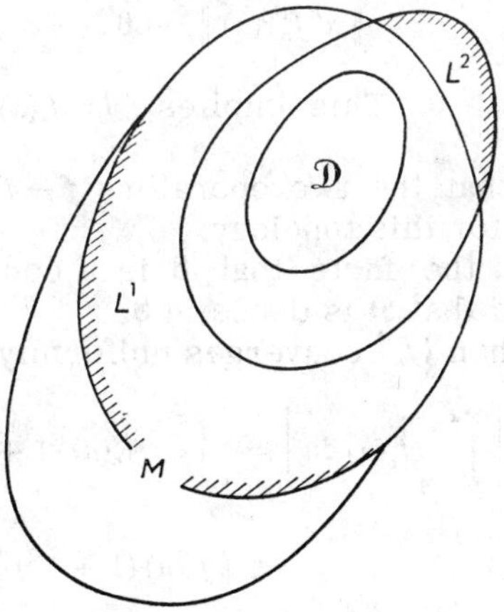

FIG. 1.

From the above, we can derive the following theorem:

Theorem 10 (Plancherel): *If* $f \in \mathbf{L}^1 \cap \mathbf{L}^2$, *then*

$$\mathcal{F}f \in \mathbf{L}^2 \text{ and } \int |\mathcal{F}f|^2 = 2\pi \int |f|^2 .$$

If $f, g \in \mathbf{L}^1 \cap \mathbf{L}^2$, *then*

$$\int \overline{\mathcal{F}f} \,.\, \mathcal{F}g = 2\pi \int \bar{f}g .$$

7. THE FOURIER TRANSFORM OF INFINITELY DIFFERENTIABLE FUNCTIONS OF RAPID DECREASE

Let us consider the space $\mathcal{S}$ of infinitely differentiable functions f all derivatives of which are of rapid decrease, that is, functions such that $x^k f^{(h)}(x) \to 0$ as $(x) \to +\infty$ for all nonnegative h and k.

One can immediately verify that, if $f \in \mathcal{S}$, then the same is true of f' and $xf(x)$ and, more generally, of $x^k f^{(h)}(x)$ or $(x^k f(x))^{(h)}$.

Obviously,

$$\mathcal{D} \subset \mathcal{S} \subset \mathbf{L}^1 \cap \mathbf{L}^2 .$$

Let us consider on $\mathcal{S}$ the metrizable topology defined by the (double) sequence of seminorms

$$f \to \| x^k f^{(h)}(x) \|_2$$

(where $\| \cdot \|_2$ denotes the norm in $\mathbf{L}^2$).

A sequence $\{f_n\}$ of functions $f_n \in \mathcal{S}$ tends to zero if

$$\| x^k f_n^{(h)}(x) \|_2 \to 0 .$$

for every h and $k \geqslant 0$. This implies $\| (x^k f_n(x))^{(h)} \|_2 \to 0$, for every h and $k \geqslant 0$.

It is obvious that the two operators $f \to f'$ and $f \to xf$ are continuous operators for this topology.

Let us accept the facts that $\mathcal{S}$ is a complete space, hence a Fréchet space, and that $\mathcal{D}$ is dense in $\mathcal{S}$.

If $f_n \to 0$ in $\mathcal{S}$, then $\{f_n\}$ converges uniformly to 0 since

$$|f_n(x)| = \left| \int_{-\infty}^{x} f_n'(u)\, du \right| = \left| \int_{-\infty}^{x} f_n'(u)(1 + u^2) \frac{du}{1 + u^2} \right|$$

$$\leqslant \| f_n'(u)(1 + u^2) \|_2 \left\| \frac{1}{1 + u^2} \right\|_2 .$$

From this we conclude that $x^k f_n^{(h)}(x) \to 0$ uniformly for every h and k. Conversely, one can show that, if $x^k f_n^{(k)}(x) \to 0$ uniformly for every h and k, then $\{f_n\}$ tends to 0 in $\mathcal{S}$.

Theorem 11. *If $f \in \mathcal{S}$, then $\mathcal{F}f$ exists and belongs to $\mathcal{S}$, and $\mathcal{F}$ is an isomorphism on $\mathcal{S}$; the inverse of $\mathcal{F}$ is*

$$\frac{1}{2\pi} \mathcal{F}^* .$$

Proof. If $f \in \mathcal{S}$, we have $f \in \mathbf{L}^1$. Therefore, $\mathcal{F}f$ exists and is given by the formula

$$(\mathcal{F}f)(\omega) = \int_{-\infty}^{+\infty} f(x)\, e^{i\omega x}\, dx\, .$$

Since, in addition, $x^k f \in \mathbf{L}^1$ for all k, it follows that $\mathcal{F}f$ is infinitely differentiable.

Since $(x^k f)^{(h)}$ exists for all h and belongs to $\mathbf{L}^1$ and since

$$\mathcal{F}[(x^k f)^{(h)}](\omega) = (-i\omega)^h \mathcal{F}[(x^k f)](\omega) = (-i\omega)^h (-i)^k (\mathcal{F}f)^{(k)}(\omega)\, ,$$

we can see that $\omega^h(\mathcal{F}f)^{(k)}(\omega)$ bounded for every h and k and hence $(\mathcal{F}f)^{(k)}$ is of rapid decrease for all k. Therefore, $\mathcal{F}f \in \mathbf{S}$.

Furthermore, if $f_n \to 0$ in $\mathbf{S}$, then $\mathcal{F}f_n \to 0$ in $\mathbf{S}$ since, by virtue of Plancherel's theorem,

$$\| \omega^h(\mathcal{F}f_n)^{(k)} \|_2 = \sqrt{2\pi}\, \| (x^k f_n)^{(h)} \|_2\, .$$

Furthermore, since $\mathcal{F}$ and $\frac{1}{2\pi}\mathcal{F}^*$ are inverses of each other in $\mathbf{L}^2$, they are also inverses of each other in $\mathbf{S}$.

8. TEMPERED DISTRIBUTIONS

A *tempered distribution* is defined as any continuous linear form on $\mathbf{S}$. We denote by $\mathbf{S}'$ the space of tempered distributions ($\mathbf{S}$ is the dual of $\mathbf{S}'$).

The restriction to $\mathcal{D}$ of a tempered distribution is a distribution.

Notation. The value of a tempered distribution T for the function $f \in \mathbf{S}$ will be denoted by $\int f \cdot T$. We set

$$\langle f, T \rangle = \int \bar{f} \cdot T\, .$$

If a distribution T is continuous on $\mathcal{D}$ for the topology induced by $\mathbf{S}$, we can extend it by virtue of the continuity, to $\mathbf{S}$, and this extension is a tempered distribution. For convenience, we shall say that T is a tempered distribution and we shall also denote its extension to $\mathbf{S}$ by T.

Similarly, if H is a Banach space, we define a tempered distribution with range in H as a continuous mapping of $\mathbf{S}$ into H. We shall not use this concept until we get to Chapter 7.

Examples. I. Every distribution with compact support is a tempered distribution.

II. Every bounded measure μ defines (by restriction to $\mathbf{S}$) a tempered distribution. Specifically, the mapping $f \to \int f \cdot \mu$ is defined on $\mathcal{S}$ and if $f_n \to 0$ in $\mathcal{S}$, then $\{f_n\}$ converges uniformly to 0 and hence $\left\{\int f_n \cdot \mu\right\}$ converges to 0. Therefore, we may write $\mathbf{M} \subset \mathcal{S}'$. In particular, $\mathbf{L}^1$, $\mathcal{S} \subset \mathcal{S}'$.

III. More generally, a measure μ defines a tempered distribution if the mapping $f \to \int f \cdot \mu$ is continuous on $\mathcal{D}$ for the topology induced on $\mathcal{D}$ by the topology of $\mathcal{S}$. Therefore, we can extend the above mapping to $\mathcal{S}$ by virtue of the continuity and we can show that, if $f \in \mathcal{S}$ is integrable with respect to μ, the integral $\int f \cdot \mu$ is simply the value of the extension of $\mathcal{S}$ of the mapping $f \to \int f \cdot \mu$ defined on $\mathcal{D}$. In particular, a locally integrable function φ defines a distribution in $\mathcal{S}'$ if the mapping $f \to \int f \cdot \varphi$ is continuous on $\mathcal{D}$ for the topology of $\mathcal{S}$. This will be the case in particular if φ is a continuous function of slow decrease, that is, if there exists an integer k such that $\varphi(x) = O(x^k)$ as $x \to \infty$. In particular, every polynomial is a tempered distribution.

Topology in $\mathcal{S}'$. We shall confine ourselves to the weak topology or the topology of simple convergence in $\mathcal{S}$. A sequence $\{T_n\}$ of tempered distributions converges to 0 if $\int f \cdot T_n \to 0$ for every $f \in \mathcal{S}$, in other words, if $\langle f, T_n \rangle \to 0$ for every $f \in \mathcal{S}$.

9. DIFFERENTIATION OF TEMPERED DISTRIBUTIONS

The derivative of T' of a tempered distribution T is defined just as for any distribution by

$$\int f \cdot T' = - \int f' \cdot T, \qquad \forall f \in \mathcal{D}.$$

Since the right-hand member of this equation is continuous with respect to f for the topology induced by $\mathcal{S}$ on $\mathcal{D}$, so is the left-hand member, which shows that T' is a tempered distribution. This equation remains valid by virtue of the continuity with respect to $f \in \mathcal{S}$. Thus, we have the following result:

Theorem 12. *The derivative T' of a tempered distribution is a tempered distribution, and*

$$\int f \cdot T' = - \int f' \cdot T, \qquad \forall f \in \mathcal{S},$$

or, in other words,

$$\langle f, T' \rangle = -\langle f', T \rangle \qquad \forall f \in \mathbf{S}.$$

We note that the derivative of a function of slow increase is not necessarily of slow increase, but its derivative nonetheless defines a tempered distribution. Also, a continuous function of slow increase has a derivative $\mathbf{S}'$ even if it is not differentiable in the elementary sense. Conversely, one can show that every tempered distribution is the derivative of some order of a continuous function of slow increase.

Theorem 13. *Differentiation is a continuous operation in* $\mathbf{S}'$.

Proof. This follows from the fact that differentiation in $\mathbf{S}'$ is, up to sign, the adjoint of differentiation in $\mathbf{S}$. The general proof, applied to the present case, is given in the following fashion. Let $\mathcal{U}$ denote a neighborhood of 0 in $\mathbf{S}'$ and defined in terms of a finite number $g_1, \ldots, g_k$ of elements of $\mathbf{S}$ by

$$S \in \mathcal{U} \quad \Leftrightarrow \quad \left| \int g_i \,.\, S \right| \leqslant 1 .$$

In particular, we have

$$T' \in \mathcal{U} \quad \Leftrightarrow \quad \left| \int g_i \,.\, T' \right| \leqslant 1 ,$$

that is,

$$T' \in \mathcal{U} \quad \Leftrightarrow \quad \left| \int g_i' \,.\, T \right| \leqslant 1 .$$

Since the set of those $T \in \mathbf{S}$ such that $\left| \int g_i' \,.\, T \right| \leqslant 1$ is a neighborhood $\mathcal{U}'$ of 0, we have thus assigned to every neighborhood $\mathcal{U}$ of 0 a neighborhood $\mathcal{U}'$ of 0 such that

$$T \in \mathcal{U}' \quad \Leftrightarrow \quad T' \in \mathcal{U}$$

which shows that the mapping $T \to T'$ is continuous.

10. THE FOURIER TRANSFORM OF TEMPERED DISTRIBUTIONS

Definition. If T is a tempered distribution, we define its Fourier transform $\mathcal{F}T$ by the formula

$$\langle f, \mathcal{F}T \rangle = \langle \mathcal{F}^* f, T \rangle , \qquad \forall f \in \mathbf{S}.$$

This definition is compatible with the definitions given in advance for $T \in \mathbf{M}$ and $T \in \mathbf{L}^2$ since the preceding formula is valid in these two cases. Thus, we have an extension to $\mathbf{S}'$ of the Fourier transform on $\mathbf{M}$ and on $\mathbf{L}^2$.

Similarly, the conjugate Fourier transform $\mathcal{F}^\star$ is defined on $\mathbf{S}'$ by

$$\langle f, \mathcal{F}^\star T \rangle = \langle \mathcal{F} f, T \rangle \qquad \forall f \in \mathbf{S}\,.$$

Remark. We shall often use the following formulas, which are equivalent to those in the preceding definition:

$$\int f.\mathcal{F}T = \int \mathcal{F}f.T\,, \qquad \int f.\mathcal{F}^\star T = \int \mathcal{F}^\star f.T. \qquad (\forall f \in \mathbf{S},\ T \in \mathbf{S}').$$

Theorem 14. *The Fourier transforms $\mathcal{F}$ and $\mathcal{F}^\star$ defined on $\mathbf{S}'$ are continuous isomorphisms, and*

$$\mathcal{F}\mathcal{F}^\star = \mathcal{F}^\star \mathcal{F} = 2\pi\,.$$

Proof. The transform $\mathcal{F}$ on $\mathbf{S}'$ is the transpose of $\mathcal{F}^\star$ on $\mathbf{S}$. Therefore, it is continuous (as can be shown directly more or less in the proof for differentiation). Similarly, $\mathcal{F}^\star$ defined on $\mathbf{S}'$ is the adjoint of $\mathcal{F}$ defined on $\mathbf{S}$; hence, it is continuous.

Furthermore, since $\mathcal{F}^\star$ and $\mathcal{F}/2\pi$ are inverses of each other on $\mathbf{S}$, their adjoints $\mathcal{F}$ and $\mathcal{F}^\star/2\pi$ are inverses of each other on $\mathbf{S}'$. Therefore, $\mathcal{F}\mathcal{F}^\star = \mathcal{F}^\star\mathcal{F} = 2\pi$ (on $\mathbf{S}'$). This can be verified directly: For all $f \in \mathbf{S}$ and all $T \in \mathbf{S}'$,

$$\langle f, \mathcal{F}\mathcal{F}^\star T \rangle = \langle \mathcal{F}^\star f, \mathcal{F}^\star T \rangle = \langle \mathcal{F}\mathcal{F}^\star f, T \rangle = 2\pi \langle f, T \rangle\,.$$

Therefore,

$$\mathcal{F}\mathcal{F}^\star T = 2\pi T\,, \qquad (\forall T \in \mathbf{S}')\,.$$

Corollary. *If a sequence $\{T_n\}$ of distributions $T_n \in \mathbf{S}'$ is such that $\mathcal{F}T_n \to 0$ as $n \to \infty$, then $T_n \to 0$ as $n \to \infty$.*

Example. Let us calculate the Fourier transform $\mathcal{F}\delta'_a$ of δ'_a. It is defined by the formula

$$\langle f, \mathcal{F}\delta'_a \rangle = \langle \mathcal{F}^\star f, \delta'_a \rangle = -\int \overline{(\mathcal{F}^\star f)'}\,.\,\delta_a = -\overline{(\mathcal{F}^\star f)'}(a)\,.$$

Now,

$$(\mathcal{F}^\star f)(x) = \int f(\omega)\,\mathrm{e}^{-\mathrm{i}\omega x}\,\mathrm{d}\omega\,, \qquad (\mathcal{F}^\star f)'(x) = -\mathrm{i}\int \omega f(\omega)\,\mathrm{e}^{-\mathrm{i}\omega x}\,\mathrm{d}\omega\,,$$

from which we get

$$(\mathcal{F}^\star f)'(a) = -\mathrm{i}\int \omega f(\omega)\,\mathrm{e}^{-\mathrm{i}a\omega}\,\mathrm{d}\omega\,, \qquad \overline{(\mathcal{F}^\star f)'}(a) = \mathrm{i}\int \omega f(\omega)\,\mathrm{e}^{\mathrm{i}a\omega}\,\mathrm{d}\omega\,,$$

and

$$\mathcal{F}\delta_a'(\omega) \quad = -\,\mathrm{i}\,\omega\,\mathrm{e}^{\mathrm{i}a\omega}\,.$$

Similarly,

$$\mathcal{F}\delta_a^{(k)}(\omega) \quad = (-\,\mathrm{i}\,\omega)^k\,\mathrm{e}^{\mathrm{i}a\omega},$$

and in particular,

$$\mathcal{F}\delta^{(k)}(\omega) \quad = (-\,\mathrm{i}\,\omega)^k,$$

and similarly,

$$\mathcal{F}^\star\,\delta^{(k)}(\omega) = (\mathrm{i}\,\omega)^k,$$

that is,

$$\delta^{(k)} \quad = \frac{1}{2\,\pi}\,\mathcal{F}\big((\mathrm{i}\,\omega)^k\big),$$

and hence

$$\mathcal{F}(\omega^k) = 2\,\pi(-\,\mathrm{i})^k\,\delta^{(k)}\,.$$

11. MULTIPLIERS IN $\mathbf{S}$ AND $\mathbf{S}'$

We denote by $\mathbf{O}_m$ the space of functions φ such that $\varphi f \in \mathbf{S}$ for every $f \in \mathbf{S}$. The elements of $\mathbf{O}_m$ are called *multipliers* on $\mathbf{S}$. They are necessarily infinitely differentiable functions.

We shall say that a function φ is of slow increase if there exists a nonnegative k such that $\varphi(x) = \mathrm{O}(x^k)$ as $x \to \infty$. Obviously, if a function φ and all its derivatives are of slow increase, then φ is a multiplier on $\mathbf{S}$. Let us accept the converse of this.

We note that $\mathbf{O}_m$ is an algebra.

If $\varphi \in \mathbf{O}_m$ and $T \in \mathbf{S}'$, we define φT by

$$\int f\,.\,(\varphi T) = \int f\varphi\,.\,T$$

or, equivalently,

$$\langle f, \varphi T\rangle = \langle \overline{\varphi} f, T\rangle\,.$$

Conversely, if φ is an infinitely differentiable function such that $\varphi T \in \mathbf{S}'$ for every $T \in \mathbf{S}'$, we can show that $\varphi \in \mathbf{O}_m$.

Thus, $\mathbf{O}_m$ is the algebra of the multipliers on $\mathbf{S}'$ as well as on $\mathbf{S}$. We note that $\mathbf{O}_m \subset \mathbf{S}'$.

Example. Every polynomial belongs to $\mathbf{O}_m$.

Theorem 15. *Suppose that* $T \in \mathbf{S}$. *If* T *is considered as operating on the functions of the variable* ω, *we have*

$$(\mathcal{F}T)' = \mathcal{F}(\mathrm{i}\,\omega T) \quad \text{and} \quad (\mathcal{F}^\star T)' = \mathcal{F}^\star(-\mathrm{i}\,\omega T).$$

Proof. We have

$$\langle f, (\mathcal{F}T)' \rangle = -\langle f', \mathcal{F}T \rangle = -\langle \mathcal{F}^\star f', T \rangle = -\langle \mathrm{i}\,\omega \mathcal{F}^\star f, T \rangle$$
$$= \langle \mathcal{F}^\star f, \mathrm{i}\,\omega T \rangle = \langle f, \mathcal{F}(\mathrm{i}\,\omega T) \rangle .$$

Corollary. *Suppose that* $T \in \mathbf{S}'$. *If* $\mathcal{F}T$ *is considered as operating on the functions of the variable* ω, *we have*

$$\mathcal{F}^\star(T') = \mathrm{i}\omega \mathcal{F}^\star T \quad \text{and} \quad \mathcal{F}(T') = -\mathrm{i}\omega \mathcal{F}T.$$

Proof. In the first formula of the preceding theorem, apply $\mathcal{F}^\star$ to both sides and set

$$\begin{cases} \mathcal{F}T = U \\ T = \dfrac{1}{2\pi}\mathcal{F}^\star U . \end{cases}$$

Then

$$\mathcal{F}^\star U' = 2\pi \, . \, \mathrm{i}\,\omega \frac{1}{2\pi}\mathcal{F}^\star U,$$

that is,

$$\mathcal{F}^\star U' = \mathrm{i}\omega\, \mathcal{F}^\star U .$$

12. INVOLUTIONS. THE CONVOLUTION PRODUCT

The mappings $f \to \bar{f}$ and $f \to \tilde{f}$ are involutions on $\mathbf{S}$ and the mappings $T \to \bar{T}$ and $T \to \tilde{T}$ are involutions on $\mathbf{S}'$.

For $f \in \mathbf{S}$ and $T \in \mathbf{S}'$, we always have

$$\int f \, . \, \bar{T} = \overline{\int \bar{f} \, . \, T} \quad \text{or} \quad \langle f, \bar{T} \rangle = \overline{\langle \bar{f}, T \rangle},$$

$$\int f \, . \, \tilde{T} = \overline{\int \tilde{f} \, . \, T} \quad \text{or} \quad \langle f, \tilde{T} \rangle = \overline{\langle \tilde{f}, T \rangle}.$$

Theorem 16. *We have*

$$\begin{cases} \mathcal{F}\tilde{f} = \overline{\mathcal{F}f} \\ \mathcal{F}^\star\tilde{f} = \overline{\mathcal{F}^\star f} \end{cases} \quad \textit{for } f \in \mathbf{S} \textit{ and } \begin{cases} \mathcal{F}\tilde{T} = \overline{\mathcal{F}T} \\ \mathcal{F}^\star\tilde{T} = \overline{\mathcal{F}^\star T} \end{cases} \quad \textit{for } T \in \mathbf{S}'.$$

In other words, the involutions $\sim$ and $-$ are interchanged by $\mathcal{F}$ and $\mathcal{F}^\star$.

Proof. If $f \in \mathbf{S}$, we have

$$\mathcal{F}\tilde{f}(\omega) = \int \tilde{f}(x) e^{i\omega x}\, dx = \int \overline{f(-x)}\, e^{i\omega x}\, dx$$

$$= \overline{\int f(-x)\, e^{-i\omega x}\, dx} = \overline{\int f(x) e^{i\omega x}\, dx} = \overline{\mathcal{F}f(\omega)},$$

so that

$$\mathcal{F}\tilde{f} = \overline{\mathcal{F}f}.$$

Let us define $\mathcal{F}f = g$ and let us apply $\mathcal{F}^\star$ to both sides. We get

$$2\pi\tilde{f} = \mathcal{F}^\star \bar{g},$$

that is,

$$(\mathcal{F}^\star g)^{\sim} = \mathcal{F}^\star \bar{g}.$$

If $T \in \mathbf{S}'$, then, for every $f \in \mathbf{S}$,

$$\langle f, \mathcal{F}\tilde{T}\rangle = \langle \mathcal{F}^\star f, \tilde{T}\rangle = \overline{\langle (\mathcal{F}^\star f)^{\sim}, T\rangle}$$

$$= \overline{\langle \mathcal{F}^\star \bar{f}, T\rangle} = \overline{\langle \bar{f}, \mathcal{F}T\rangle} = \langle f, \overline{\mathcal{F}T}\rangle,$$

so that

$$\mathcal{F}\tilde{T} = \overline{\mathcal{F}T}.$$

If $f \in \mathbf{S}$ and $U \in \mathbf{S}'$, we can define the convolution product $U * f$ by the following formula (already proven in the case in which either f or U has compact support or if both f and U have positive support):

$$(U * f)(x) = \int f(x - y)\, U^y.$$

The function $U * f$ is an infinitely differentiable function. We verify first that, for every $f \in \mathbf{S}$, the function

$$x \to \frac{1}{h}[f(x + h) - f(x)],$$

tends to f' in $\mathbf{S}$ as $h \to 0$. We have

$$\frac{1}{h}[(U*f)(x+h)-(U*f)(x)] = \int \frac{1}{h}[f(x+h-y)-f(x-y)] \,.\, U^y,$$

and, consequently,

$$\lim \frac{1}{h}[(U*f)(x+h)-(U*f)(x)] = \int f'(x-y) \,.\, U^y,$$

so that

$$(U*f)'(x) = \int f'(x-y) \,.\, U^y,$$

and similarly

$$(U*f)^{(m)}(x) = \int f^{(m)}(x-y) \,.\, U^y.$$

We denote by $\mathbf{O}'_c$ the subset of $\mathbf{S}'$ consisting of tempered distributions U such that $U*f \in \mathbf{S}$ for every $f \in \mathbf{S}$. The elements of $\mathbf{O}'_c$ will be called the convolvers on $\mathbf{S}$. The set $\mathbf{O}'_c$ includes all continuous functions of rapid decrease and all the derivatives in the distribution sense of such functions. Every distribution with compact support belongs to $\mathbf{O}'_c$.

Theorem 17. *If $U \in \mathbf{O}'_c$, the operator $f \to U*f$ defined in $\mathbf{S}$ is continuous.*

Proof. This operator has a closed graph. To see this, note that, if $f_n \to 0$ (in $\mathbf{S}$) and $U*f_n \to g$ (in $\mathbf{S}$), we have $g=0$ since $\{U*f_n\}$ converges to 0 simply. Furthermore, $\mathbf{S}$ is a Fréchet space and therefore, every operator on $\mathbf{S}$ possessing a closed graph is continuous.

Definition. If $U \in \mathbf{O}'_c$ and $T \in \mathbf{S}'$, we define $U*T$ by

$$\langle f, U*T \rangle = \langle \tilde{U}*f, T \rangle \qquad \forall f \in \mathbf{S}.$$

We verify the legitimacy of this definition by remarking that the right-hand member is well defined ($\tilde{U}*f \in \mathbf{S}$) and that, if a sequence $\{f_n\}$ converges to 0 in $\mathbf{S}$, then the sequence $\{\tilde{U}*f_n\}$ converges to 0 in $\mathbf{S}$ and hence $\langle \tilde{U}*f_n, T \rangle$ converges to 0. This definition is compatible with the definition given earlier for $U*T$ (see Chapter 2, Theorem 15).

Furthermore, the mapping $T \to U*T$ is continuous in $\mathbf{S}'$. This is true because, if $T_n \to 0$ in $\mathbf{S}'$, we have

$$\lim_{n\to+\infty} \langle f, U*T_n \rangle = \lim_{n\to+\infty} \langle \tilde{U}*f, T_n \rangle = 0, \qquad \forall f \in \mathbf{S}.$$

The preceding justifications can be summarized in the fact that the mapping $T \to U * T$ is defined in $\mathbf{S}'$ as the adjoint of the mapping $f \to \tilde{U} * f$ in $\mathbf{S}$.

By virtue of the preceding definition, $\mathbf{O}_c'$ will also be called both the set of convolvers of $\mathbf{S}'$ and the set of convolvers of $\mathbf{S}$.

We note that $\mathbf{O}_c' \subset \mathbf{S}'$ and, hence that $U * V$ is defined for U, $V \in \mathbf{O}_c'$. We shall see that $\mathbf{O}_c'$ is a commutative algebra under convolution.

Theorem 18. *If $U \in \mathbf{O}_c'$ and $f, g \in \mathbf{S}$, then*

$$\langle U * f, g \rangle = \langle U, g * \tilde{f} \rangle .$$

Proof. The formula is valid for $f, g \in \mathfrak{D}$. By virtue of the continuity, it is valid for $f \in \mathbf{S}$ and $g \in \mathfrak{D}$ and hence for $f, g \in \mathbf{S}$.

Theorem 19.

$$\begin{array}{lll} \textit{If } \mu, \nu \in \mathbf{M}, & \textit{then } \mathcal{F}(\mu * \nu) = \mathcal{F}\mu \,.\, \mathcal{F}\nu . & \\ \textit{If } U \in \mathbf{O}_c' \textit{ and } f \in \mathbf{S}, & \textit{then } \mathcal{F}(U * f) = \mathcal{F}U \,.\, \mathcal{F}f & \rbrace \textit{ with } \mathcal{F}U \in \mathbf{O}_m . \\ \textit{If } U \in \mathbf{O}_c' \textit{ and } T \in \mathbf{S}', & \textit{then } \mathcal{F}(U * T) = \mathcal{F}U \,.\, \mathcal{F}T & \end{array}$$

Proof.

(a)
$$\begin{aligned} [\mathcal{F}(\mu * \nu)](\omega) &= \int e^{i\omega z}(\mu * \nu)^z \\ &= \int e^{i\omega(x+y)} \,.\, \mu^x \nu^y \\ &= \int e^{i\omega x} \,.\, \mu^x \,.\, \int e^{i\omega y} \,.\, \nu^y \\ &= (\mathcal{F}\mu)(\omega) \,.\, (\mathcal{F}\nu)(\omega) . \end{aligned}$$

(b) Suppose that $g \in \mathbf{S}$. Then,

$$\begin{aligned} \langle \mathcal{F}(U * f), g \rangle &= \langle U * f, \mathcal{F}^\star g \rangle \\ &= \langle U, \mathcal{F}^\star g * \tilde{f} \rangle \\ &= \frac{1}{2\pi} \langle \mathcal{F}U, \mathcal{F}(\mathcal{F}^\star g * \tilde{f}) \rangle \\ &= \langle \mathcal{F}U, g.\overline{\mathcal{F}f} \rangle \\ &= \langle \mathcal{F}U.\mathcal{F}f, g \rangle , \end{aligned}$$

from which the formula follows. For all $U \in \mathbf{O}_c'$, we have

$$\mathcal{F}U \,.\, \mathcal{F}f = \mathcal{F}(U * f) \in \mathbf{S}$$

for every $f \in \mathbf{S}$. Therefore, $\mathcal{F}U \,.\, g \in \mathbf{S}$ for every $g \in \mathbf{S}$. Hence, $\mathcal{F}U \in \mathbf{O}_m$.

Suppose that $g \in \mathbf{S}$. We have

$$\begin{aligned} \langle \mathcal{F}(U * T), g \rangle &= \langle U * T, \mathcal{F}^\star g \rangle \\ &= \langle T, \tilde{U} * \mathcal{F}^\star g \rangle \\ &= \frac{1}{2\pi} \langle \mathcal{F}T, \mathcal{F}(\tilde{U} * \mathcal{F}^\star g) \rangle \\ &= \langle \mathcal{F}T, \overline{\mathcal{F}U} \,.\, g \rangle \\ &= \langle \mathcal{F}T \,.\, \mathcal{F}U, g \rangle , \end{aligned}$$

from which the formula follows.

Corollaries. I. *If $U, V \in \mathbf{O}_c'$, we have*

$$U * V = V * U .$$

This is true because this relation can be written $\mathcal{F}(U * V) = \mathcal{F}(V * U)$, that is,

$$\mathcal{F}U \,.\, \mathcal{F}V = \mathcal{F}V \,.\, \mathcal{F}U .$$

II. *If $U, V \in \mathbf{O}_c'$ and $T \in \mathbf{S}'$, then*

$$(U * V) * T = U * (V * T) .$$

Proof. This relation can be written

$$\mathcal{F}\left[(U * V) * T\right] = \mathcal{F}\left[U * (V * T)\right] ,$$

that is,

$$\mathcal{F}(U * V)\, \mathcal{F}T = \mathcal{F}U\, \mathcal{F}(V * T) ,$$

or

$$[\mathcal{F}U \,.\, \mathcal{F}V]\, \mathcal{F}T = \mathcal{F}U\, [\mathcal{F}V \,.\, \mathcal{F}T].$$

In particular, convolution is associative over $\mathbf{O}_c'$, which is therefor a convolution algebra.

Examples. I. We have

$$\mathcal{F}\delta' = -\, \mathrm{i}\, \omega ,$$

so that

$$\mathcal{F}(T') = \mathcal{F}(\delta' * T) = \mathcal{F}\delta' \,.\, \mathcal{F}T = -\, \mathrm{i}\, \omega \mathcal{F}T .$$

II. We have

$$\mathcal{F}(T * \delta_a) = \mathcal{F}T \,.\, \mathcal{F}\delta_a = \mathcal{F}T \,.\, \mathrm{e}^{\mathrm{i}a\omega} .$$

13. APPLICATIONS TO THE FOURIER TRANSFORM OF A BOUNDED MEASURE

Theorem 20. *On the unit ball* $\mathbf{M}_1$ *in* $\mathbf{M}$ *(that is, on the set of measures of norm* $\leqslant 1$*), the weak topology coincides with the vague topology and with the topology induced by the weak topology of* $\mathbf{S}'$*. For this topology,* $\mathbf{M}_1$ *is compact and closed in* $\mathbf{S}'$.

Proof. On the unit ball in $\mathbf{M}$, the weak topology coincides, on every dense subset of $\mathbf{L}_\infty$, with the topology of simple convergence, in particular, with the topology of simple convergence on $\mathfrak{D}^0$ (the vague topology) and the topology of simple convergence on $\mathbf{S}$ (the topology induced by the weak topology of $\mathbf{S}'$). Furthermore, $\mathbf{M}_1$ is compact for that topology since $\mathbf{M}_1$ is the unit ball in the dual of a Banach space. Therefore, $\mathbf{M}_1$ is closed in $\mathbf{S}'$.

Remark. $\mathbf{M}$ is not closed in $\mathbf{S}'$.

If a sequence $\{\mu_n\}$ of measures μ_n converges narrowly to a measure μ, then the sequence $\{\mathcal{F}\mu_n\}$ converges simply to $\mathcal{F}\mu$ (according to the definition of narrow convergence). The purpose of the following theorems is to study the convergence of measures, beginning with the convergence of their Fourier transforms.

Theorem 21. *If a sequence of measures* μ_n *is such that*

(a) $\| \mu_n \| \leqslant 1$

and

(b) $\mathcal{F}\mu_n$ converges almost everywhere to a function φ,

then $\{\mu_n\}$ *converges weakly to a measure* μ *of norm* $\leqslant 1$ *such that* $\mathcal{F}\mu = \varphi$ *almost everywhere.*

Proof. We have $|\mathcal{F}\mu_n(\omega)| \leqslant 1$ for every ω. Consequently, $|\varphi(\omega)| \leqslant 1$ almost everywhere. Therefore, φ defines a tempered distribution. According to Lebesgue's theorem,

$$\int f \cdot \mathcal{F}\mu_n \to \int f \cdot \varphi,$$

for every $f \in \mathbf{S}$ and hence $\mathcal{F}\mu_n \to \varphi$ in $\mathbf{S}'$ and

$$\mu_n \to \overset{-1}{\mathcal{F}} \varphi = \frac{1}{2\pi} \mathcal{F}^\star \varphi \quad \text{in} \quad \mathbf{S}'.$$

Since the unit ball of $\mathbf{M}$ is closed in $\mathbf{S}'$, it follows that $\overset{-1}{\mathcal{F}} \varphi$ is a measure μ of norm $\leqslant 1$ (such that $\mathcal{F}\mu = \varphi$ almost everywhere).

Theorem 22. *In addition to the hypotheses of the preceding theorem, suppose that*

(c) $\mu_n \geqslant 0$;
(d) $\{\mathcal{F}\mu_n\}$ *converges uniformly to* φ *on a neighborhood of 0.*

Then, $\{\mu_n\}$ *converges narrowly to a positive measure* μ *such that* $\mathcal{F}\mu = \varphi$ *almost everywhere.*

Proof. The function φ is continuous at 0 and

$$\int \mu = \varphi(0) = \lim_{n\to+\infty} (\mathcal{F}\mu_n)(0) = \lim_{n\to+\infty} \int \mu_n,$$

Therefore, $\{\mu_n\}$ converges narrowly to μ. Furthermore, μ is positive since every vague limit of positive measures is positive.

The following is another theorem that we shall have occasion to use later:

Theorem 23. *Let* $\{\mu_n\}$ *denote a sequence of positive measures* μ_n *of mass 1 and let* μ *denote a measure of mass 1. If* $\{\mathcal{F}\mu_n\}$ *converges almost everywhere to* $\mathcal{F}\mu$, *then* $\{\mu_n\}$ *converges narrowly to* μ.

Proof. $\{\mu_n\}$ converges weakly to μ by virtue of Theorem 21 and it converges narrowly to μ because $\left\{\int \mu_n\right\}$ converges to $\int \mu$.

14. THE FOURIER TRANSFORM IN SEVERAL VARIABLES

The theory of the Fourier transform is developed in the same way for functions, measures, or distributions on a real finite-dimensional vector space E. The concepts of measure, bounded measure, and distribution on E are defined without difficulty by using any base of E and thereby referring the discussion to the case of a space $\mathbf{R}^n$. With regard to measures, we shall come back to this question at the beginning of Chapter 5.

If μ is a bounded measure on E, we shall define $\mathcal{F}\mu$ as a function on the dual E' of E by

$$(\mathcal{F}\mu)(U) = \int e^{i\langle X,U\rangle} \,.\, \mu^X.$$

We can replace E' with any space in duality with E, in particular, with E itself if E is Euclidean.

If $E = \mathbf{R}^n$, we set

$$X = \begin{bmatrix} X^1 \\ \vdots \\ X^n \end{bmatrix} \text{ and } U = [U_1, \ldots, U_n].$$

We then have

$$(\mathcal{F}\mu)(U_1, \ldots, U_n) = \int e^{i(U_1X^1 + \cdots + U_nX^n)} \,.\, \mu^X.$$

We shall use the formula $\mathcal{F}(\mu * \nu) = \mathcal{F}\mu \,.\, \mathcal{F}\nu$, which can be generalized without any difficulty.

Let us accept the fact that the Fourier transform of bounded measures is one-to-one. We shall also use natural generalizations of the results of section 13.

FORMULAS

One finds six different definitions of a Fourier transform in the literature. The one chosen depends on the author and the field of use. These definitions are

$$(\mathcal{F}f)(\omega) = \int f(x)\, e^{i\omega x}\, dx$$

$$(\mathcal{F}f)(\omega) = \frac{1}{\sqrt{2\pi}} \int f(x)\, e^{i\omega x}\, dx$$

$$(\mathcal{F}f)(\omega) = \int f(x)\, e^{2i\pi\omega x}\, dx$$

and the three definitions obtained from these by replacing i with −i, that is, by exchanging the definitions of $\mathcal{F}$ and $\mathcal{F}^*$. Here, we give a list of pairs of formulas for the first and third definitions:

First definition	Third definition
$(\mathcal{F}f)(\omega) = \int f(x)\, e^{i\omega x}\, dx$	$(\mathcal{F}f)(\omega) = \int f(x)\, e^{2i\pi\omega x}\, dx$
$(\mathcal{F}^*f)(\omega) = \int f(x)\, e^{-i\omega x}\, dx$	$(\mathcal{F}^*f)(\omega) = \int f(x)\, e^{-2i\pi\omega x}\, dx.$
Plancherel's theorem:	
$\int \lvert\mathcal{F}f\rvert^2 = 2\pi \int \lvert f\rvert^2$	$\int \lvert\mathcal{F}f\rvert^2 = \int \lvert f\rvert^2$
$(\mathcal{F}T)' = \mathcal{F}(i\,xT)$	$(\mathcal{F}T)' = \mathcal{F}(2\,i\,\pi xT)$
$(\mathcal{F}^*T)' = \mathcal{F}^*(-i\,xT)$	$(\mathcal{F}^*T)' = \mathcal{F}^*(-2\,i\,\pi xT)$
$\mathcal{F}T' = -i\,\omega\mathcal{F}T$	$\mathcal{F}T' = -2\,i\,\pi\omega\mathcal{F}T$
$\mathcal{F}^*T' = i\,\omega\mathcal{F}^*T$	$\mathcal{F}^*T' = 2\,i\,\pi\omega\,\mathcal{F}^*T.$

$\mathcal{F}(\delta^{(h)}) = (-\mathrm{i}\,\omega)^h$	$\mathcal{F}(\delta^{(h)}) = (-2\,\mathrm{i}\,\pi\omega)^h$
$\mathcal{F}^\star(\delta^{(h)}) = (\mathrm{i}\,\omega)^h$	$\mathcal{F}^\star(\delta^{(h)}) = (2\,\mathrm{i}\,\pi\omega)^h$
$\mathcal{F}(x^h) = \dfrac{2\pi}{\mathrm{i}^h}\,\delta^{(h)}$	$\mathcal{F}(x^h) = \dfrac{1}{(2\mathrm{i}\pi)^h}\,\delta^{(h)}$
$\mathcal{F}^\star(x^h) = \dfrac{2\pi}{(-\mathrm{i})^h}\,\delta^{(h)}$	$\mathcal{F}^\star(x^h) = \dfrac{1}{(-2\mathrm{i}\pi)^h}\,\delta^{(h)}$
$\overset{-1}{\mathcal{F}} = \dfrac{1}{2\pi}\,\mathcal{F}^\star$	$\overset{-1}{\mathcal{F}} = \mathcal{F}^\star$

4

The Laplace Transform

1. THE LAPLACE TRANSFORM OF MEASURES WITH POSITIVE SUPPORT

Definition. The *Laplace transform* of a measure μ with positive support is the function $\mathcal{L}\mu$ of a complex variable p defined by

$$(\mathcal{L}\mu)(p) = \int e^{-pt} \,.\, \mu^t \,.$$

The function $\mathcal{L}\mu$ is initially defined for values of p such that e^{-pt} is integrable with respect to μ^t. Afterwards, in view of the analytical properties of $\mathcal{L}\mu$ in its initial domain of definition, we can extend this definition analytically.

Examples. I. We have

$$(\mathcal{L}\delta_a)(p) = \int e^{-pt} \,.\, \delta_a^t = e^{-pa}.$$

Note that $\mathcal{L}\delta_a$ is an entire function of the complex variable p.

II. If μ has a density f with respect to Lebesgue measure, we adopt the convention that $\mathcal{L}f = \mathcal{L}\mu$, that is,

$$(\mathcal{L}f)(p) = \int_0^{+\infty} f(t)\, e^{-pt}\, dt \,.$$

The preceding definition applies to every locally integrable function of positive support. In particular, let

$$f(t) = \mathcal{Y}(t)\, e^{\lambda t} \qquad (\lambda \in \mathbf{C}),$$

where

$$\mathcal{Y}(t) = \begin{cases} +1 \text{ for } t \geqslant 0, \\ 0 \text{ for } t < 0. \end{cases}$$

We have

$$(\mathcal{L}f)(p) = \int_0^{+\infty} e^{\lambda t}\, e^{-pt}\, dt = \int_0^{+\infty} e^{(\lambda - p)t}\, dt .$$

The integral on the right is absolutely convergent for $\mathcal{R}(p) > \mathcal{R}(\lambda)$. We then have

$$(\mathcal{L}f)(p) = \frac{1}{p - \lambda}.$$

In particular,

$$(\mathcal{L}\mathcal{Y})(p) = \frac{1}{p}.$$

Thus, the Laplace transform of $f(t) = \mathcal{Y}(t)\, e^{\lambda t}$ can be extended as a rational fraction.

Remark. Two functions with positive support f_1 and f_2 that are locally integrable and equal to each other almost everywhere (with respect to Lebesgue measure) have the same Laplace transform.

We shall now give certain general properties of the set of values of p for which

$$(\mathcal{L}\mu)(p) = \int e^{-pt} \,.\, \mu^t$$

is defined. This set is the set of those p for which one of the following equivalent properties is satisfied:

$$\begin{cases} e^{-pt} \text{ is integrable with respect to } \mu^t; \\ e^{-pt} \,.\, \mu^t \in \mathbf{M} \text{ (the set of bounded measures).} \end{cases}$$

Let us set $p = \xi + i\eta$. We have

$$e^{-pt} = e^{-\xi t}\, e^{-i\eta t}.$$

Consequently, a necessary and sufficient condition for e^{-pt} to be integrable with respect to μ^t is that $e^{-\xi t}$ be integrable with μ^t. In other words, the set of those p such that e^{-pt} is integrable with respect to μ^t is invariant under purely imaginary translations.

This set is a strip representing the union of lines parallel to the imaginary axis.

Furthermore, if $e^{-\xi_1 t}$ is integrable with respect to μ^t for $\xi_1 \in \mathbf{R}$, then $e^{-\xi t}$ will also be integrable with respect to μ^t if $\xi \geqslant \xi_1$ since

$$e^{-\xi t} = e^{-\xi_1 t} \,.\, e^{-(\xi - \xi_1)t}.$$

Consequently, the set of those ξ such that $e^{-\xi t}$ is integrable with respect to μ^t is an open ray $]\xi_0, +\infty[$ or a closed ray $[\xi_0, +\infty[$, or the entire real line or the empty set. Thus, we have the following theorem.

Theorem 1. *There exists a ξ_0 (which may be equal to $+\infty$ or $-\infty$) in the extended real number system such that the integral*

$$(\mathfrak{L}\mu)(p) = \int e^{-pt} \,.\, \mu^t,$$

is defined for $\mathfrak{R}(p) > \xi_0$ but not for $\mathfrak{R}(p) < \xi_0$.

The number ξ_0 is called the ***abscissa of integrability.*** We note that if μ has a density f, the above theorem means that the integral

$$(\mathfrak{L}f)(p) = \int_0^{+\infty} f(t)\, e^{-pt}\, dt,$$

is absolutely convergent for $\mathfrak{R}(p) > \xi_0$ but not absolutely convergent for $\mathfrak{R}(p) < \xi_0$. It can be semiconvergent for values of p such that $\mathfrak{R}(p) < \xi_0$. Thus, the situation is not completely parallel to the situation that exists in the study of sets of absolute convergence of power series.

In no case, shall we use the determination of $(\mathfrak{L}f)(p)$ by the integral

$$\int_0^{+\infty} f(t)\, e^{-pt}\, dt,$$

unless this integral is absolutely convergent. In the case of a function f, the abscissa of integrability is generally called the ***abscissa of absolute convergence.***

Examples. I. The abscissa of integrability of $(\mathfrak{L}\delta_a)(p) = \int e^{-pt} \,.\, \delta_a = e^{-ap}$ is $-\infty$.

II. For $f(t) = \mathcal{Y}(t)\, e^{\lambda t}$, the abscissa of absolute convergence of

$$(\mathfrak{L}f)(p) = \int_0^{+\infty} e^{(\lambda - p)t}\, dt = \frac{1}{p - \lambda}$$

is $\mathfrak{R}(\lambda)$.

III. If μ or f has compact support, the abscissa of integrability of $\mathfrak{L}\mu$ or $\mathfrak{L}f$ is equal to $-\infty$.

Theorem 2. *If* $|f(t)| \leqslant |g(t)|$, *the abscissa of absolute convergence of* $\mathfrak{L}f$ *does not exceed the abscissa of absolute convergence of* $\mathfrak{L}g$.

This follows immediately from the definitions.

An application. If $|f(t)| \leqslant A\,e^{\lambda t}$ (where $\lambda \in \mathbf{R}$), the abscissa of absolute convergence of $\mathfrak{L}f$ does not exceed λ.

Notations. In the theory, we can denote by $\mathfrak{L}f$ the Laplace transform of f. In practice, one often denotes f and $\mathfrak{L}f$ by their analytic expressions as functions of t and p, respectively.

If $F = \mathfrak{L}f$, we may write

$$f(t) \sqsupset F(p)\,.$$

Examples.

$$\mathcal{Y}(t)\,e^{\lambda t} \sqsupset \frac{1}{p-\lambda}\,.$$

$$\mathcal{Y}(t) \sqsupset \frac{1}{p}\,.$$

Some common formulas. The Laplace transform is a linear transformation: If $\mathfrak{L}\mu$ and $\mathfrak{L}\nu$ have abscissas of integrability ξ_0 and ξ_1, respectively, the abscissa of integrability of $\mathfrak{L}(a\mu + b\nu)$ does not exceed max (ξ_0, ξ_1) and we have

$$[\mathfrak{L}(a\mu + b\nu)]\,(p) = a(\mathfrak{L}\mu)\,(p) + b(\mathfrak{L}\nu)\,(p)$$

for $\mathcal{R}(p) >$ max (ξ_0, ξ_1).

The following table gives several Laplace transforms:

	$f(t)$	$F(p)$
$\mathcal{Y}(t)\,\times$	$\frac{1}{\omega}\sin \omega t$	$\frac{1}{p^2+\omega^2}$
	$\cos \omega t$	$\frac{p}{p^2+\omega^2}$
	$\frac{1}{\omega}\sinh \omega t$	$\frac{1}{p^2-\omega^2}$
	$\cosh \omega t$	$\frac{p}{p^2-\omega^2}$

These formulas are derived from the formula

$$\mathcal{Y}(t)\, e^{\lambda t} \sqsupset \frac{1}{p-\lambda}.$$

We also note the following rule: If $f \sqsupset F$, then

$$f\left(\frac{t}{\lambda}\right) \sqsupset \lambda F(\lambda p) \quad \text{and} \quad \frac{1}{\lambda} f\left(\frac{t}{\lambda}\right) \sqsupset F(\lambda p).$$

Theorem 3 (holomorphicity of the Laplace transform). *If*

$$(\mathfrak{L}\mu)(p) = \int e^{-pt} \,.\, \mu^t$$

has abscissa of integrability ξ_0, ***then*** $(\mathfrak{L}\mu)(p)$ ***is a holomorphic function of*** p ***for*** $\mathcal{R}(p) > \xi_0$ ***and***

$$(\mathfrak{L}\mu)^{(k)}(p) = (-1)^k \left[\mathfrak{L}(t^k \mu)\right](p).$$

In other words, if $f(t) \sqsupset F(p)$, ***then*** $(-t)^k f(t) \sqsupset F^{(k)}(p)$.

Proof. Let ξ_1 denote the abscissa of integrability of $\mathfrak{L}(t\mu^t)$. Let us show that $\xi_1 \geqslant \xi_0$. For every $\xi > \xi_1$, we have

$$t\, e^{-\xi t} \mu^t \in \mathbf{M}, \quad \text{so that} \quad e^{-\xi t} \mu^t \in \mathbf{M} \quad \text{and} \quad \xi \geqslant \xi_0 .$$

(b) Let us show that $\xi_1 \leqslant \xi_0$. Suppose that $\xi > \xi_0$ and $\xi' \in]\xi_0, \xi[$. Then,

$$e^{-\xi' t} \mu^t \in \mathbf{M},$$

so that

$$t\, e^{-\xi t} \,.\, \mu^t = t\, e^{-(\xi - \xi')t} e^{-\xi' t} \,.\, \mu^t \in \mathbf{M},$$

and, consequently, $\xi \geqslant \xi_1$.

It follows from (a) and (b) that $\xi_0 = \xi_1$.

(c) Suppose that $\xi' > \xi_0$. Suppose that $\xi > \xi'$. Consider

$$(\mathfrak{L}\mu)(\xi + i\eta) = \int e^{-(\xi + i\eta t)} \,.\, \mu^t$$

as a function of two real variables ξ and η.

Subject to the applicability of the rule for differentiating an integral depending on a parameter, we shall have

$$\frac{\partial}{\partial \xi}(\mathfrak{L}\mu)(\xi + i\eta) = -\int t\, e^{-(\xi + i\eta)t} \,.\, \mu^t$$

$$\frac{\partial}{\partial \eta}(\mathfrak{L}\mu)(\xi + i\eta) = -i \int t\, e^{-(\xi + i\eta)t} \,.\, \mu^t$$

We have

$$|t\,e^{-(\xi+i\eta)t}| \leqslant |t|\,e^{-\xi' t}\ (\in \mathbf{L}^1(\mu^t)).$$

Thus, the rule for differentiality is applicable. Therefore,

$$d[\mathfrak{L}\mu(\xi + i\,\eta)] = -(d\xi + i\,d\eta)\int t\,e^{-(\xi+i\eta)t}\cdot\mu^t,$$

which shows that $(\mathfrak{L}\mu)(\xi + i\eta)$ is holomorphic in the half-plane $\xi > \xi'$, and hence in the half-plane $\xi > \xi_0$, and that

$$(\mathfrak{L}\mu)'(p) = -\int t\,e^{-pt}\,.\,\mu^t = [\mathfrak{L}(-t\mu^t)](p)\,.$$

From this we immediately derive the rule for forming the kth derivative.

Corollary. *If μ has compact support, $\mathfrak{L}\mu$ is an entire function.*

An application. Suppose that

$$f(t) = \mathcal{Y}(t)\frac{t^{\alpha-1}}{\Gamma(\alpha)}, \quad \text{with} \quad \alpha > 0\,.$$

For real positive p,

$$(\mathfrak{L}f)(p) = \int_0^{+\infty}\frac{t^{\alpha-1}}{\Gamma(\alpha)}e^{-pt}\,dt$$

$$= \frac{1}{\Gamma(\alpha)}\cdot\frac{1}{p^\alpha}\int_0^{+\infty}u^{\alpha-1}\,.\,e^{-u}\,du = \frac{1}{p^\alpha}\,.$$

$(\mathfrak{L}f)(p)$ is a holomorphic function of p for $\mathfrak{R}(p) > 0$. We have necessarily

$$(\mathfrak{L}f)(p) = \frac{1}{p^\alpha}$$

(real determination for $p \in \mathbf{R}_+$).

Theorem 4. *If the abscissa of integrability of $\mathfrak{L}\mu$ is ξ_0, then the abscissa of integrability of $\mathfrak{L}(e^{\lambda t}\mu^t)$ is $\xi_0 + \lambda$ and we have*

$$[\mathfrak{L}(e^{\lambda t}\mu^t)](p) = [\mathfrak{L}(\mu)](p - \lambda)\,, \quad (\textit{for } \mathfrak{R}(p) > \xi_0 + \lambda).$$

Proof. We have

$$e^{-\xi t}\,e^{\lambda t}\,.\,\mu^t = e^{-(\xi-\lambda)t}\,.\,\mu^t.$$

Thus, this measure is bounded if $\xi - \lambda$ belongs to the set of integrability of $\mathfrak{L}\mu$, from which the first part of the theorem follows. We then have

$$[\mathfrak{L}(e^{\lambda t} . \mu^t)](p) = \int e^{-(p-\lambda)t} . \mu^t = [\mathfrak{L}(\mu)](p - \lambda).$$

An application. We have

$$e^{\lambda t} \frac{t^{\alpha-1}}{\Gamma(\alpha)} \sqsupset \frac{1}{(p-\lambda)^\alpha}, \quad (\alpha > 0).$$

A relationship between the Laplace and Fourier transforms. Let ξ_0 be the abscissa of integrability of

$$(\mathfrak{L}\mu)(p) = \int e^{-pt} . \mu^t .$$

Let us set $p = \xi + i\eta$. For $\xi > \xi_0$ (and also for $\xi = \xi_0$ if the straight line $\xi = \xi_0$ is part of the set of integrability), we have

$$(\mathfrak{L}\mu)(\xi + i\eta) = \int e^{-(\xi+i\eta)t} . \mu^t = \int e^{-i\eta t} e^{-\xi t} . \mu^t,$$

that is,

$$(\mathfrak{L}\mu)(\xi + i\eta) = [\mathcal{F}^\star(e^{-\xi t} . \mu^t)](\eta).$$

This formula can be written

$$(\mathfrak{L}\mu)(\xi + i\eta) = [\mathcal{F}^\star(e^{-\xi \cdot} . \mu)](\eta),$$

where $e^{-\xi \cdot}$ denotes the function $t \to e^{-\xi t}$.

In particular, if f is locally integrable and if $\xi > \xi_0$ (or $\xi = \xi_0$ under the condition stated above),

$$(\mathfrak{L}f)(\xi + i\eta) = [\mathcal{F}^\star(e^{-\xi t} f(t))](\eta).$$

We can rewrite this formula

$$(\mathfrak{L}f)(\xi + i\eta) = [\mathcal{F}^\star(e^{-\xi \cdot} f)](\eta).$$

If μ is a bounded measure or if $f \in \mathbf{L}^1$, we have

$$(\mathfrak{L}\mu)(i\eta) = (\mathcal{F}^\star \mu)(\eta),$$
$$(\mathfrak{L}f)(i\eta) = (\mathcal{F}^\star f)(\eta),$$

that is,

$$(\mathcal{F}\mu)(\eta) = \mathfrak{L}\mu(-i\eta),$$
$$(\mathcal{F}f)(\eta) = \mathfrak{L}\mu(-i\eta).$$

Theorem 5. (on the transform of a convolution product). *Let μ and ν denote two measures with positive support and whose Laplace transforms have abscissas of integrability ξ_0 and ξ_1, respectively. If $\mathcal{R}(p) > \sup(\xi_0, \xi_1)$, then*

$$[\mathcal{L}(\mu * \nu)](p) = (\mathcal{L}\mu)(p) \, . \, (\mathcal{L}\nu)(p) \, .$$

Proof. We have

$$(e^{-p\cdot}\mu) * (e^{-p\cdot}\nu) = e^{-p\cdot}(\mu * \nu) \qquad (\forall p \in \mathbf{C}) \, .$$

Therefore, for $\xi > \sup(\xi_0, \xi_1)$, the measure $e^{-\xi\cdot}(\mu * \nu)$ is bounded and we have, for $\mathcal{R}(p) > \sup(\xi_0, \xi_1)$,

$$\int e^{-pt}(\mu * \nu)^t = \int e^{-pt} \, . \, \mu^t \, . \int e^{-pt} \, . \, \nu^t \, ,$$

from which the theorem follows.

Corollaries. I. *Let μ denote a measure with positive support and let ξ_0 denote the abscissa of integrability of $\mathcal{L}\mu$. For $\mathcal{R}(p) > \sup(\xi_0, 0)$,*

$$[\mathcal{L}(\mathcal{Y} * \mu)](p) = \frac{1}{p}(\mathcal{L}\mu)(p) \, .$$

We note that

$$(\mathcal{Y} * \mu)(t) = \int_{[0,t[} \mu \, ,$$

almost everywhere. In other words, $\mathcal{Y} * \mu$ is equal almost everywhere to the cumulative function of μ that vanishes for $t < 0$.

We recall that the equation $f = \mathcal{Y} * \mu$ can also be written $\mu = \delta' * f$, from which we get the following:

II. *If $\mu = \delta' * f$ and $\mathcal{R}(p)$ exceeds the abscissas of integrability of $\mathcal{L}f$ and $\mathcal{L}\mu$, then*

$$(\mathcal{L}\mu)(p) = p(\mathcal{L}f)(p) \, .$$

Of course, in applying this result, we can replace μ with a locally integrable function.

2. THE LAPLACE TRANSFORM OF A DISTRIBUTION

We shall confine ourselves to distributions T with positive support, that is, to distributions in $\mathcal{D}'_+$ that satisfy the following supplementary property:

(A) T is the derivative of some order of a function with positive support and whose Laplace transform has an abscissa of absolute convergence other than $+\infty$.

A deeper study shows that this restriction does not exclude any case in which the Laplace transform could be defined by more general procedures than those that we shall use here.

Definition. For $T = D^k f \in \mathfrak{D}'_+$, where f is such that the abscissa of absolute convergence of $\mathfrak{L}f$ is ξ_0, we define

$$(1) \qquad (\mathfrak{L}T)(p) = p^k(\mathfrak{L}f)(p) \qquad (\text{for } \mathfrak{R}(p) > \xi_0).$$

Then, $\mathfrak{L}T$ is a holomorphic function for $\mathfrak{R}(p) > \xi_0$.

We should point out that this definition constitutes an extension of the definition given in the preceding section and that it is independent of the choice of k.

It should be understood that if we define $\mathfrak{L}T$ by two different methods, namely, as a holomorphic function for $\mathfrak{R}(p) > a$ in one case and as a holomorphic function for $\mathfrak{R}(p) > b$ in the other case, it is sufficient to verify that the two definitions coincide for $\mathfrak{R}(p) > \max(a, b)$. To do this, it suffices to verify that this is the case for $\mathfrak{R}(p) > c$, where $c > \max(a, b)$.

Let T denote a measure. If $T = D^k f$, we have formula (1) for $\mathfrak{R}(p)$ greater than the abscissas of integrability of f and T. Furthermore, if $T = D^k f = D^h g$, where $k < h$, it necessarily follows that $f = D^{h-k} g$, from which we conclude

$$(\mathfrak{L}f)(p) = p^{h-k}(\mathfrak{L}g)(p),$$

and

$$p^k(\mathfrak{L}f)(p) = p^h(\mathfrak{L}g)(p)$$

for $\mathfrak{R}(p)$ greater than the abscissas of absolute convergence of f and g. Thus, the definition is justified.

Example. Suppose that $\delta^{(k)} = D^{k+1}\mathcal{Y}$. We have

$$(\mathfrak{L}\delta^{(k)})(p) = p^{k+1}(\mathfrak{L}\mathcal{Y})(p) = p^{k+1} \cdot \frac{1}{p} = p^k.$$

It follows from the definition of the Laplace transform of a distribution T that $\mathfrak{L}T$ is a holomorphic function in a half-plane of the form $\mathfrak{R}(p) > \alpha$.

Theorem 6. $(\mathfrak{L}T)' = -\mathfrak{L}(tT)$.

Proof. Suppose that

$$T = D^k f, \; (\mathfrak{L}T)(p) = p^k(\mathfrak{L}f)(p).$$

We have

$$(\mathfrak{L}T)'(p) = kp^{k-1}(\mathfrak{L}f)(p) - p^k[\mathfrak{L}(tf)](p),$$
$$\mathfrak{L}(tT) = \mathfrak{L}(tD^k f) = \mathfrak{L}[D^k(tf) - kD^{k-1}f],$$

and, consequently,

$$[\mathfrak{L}(tT)](p) = p^k[\mathfrak{L}(tf)](p) - kp^{k-1}(\mathfrak{L}f)(p).$$

Theorem 7. *Let T and U denote two distributions (in $\mathfrak{D}'_+$) satisfying property* (A). *Then,*

$$\mathfrak{L}(T * U) = \mathfrak{L}(T) \,.\, \mathfrak{L}(U).$$

Proof. Assume that $T = D^k f$ and $U = D^h g$. Then,

$$T * U = D^{k+h}(f * g),$$

and, consequently,

$$(\mathfrak{L}T)(p) = p^k(\mathfrak{L}f)(p),$$
$$(\mathfrak{L}U)(p) = p^h(\mathfrak{L}g)(p),$$

and

$$[\mathfrak{L}(T * U)](p) = p^{k+h}[\mathfrak{L}(f * g)](p)$$
$$= p^{k+h}(\mathfrak{L}f)(p) \,.\, (\mathfrak{L}g)(p).$$

Corollary. *If $T = D^k U$, then* $(\mathfrak{L}T)(p) = p^k(\mathfrak{L}U)(p)$.

This formula can also be proven directly from the definition of $\mathfrak{L}T$ and $\mathfrak{L}U$.

Theorem 8. $[\mathfrak{L}(e^{\lambda t}T)](p) = (\mathfrak{L}T)(p - \lambda)$.

Proof. Let us suppose the theorem true for every distribution T_1 of the form $D^k f$ and let us show that it is true for a distribution of the form $D^{k+1}f$. Suppose that $T = D^{k+1}f = DT_1$, where $T_1 = D^k f$. We have

$$e^{\lambda t}T = e^{\lambda t}DT_1 = D(e^{\lambda t}T_1) - \lambda e^{\lambda t}T_1,$$

from which we get

$$[\mathfrak{L}(e^{\lambda t}DT_1)](p) = p\mathfrak{L}(e^{\lambda t}T_1)(p) - \lambda\mathfrak{L}(e^{\lambda t}T_1)$$
$$= (p - \lambda) \times (\mathfrak{L}T_1)(p - \lambda).$$

Since

$$[\mathfrak{L}(DT_1)](p) = p(\mathfrak{L}T_1)(p),$$

the formula is also proven for T.

Theorem 9. *Suppose that $T = D^k f$ and that the abscissa of absolute convergence of $\mathcal{L}f$ is ξ_0. For every $\xi > \xi_0$, we have*

$$(\mathcal{L}T)(\xi + i\,\eta) = [\mathcal{F}^\star(e^{-\xi t}\,T)](\eta)\,.$$

Proof. Let us suppose that $\xi_0 > 0$. (We can reduce the opposite case $\xi_0 \leq 0$ to this this one by setting $T = e^{\alpha t}\,T_0$ with $\alpha + \xi_0 > 0$.) Then,

$$\begin{aligned}(\mathcal{L}T)(\xi + i\,\eta) &= (\xi + i\,\eta)^k \times (\mathcal{L}f)(\xi + i\eta)\\ &= (\xi + i\,\eta)^k\,[\mathcal{F}^\star(e^{-\xi t} f(t))](\eta)\\ &= (\xi + i\,\eta)^k\,[\mathcal{F}^\star(e^{-\xi t}\,.\,\mathcal{Y} * \cdots * \mathcal{Y} * T)]\\ &= (\xi + i\,\eta)^k\,[\mathcal{F}^\star(e^{-\xi t}\,\mathcal{Y})]^k(\eta)\,[\mathcal{F}^\star(e^{-\xi t}\,T)](\eta)\\ &= (\xi + i\,\eta)^k\,\frac{1}{(\xi + i\,\eta)^k}\,[\mathcal{F}^\star(e^{-\xi t}\;T)](\eta)\\ &= [\mathcal{F}^\star(e^{-\xi t}\,T)](\eta)\,.\end{aligned}$$

3. INVERSE-TRANSFORM FORMULAS

The relation between the Laplace and Fourier transforms shows that if $\mathcal{L}T$ vanishes $\mathcal{R}(p) > \alpha$, then $T = 0$.

The formula for the inverse of the Fourier transform yields the following results:

Theorem 10. *Let $F = \mathcal{L}f$ and let ξ_0 denote the abscissa of absolute convergence of $\mathcal{L}f$. Let ξ denote a number greater than ξ_0. If the function $\eta \to F(\xi + i\,\eta)$ is integrable, then*

$$f(t) = \frac{1}{2\pi i}\int_{\xi - i\infty}^{\xi + i\infty} F(p)\,e^{pt}\,dp\,,$$

almost everywhere with respect to Lebesgue measure, where the integral on the right is taken along the line of abscissa ξ in the direction of increasing η.

Proof. We have

$$F(\xi + i\eta) = [\mathcal{F}^\star(e^{-\xi t}\,f(t))](\eta)\,,$$

so that

$$e^{-\xi t} f(t) = \frac{1}{2\pi}\int_{-\infty}^{+\infty} F(\xi + i\,\eta)\,e^{it\eta}\,d\eta, \qquad \text{(a. e.)}$$

and

$$f(t) = \frac{1}{2\pi}\int_{-\infty}^{+\infty} F(\xi + i\eta)\,e^{t(\xi + i\eta)}\,d\eta\,; \qquad \text{(a. e.)}$$

that is,

$$f(t) = \frac{1}{2\pi i}\int_{\xi - i\infty}^{\xi + i\infty} F(p)\, e^{pt}\, dp . \qquad \text{(a. e.)}$$

Theorem 11. *Let $F(p)$ denote a function that is holomorphic for $\mathcal{R}(p) > 0$ and continuous for $\mathcal{R}(p) \geqslant 0$. Suppose that F satisfies a condition of the form*

$$|F(p)| \leqslant \frac{C}{|p^2|}.$$

for $\mathcal{R}(p) \geqslant 0$. Then there exists a continuous function f with positive support that approaches 0 as $t \to +\infty$ and such that $F = \mathfrak{L}f$ with abscissa of absolute convergence not exceeding 0. Therefore,

$$f(t) = \frac{1}{2\pi}\int_{-\infty}^{+\infty} F(i\eta)\, e^{i\eta t}\, d\eta .$$

Proof. Suppose that

$$f_\xi(t) = \frac{1}{2i\pi}\int_{\xi - i\infty}^{\xi + i\infty} F(p)\, e^{pt}\, dp, \quad \xi \geqslant 0$$

(1)
$$= \frac{e^{\xi t}}{2\pi}\int_{-\infty}^{+\infty} F(\xi + i\eta)\, e^{i\eta t}\, d\eta .$$

For fixed ξ, the function $F(\xi + i\eta)$ is bounded and

$$F(\xi + i\eta) = O\left(\frac{1}{|\eta|^2}\right),$$

as $\eta \to \infty$. Consequently, the integral

$$\int_{-\infty}^{+\infty} F(\xi + i\eta)\, e^{i\eta t}\, d\eta,$$

is absolutely convergent for every ξ and t. We have

$$|F(p)| \leqslant \frac{C}{\eta^2},$$

which implies that the preceding integral is uniformly convergent with respect to ξ and t. Consequently, $f_\xi(t)$ is continuous with respect to ξ and t.

Let us show that $f_\xi(t)$ is independent of ξ. Since

$$f_0(t) = \lim_{\xi \to 0} f_\xi(t),$$

we need only to show that $f_{\xi_1}(t) = f_{\xi_2}(t)$ for $0 < \xi_2 < \xi_1$. Suppose that

$$\varepsilon = f_{\xi_1}(t) - f_{\xi_2}(t)$$
$$= \frac{1}{2i\pi}\int_{\xi_1-i\infty}^{\xi_1+i\infty} F(p)\, e^{pt}\, dp - \frac{1}{2i\pi}\int_{\xi_2-i\infty}^{\xi_2+i\infty} F(p)\, e^{pt}\, dp .$$

Let $\gamma = \gamma_1 + \gamma_2 + \gamma_3 + \gamma_4$ denote the rectangular contour defined by the points $\xi_1 \pm i\, R$ and $\xi_2 \pm i\, R$. Then,

$$\varepsilon = \lim_{R\to\infty} \frac{1}{2i\pi}\int_{\gamma_1+\gamma_3} F(p)\, e^{pt}\, dp ,$$

where the integration is in the positive direction (see Fig. 1). Since $F(p)$ is holomorphic for $\Re(p) > 0$,

$$\varepsilon = \lim_{R\to\infty} \frac{-1}{2i\pi}\int_{\gamma_4+\gamma_2} F(p)\, e^{pt}\, dp$$
$$= \lim_{R\to\infty} \frac{1}{2i\pi}\left[\int_{\xi_2}^{\xi_1} F(\xi + i\, R)\, e^{(\xi+iR)t}\, d\xi - \int_{\xi_2}^{\xi_1} F(\xi - i\, R)\, e^{(\xi-iR)t}\, d\xi\right] ,$$

and, consequently,

$$|\varepsilon| \leqslant \frac{1}{2\pi}\cdot 2 \cdot \frac{C}{R^2} | \xi_1 - \xi_2 |\, e^{\xi_1 t} .$$

Since the right-hand member tends to 0 as $R \to \infty$, it follows that $\varepsilon = 0$.

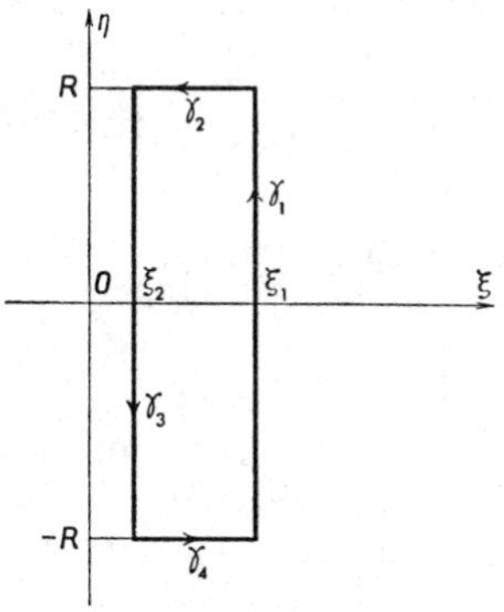

FIG. 1.

Let us now denote by $f(t)$ the value (independent of ξ) of $f_\xi(t)$. We have, in particular,

$$f(t) = \frac{1}{2\pi}\int_{-\infty}^{+\infty} F(i\eta)\, e^{i\eta t}\, d\eta .$$

Thus, f is continuous and bounded and it approaches 0 as $t \to \infty$.

Furthermore,

$$|f(t)| \leqslant \frac{e^{\xi t}}{2\pi}\int_{-\infty}^{+\infty}\frac{C}{\xi^2+\eta^2}\,d\eta \qquad (\forall \xi > 0);$$

that is,

$$|f(t)| \leqslant \frac{C e^{\xi t}}{2\xi}.$$

Suppose that $t < 0$. By letting ξ approach $+\infty$, we see that $f(t) = 0$.

We have seen that f was bounded; therefore $\mathfrak{L}f$ has a nonpositive abscissa of absolute convergence.

Finally, the formula for the inverse Fourier transform applied to (1) yields, for $\xi > 0$,

$$\begin{aligned} F(\xi + i\eta) &= \int_{-\infty}^{+\infty} f(t)\, e^{-\xi t}\, e^{-i\eta t}\, dt \\ &= (\mathfrak{L}f)(\xi + i\eta), \end{aligned}$$

which completes the proof.

Remark. The hypotheses of the preceding theorem do not enable us to show that $f \in \mathbf{L}^1$. We can show this fact in certain particular cases by using the differentiability properties that the function $\eta \to F(i\eta)$ may have and the fact that $2\pi f$ is the Fourier transform of that function.

Corollary. *If $F(p)$ is holomorphic for $\mathfrak{R}(p) > c$ and if*

$$F(p) \leqslant \frac{C}{|p|^2} \qquad (\textit{for } \mathfrak{R}(p) > c),$$

then $F(p)$ is the Laplace transform of a function f with positive support and such that the abscissa of absolute convergence of $\mathfrak{L}f$ does not exceed c.

Proof. For every $c_1 > c$, the function $F(p - c_1) = F_1$ satisfies the hypotheses of the preceding theorem. There exists a function f_1 such that $F_1 = \mathfrak{L}f_1$ and the abscissa of convergence of $\mathfrak{L}f_1$ is nonpositive. Thus,

$$F = \mathfrak{L}f \quad \text{with} \quad f = e^{c_1 t} f_1,$$

the abscissa of convergence of $\mathfrak{L}f$ not exceeding c_1. Since c_1 is arbitrarily close to (but exceeds) c, the abscissa of absolute convergence does not exceed c.

Theorem 12. *If $F(p)$ is holomorphic for $\mathfrak{R}(p) > c$ and if*

$$F(p) \leqslant C\,|p|^m \qquad (\textit{for } \mathfrak{R}(p) > c),$$

then $F(p)$ is the Laplace transform of a distribution T with positive support.

Proof. The function $F(p)/p^{m+1}$ satisfies the hypotheses of the preceding corollary and hence is the Laplace transform of a function f. Consequently, $F(p)$ is the Laplace transform of $D^{m+2}f$.

Remark. In the notations of the preceding theorem, one can can evaluate T in the following manner: $T = D^{m+2}f$ with

$$f(t) = \frac{1}{2i\pi}\int_{\xi-i\infty}^{\xi+i\infty}\frac{F(p)}{p^{m+2}}\,e^{pt}\,dp$$

(for ξ positive and greater then the abscissa of absolute convergence of $\mathcal{L}f$).

4. INITIAL-VALUE AND FINAL-VALUE THEOREMS

In order to interpret more easily the theorems of this section, we shall first make some additional remarks regarding measures. The set of real numbers together with the points $+\infty$ and $-\infty$ is called the ***extended real line***, and we denote it by $\overline{\mathbf{R}}$. We equip $\overline{\mathbf{R}}$ with the topology defined by the following base of neighborhoods: The basic neighborhoods of an $x \neq \pm\infty$ are its neighborhoods in $\mathbf{R}$, and the basic neighborhoods of $+\infty$ (resp. $-\infty$) are the unions of $+\infty$ (resp. $-\infty$) and the complements of subsets of $\mathbf{R}$ that are bounded from above (resp. bounded from below).

The mapping Arctan with the conventions Arctan $(+\infty) = \pi/2$ and Arctan $(-\infty) = -\pi/2$ is a homeomorphism of $\overline{\mathbf{R}}$ onto $[-\pi/2, +\pi/2]$. From this we conclude that $\overline{\mathbf{R}}$ is compact.

Let Φ denote the set of continuous functions defined on $\mathbf{R}$ that approach limits as their arguments approach $+\infty$ and $-\infty$. Every $f \in \Phi$ will be identified with its (continuous) extension to $\overline{\mathbf{R}}$ defined by the conventions

$$f(+\infty) = \lim_{t\to+\infty} f(t), \qquad f(-\infty) = \lim_{t\to-\infty} f(t).$$

The mapping $f \to f \circ \tan$ is an isomorphism of Φ onto the space of continuous functions defined on $[-\pi/2, +\pi/2]$. For the topology of uniform convergence, we have

$$\|f\| = \|f \circ \tan\|.$$

We shall call every continuous linear form μ defined on Φ a ***measure*** on $\overline{\mathbf{R}}$. We can associate with every measure on $\overline{\mathbf{R}}$ a measure μ_1 on $\mathbf{R}$ with support in $[-\pi/2, +\pi/2]$ such that

$$\int f \cdot \mu = \int f_1 \cdot \mu_1, \quad \text{with} \quad f_1 = f \circ \tan.$$

The measure μ_1 will be called the ***image*** of μ under the mapping Arctan.

In particular, we shall consider the measures defined by

$$\left.\begin{aligned}\int f\cdot\delta_{+\infty} &= f(+\infty)\\ \int f\cdot\delta_{-\infty} &= f(-\infty)\end{aligned}\right\}\quad \text{for} \quad f\in\Phi,$$

the images of which under the Arctan mapping are $\delta_{+\pi/2}$ and $\delta_{-\pi/2}$. Furthermore, every bounded measure on $\mathbf{R}$ defines a measure on $\overline{\mathbf{R}}$ since every function $f\in\Phi$ is integrable with respect to a bounded measure on $\mathbf{R}$.

We equip the measures on $\overline{\mathbf{R}}$ with the topology of simple convergence on Φ. We then have

$$\mu_n \to \mu \quad \Leftrightarrow \quad \int f\cdot\mu_n \to \int f\cdot\mu, \qquad \forall f\in\Phi.$$

For $\{\mu_n\}$ to converge to μ, it is necessary and sufficient that the sequence of images of the μ_n converge narrowly (or vaguely) to the image of μ.

Lemma 1. *Let μ_n denote a positive measure on $\mathbf{R}$ of total mass equal to 1. Suppose that, for every closed neighborhood V of a point a, we have $\lim_{n\to\infty}\mu_n(V) = 1$. Then, $\{\mu_n\}$ converges narrowly to δ_a.*

Proof. From the relation $\mu_n(V) + \mu_n(\complement V) = 1$, we conclude that $\mu_n(\complement V) \to 0$. Let f denote a member of $\mathfrak{D}^0$. We have

$$\int f\cdot\mu_n - f(a) = \int [f - f(a)]\cdot\mu_n = \int_V [f - f(a)]\cdot\mu_n + \int_{\complement V} [f - f(a)]\cdot\mu_n$$

Let ε denote a positive number. Let us choose V such that $x\in V \Rightarrow |f(x) - f(a)| \leqslant \varepsilon$. We then have

$$\left|\int_V [f - f(a)]\cdot\mu_n\right| \leqslant \varepsilon.$$

Let N denote an integer such that

$$n \geqslant N \Rightarrow \mu_n(\complement V) \leqslant \frac{\varepsilon}{2\,\|f\|_u}.$$

Then, for $n \geqslant N$, we have

$$\left|\int_{\complement V} [f - f(a)]\cdot\mu_n\right| \leqslant \varepsilon,$$

and, consequently,

$$\left| \int f \cdot \mu_n - f(a) \right| \leqslant 2\varepsilon .$$

Corollary. *Suppose that μ_n is a positive measure on* $\mathbf{R}$ *of total mass equal to 1 and that* $\{\mu_n([A, +\infty])\}$ *converges to 1 for every A. Then $\{\mu_n\}$, where each μ_n is treated as a measure on* $\overline{\mathbf{R}}$, *converges to $\delta_{+\infty}$.*

Theorem 13. *For $\lambda \in \mathbf{R}_+$, the measure $\lambda \mathcal{Y}(t)\,\mathrm{e}^{-\lambda t}\,\mathrm{d}t$ is positive with total mass 1. As $\lambda \to +\infty$, this measure tends narrowly to δ; as $\lambda \to 0$, it tends to $\delta_{+\infty}$.*

Proof. We have

$$\lambda \int_0^{+\infty} \mathrm{e}^{-\lambda t}\,\mathrm{d}t = 1 .$$

To study the limit as $\lambda \to +\infty$, we can either use the preceding lemma or note that the Fourier transform is*

$$\omega \to \frac{\lambda}{\lambda - \mathrm{i}\omega},$$

and tends simply to 1 as $\lambda \to +\infty$. Therefore, the measure $\lambda \mathcal{Y}(t)\,\mathrm{e}^{-\lambda t}\,\mathrm{d}t$ tends narrowly to δ as $\lambda \to +\infty$. Furthermore,

$$\lim_{\lambda \to 0} \lambda \int_A^{+\infty} \mathrm{e}^{-\lambda t}\,\mathrm{d}t = \lim_{\lambda \to 0} \mathrm{e}^{-\lambda A} = 1 .$$

Consequently, the measure $\lambda \mathcal{Y}(t)\,\mathrm{e}^{-\lambda t}\,\mathrm{d}t$ tends to $\delta_{+\infty}$ as $\lambda \to 0$.

Corollary. I. *For every continuous bounded function f,*

$$\lim_{\left\{\substack{p \to +\infty \\ p \in \mathbf{R}}\right.} pF(p) = f(0), \qquad \textit{with} \qquad F = \mathcal{L}(\mathcal{Y}f) .$$

II. *For every function $f \in \Phi$,*

$$\lim_{\left\{\substack{p \to 0 \\ p \in \mathbf{R}}\right.} pF(p) = f(+\infty), \qquad \textit{with} \qquad F = \mathcal{L}(\mathcal{Y}f) .$$

Let us give results improving these corollaries.

Theorem 14. *Let f denote a function of positive support with Laplace transform $\mathcal{L}f = F$.*

(a) *If f has a right-hand limit as $t \to 0$, then*

---

*The Laplace transform is $\lambda/(p + \lambda)$.

$$\lim_{t \to +0} f(t) = \lim_{\left\{\substack{p \in \mathbf{R} \\ p \to +\infty}\right.} pF(p) .$$

(b) *If* f *has a limit as* $t \to +\infty$, *then* $\mathfrak{L}f$ *has a nonpositive abscissa of absolute convergence, and*

$$\lim_{t \to +\infty} f(t) = \lim_{\left\{\substack{p \in \mathbf{R} \\ p \to 0}\right.} pF(p) .$$

Proof. (a) Let us show first that if f vanishes everywhere on an interval of the form $[0, \varepsilon]$, where $(\varepsilon > 0)$, then

$$\lim_{p \to +\infty} pF(p) = 0 .$$

Let ξ_0 denote a member of $\mathbf{R}_+$ such that

$$\int_0^{+\infty} |f(t)|\, e^{-\xi_0 t}\, dt < +\infty .$$

For $p > \xi_0$, we have

$$\int_\varepsilon^{+\infty} |f(t)|\, p\, e^{-pt}\, dt = \int_\varepsilon^{+\infty} |f(t)|\, e^{-\xi_0 t}\, p\, e^{-(p-\xi_0)t}\, dt$$
$$\leqslant p\, e^{-(p-\xi_0)\varepsilon} \int_\varepsilon^{+\infty} |f(t)|\, e^{-\xi_0 t}\, dt .$$

Since the right-hand member of this inequality tends to 0 as $p \to +\infty$, the asserted property is proven.

Suppose now that

$$\lim_{t \to +0} f(t) = 0 .$$

Let ε denote any positive number. There exists an η such that $0 \leqslant t \leqslant \eta$ implies $|f(t)| \leqslant \varepsilon$. We have

$$\int_0^{+\infty} p\, e^{-pt} f(t)\, dt = \int_0^{\eta} p\, e^{-pt} f(t)\, dt + \int_\eta^{+\infty} p\, e^{-pt} f(t)\, dt$$

with

$$\int_0^{\eta} p\, e^{-pt} f(t)\, dt \leqslant \varepsilon .$$

Therefore,

$$\lim_{p \to \infty} \int_\eta^{+\infty} p\, e^{-pt} f(t)\, dt = 0 .$$

Hence, there exists a P such that $p \geqslant P$ implies

$$\left| \int_0^{+\infty} p\,e^{-pt} f(t)\,dt \right| \leqslant 2\varepsilon,$$

so that

$$\lim_{p\to+\infty} \int_0^{+\infty} p\,e^{-pt} f(t)\,dt = 0.$$

Suppose, finally, that

$$\lim_{t\to+0} f(t) = a.$$

We have

$$\lim_{t\to+0} \left[f(t) - a\,\mathcal{Y}(t)\right] = 0,$$

so that

$$\lim_{p\to+\infty} (pF(p) - a) = 0;$$

that is,

$$\lim_{p\to+\infty} pF(p) = a.$$

(b) Let us suppose first that

$$\lim_{t\to+\infty} f(t) = 0.$$

Let ε denote any positive number. There exists an A such that $t \geqslant A$ implies $|f(t)| \leqslant \varepsilon$. We have

$$\int_0^{+\infty} p\,e^{-pt} f(t)\,dt = p \int_0^A e^{-pt} f(t)\,dt + \int_A^{+\infty} p\,e^{-pt} f(t)\,dt.$$

The function

$$p \to \int_0^A e^{-pt} f(t)\,dt,$$

is an entire function and hence,

$$\lim_{p\to 0} p \int_0^A e^{-pt} f(t)\,dt = 0.$$

Furthermore,

$$\left| \int_A^{+\infty} p\,e^{-pt} f(t)\,dt \right| \leqslant \varepsilon.$$

Consequently, there exists an η such that $0 \leqslant p \leqslant \eta$ implies

$$\left| \int_0^\infty p\, e^{-pt} f(t) dt \right| \leqslant 2\varepsilon ,$$

so that

$$\lim_{p \to 0} \int_0^{+\infty} p\, e^{-pt} f(t)\, dt = 0 .$$

Let us now suppose that

$$\lim_{t \to +\infty} f(t) = a .$$

We have

$$\lim_{t \to \infty} [f(t) - a\mathcal{Y}(t)] = 0 ,$$

so that

$$\lim_{p \to 0} [pF(p) - a] = 0 ;$$

that is

$$\lim_{p \to 0} pF(p) = a ,$$

which completes the proof.

5. APPLICATIONS OF THE LAPLACE TRANSFORM

Among the numerous applications of the Laplace transform, we shall treat its applications to convolution equations. Among the applications that we shall not treat here are the solutions of certain differential equations with variable coefficients and certain partial differential equations.

Consider a convolution equation of the form

$$a * x = b ,$$

where a, b, and x are the distributions with positive support. If a and b have Laplace transforms A and B and if the equation has a solution (necessarily unique in $\mathcal{D}'_+$) having a Laplace transform X, we have

$$A(p)\, X(p) = B(p) ,$$

so that

$$X(p) = \frac{B(p)}{A(p)} .$$

We need only find the inverse transform of $X(p)$ with the aid of a table of Laplace transforms.

This method fails if a or b does not have a Laplace transform or if the equation $a * x = b$ has a solution that does not have a Laplace transform.

Thus, we may simply state that if a and b have Laplace transforms and if $B(p)/A(p)$ is the Laplace transform of a distribution x with positive support, this distribution is the unique solution of the equation $a * x = b$ in $\mathfrak{D}'_+$.

We shall refer to the solution (unique if it exists) of the equation

$$a * x_0 = \delta ,$$

that is, the inverse of a in the convolution algebra $\mathfrak{D}'_+$, as the *elementary solution* of the equation $a * x = b$ (in $\mathfrak{D}'_+$).

If x_0 has a Laplace transform, this transform is equal to

$$X_0(p) = \frac{1}{A(p)} .$$

The general solution of $a * x = b$ is then $x = x_0 * b$.

Differential equations. The method of solutions of differential equations with the aid of the Laplace transform is as follows: We replace the original equation with an equation in $\mathfrak{D}'_+$ that is satisfied by the products of $\mathcal{Y}$ and the unknowns in the original equation. The initial values of the original unknowns provide the relations among the derivatives in the ordinary sense and the derivatives in the distribution sense of the modified unknowns.

Example I. Solve the system of differential equations

$$(1) \qquad \begin{cases} \dfrac{dx}{dt} + \dfrac{dy}{dt} = x - y + e^{2t} \\[2ex] \dfrac{d^2x}{dt^2} + \dfrac{dy}{dt} = 3\, e^{2t} . \end{cases}$$

with initial conditions

$$\begin{cases} x(0) = 0 \\ x'(0) = -1 \\ y(0) = 1 . \end{cases}$$

Let us replace this problem with the following one: Find two functions $x(t)$ and $y(t)$ such that

$x(t) = y(t) = 0$ for $t < 0$,
$x(t)$ and $y(t)$ are differentiable for $t > 0$, and their derivatives satisfy the system (1) for $t > 0$,

$$x(+0) = 0, \qquad x'(+0) = -1, \qquad y(+0) = 1\,.$$

The relationships between the ordinary derivatives dx/dt and dy/dt on the one hand and the derivatives in the distribution sense Dx and Dy are

$$Dx = \frac{dx}{dt}\,,\; D^2 x = \frac{d^2x}{dt^2} - \delta\,,$$

$$Dy = \frac{dy}{dt} + \delta\,.$$

From this we derive the following system

$$\begin{cases} Dx + Dy - \delta - x + y = e^{2t}\,, \\ D^2 x + \delta + Dy - \delta = 3\,e^{2t}\,, \end{cases}$$

that is,

$$\begin{cases} (D-1)\,x + (D+1)\,y = \delta + e^{2t} \\ D^2 x + Dy = 3\,e^{2t}\,. \end{cases}$$

We define $X = \mathfrak{L}x$ and $Y = \mathfrak{L}y$. Then

$$\begin{cases} (p-1)\,X + (p+1)\,Y = 1 + \dfrac{1}{p-2} \\ p^2\,X + pY = \dfrac{3}{p-2}\,. \end{cases}$$

Solution of this system as a system of algebraic equations yields

$$\begin{cases} X = -\dfrac{p^2 - 4p - 3}{p(p^2+1)(p-2)} \\ Y = \dfrac{p^3 - p^2 - 3p + 3}{p(p^2+1)(p-2)}\,. \end{cases}$$

Decomposing the terms on the right sides into partial fractions, we obtain

$$X = -\frac{3}{2p} + \frac{7}{10(p-2)} + \frac{4}{5}\,\frac{p-3}{p^2+1}\,,$$

$$Y = -\frac{3}{2p} + \frac{1}{10(p-2)} + \frac{4}{5}\,\frac{3p+1}{p^2+1}\,,$$

from which we get

$$x(t) = -\frac{3}{2} + \frac{7}{10}e^{2t} + \frac{4}{5}\cos t - \frac{12}{5}\sin t \qquad (\text{for } t > 0),$$

and

$$y(t) = -\frac{3}{2} + \frac{1}{10}e^{2t} + \frac{12}{5}\cos t + \frac{4}{5}\sin t \qquad (\text{for } t > 0).$$

II. Find the elementary solution in $\mathfrak{D}'_+$ of the equation

$$x'' - 3x' + 2x = 0.$$

This will be the solution of

$$D^2 x - 3Dx + 2x = \delta,$$

so that

$$(p^2 - 3p + 2)X(p) = 1 \qquad (X = \mathfrak{L}x);$$

that is,

$$X(p) = \frac{1}{p^2 - 3p + 2} = \frac{1}{(p-1)(p-2)} = \frac{-1}{p-1} + \frac{1}{p-2},$$

and

$$x(t) = -e^{t} + e^{2t} \qquad (\text{for } t > 0).$$

Convolution integral equations. *Example.* Solve the equation

(1) $$\int_0^t \sin(t-\theta)\, x(\theta)\, d\theta = t^2 \qquad \text{for } t \geqslant 0,$$

where the function x has positive support.

The equation is of the form $a * x = b$ with

$$a(t) = \mathcal{Y}(t)\sin t, \qquad b(t) = \mathcal{Y}(t)\, t^2.$$

Taking the Laplace transforms, we obtain

$$X(p) = \frac{2(p^2+1)}{p^3} = \frac{2}{p} + \frac{2}{p^3},$$

and

$$x(t) = (2 + t^2)\,\mathcal{Y}(t).$$

If, instead of equation (1), we are required to solve the equation

(2) $$\begin{cases} \displaystyle\int_0^t \sin(t-\theta)\, y(\theta)\, d\theta = 2t & (t \geqslant 0), \\ y \text{ of positive support} \end{cases}$$

we obtain in a similar way

$$\frac{1}{p^2+1}\,Y(p) = \frac{2}{p^2},$$

so that

$$Y(p) = 2 + \frac{2}{p^2}$$

and

$$y = 2\,\delta + 2\,t\mathcal{Y}(t)\,.$$

Thus, equation (2) has no solution in the form of a function. The equation $a * y = c$ with

$$a(t) = \sin t, \quad c(t) = 2\,\mathcal{Y}(t)\,t\,,$$

has the solution*

$$y = 2\,\delta + 2\,t\mathcal{Y}(t) \qquad (\text{in } \mathcal{D}'_+).$$

Naturally, $y = Dx$ since $c = Db$.

6. THE LAPLACE TRANSFORM OF VECTOR-VALUED FUNCTIONS

The Laplace transform can easily be generalized to the case of functions f with positive support and range in a finite-dimensional vector space E. By choosing a suitable basis, we can identify E with some space $\mathbf{R}^n$. Thus, we set

$$f(t) = \begin{bmatrix} f^1(t) \\ \vdots \\ f^n(t) \end{bmatrix} \quad \text{and} \quad (\mathcal{L}f)(p) = \begin{bmatrix} (\mathcal{L}f^1)(p) \\ \vdots \\ (\mathcal{L}f^n)(p) \end{bmatrix}.$$

The Laplace transform $\mathcal{L}f$ is a holomorphic functions with range in $\mathbf{R}^n$.

These considerations are applicable, in particular, if $f(t) \in L(\mathbf{R}^n, \mathbf{R}^p)$. We then have the following result:

---

*The right-hand member represents the sum of the Dirac measure δ and the function $t \to 2\,t\mathcal{Y}(t)$ or, more exactly, the equivalence class of functions that are equal to this function modulo functions that vanish almost everywhere. The notations used are far from satisfactory. However, it is quite difficult not to make compromises in our examples because the functions are almost always described in terms of analytic expressions and this necessitates the introduction of an independent variable.

Theorem 15. *The Laplace transform of the function* $t \to \mathcal{Y}(t)\,e^{tA}$ *is equal to* $(p - A)^{-1}$ (*for* $A \in L(\mathbf{R}^n, \mathbf{R}^n)$).

Proof. Let p denote a number such that $\Re(p)$ is greater than the real parts of all the eigenvalues of A. Then,

$$\int_0^{+\infty} e^{tA}\,e^{-pt}\,dt = \int_0^{+\infty} e^{-(p-A)t}\,dt = -(p - A)^{-1}\left[e^{-(p-A)t}\right]_{t=0}^{t=+\infty}$$
$$= (p - A)^{-1}\,.$$

This completes the proof of the theorem since the function $p \to (p - A)^{-1}$ is a rational fraction.

Theorem 15 can be used to evaluate e^{tA}.

Example. Suppose that

$$A = \begin{bmatrix} 0 & 2 \\ 2 & 3 \end{bmatrix}.$$

We have

$$(p - A)^{-1} = \frac{1}{(p+1)(p-4)}\begin{bmatrix} p-3 & 2 \\ 2 & p \end{bmatrix}.$$

Decomposing the right side into partial fractions, we have

$$(p - A)^{-1} = \frac{1}{5}\begin{bmatrix} 4 & -2 \\ -2 & 1 \end{bmatrix}\frac{1}{p+1} + \frac{1}{5}\begin{bmatrix} 1 & 2 \\ 2 & 4 \end{bmatrix}\frac{1}{p-4}\,.$$

Now, by using the table of Laplace transforms, we obtain

$$e^{tA} = \frac{1}{5}\begin{bmatrix} 4 & -2 \\ -2 & 1 \end{bmatrix}e^{-t} + \frac{1}{5}\begin{bmatrix} 1 & 2 \\ 2 & 4 \end{bmatrix}e^{4t}.$$

TABLE OF LAPLACE TRANSFORMS

$f(t)$	$F(p)$	$\mathcal{Y}(t)\times$	
Df	$pF(p)$		
$e^{\lambda t} f(t)$	$F(p-\lambda)$	$e^{\lambda t}$	$\frac{1}{p-\lambda}$
$f\left(\frac{t}{\lambda}\right)$	$\lambda F(\lambda p)$	$\frac{1}{\omega}\sin \omega t$	$\frac{1}{p^2+\omega^2}$
$\frac{1}{\lambda} f\left(\frac{t}{\lambda}\right)$	$F(\lambda p)$	$\cos \omega t$	$\frac{p}{p^2+\omega^2}$

$\int_0^t f$	$\frac{F(p)}{p}$	$\frac{1}{\omega}\sinh \omega t$	$\frac{1}{p^2 - \omega^2}$
$f_1 * f_2$	$F_1(p)\, F_2(p)$	$\cosh \omega t$	$\frac{p}{p^2 - \omega^2}$
$(-t)^k f(t)$	$F^{(k)}(p)$	$\frac{t^{\alpha-1}}{\Gamma(\alpha)}\ (\alpha > 0)$	$\frac{1}{p^\alpha}$
δ	1	$\frac{1}{\sqrt{t}}$	$\frac{\sqrt{\pi}}{\sqrt{p}}$
δ_a	e^{-ap}		
δ'	p	$e^{\lambda t}\, \frac{t^{\alpha-1}}{\Gamma(\alpha)}$ $(\alpha > 0)$	$\frac{1}{(p-\lambda)^\alpha}$
δ_a'	$e^{-ap}\, p$		
$\delta^{(k)}$	p^k		
$\mathcal{Y}$	$\frac{1}{p}$		

5

Elements of Probability Theory

1. MEASURES IN A VECTOR SPACE

Let E denote a vector space of finite dimension n. Let $\mathfrak{D}^0(E)$ denote the space of continuous functions defined on E and with compact support. Let $\mathfrak{D}^0_K(E)$ denote the space of continuous functions on E with support in a compact set K. We recall that any linear form μ $\left(f \to \int f \cdot \mu\right)$ defined on the space $\mathfrak{D}^0(E)$ and whose restriction to $\mathfrak{D}^0_K(E)$ is continuous for every compact set K is called a *measure* on E.

Let S denote a basis for E. For every $X \in E$, let X_S denote the n-tuple of components of X. In a basis S, every function f is represented by a function f_S defined by

$$f(X) = f_S(X_S).$$

A necessary and sufficient condition for f to be a member of $\mathfrak{D}^0(E)$ is that f_S be a member of $\mathfrak{D}^0(\mathbf{R}^n)$.

Every measure μ in E is represented by a measure μ_S on $\mathbf{R}^n$ defined by

$$\int f(X) \cdot \mu = \int f_S(X_S) \cdot \mu_S.$$

We note that μ_S is simply the image of μ under the mapping $X \to X_S$. For μ to be bounded, it is necessary and sufficient that μ_S be bounded. Conversely, if we assign to every basis S a measure μ_S, a necessary and sufficient condition for μ_S to be the representation of a measure μ on E is that

$$\int f_S(X_S) \cdot \mu_S ,$$

be independent of S, that is, that

$$\int f_S(KX_{S'}) \cdot \mu_{S'} = \int f_S(X_S) \cdot \mu_S ,$$

for every change of basis defined by $X_S = KX_{S'}$ or, in short, that μ_S be the image of $\mu_{S'}$ under the mapping K: $X_{S'} \to X_S$ (where K is the matrix of the change of basis).

Example. Lebesgue measure in a vector space.

Lebesgue measure in a vector space is defined only up to a constant of proportionality. Specifically, if, for a particular basis S, we set

$$\mu_S = \mathrm{d}X_S^1 \ldots \mathrm{d}X_S^n ,$$

we shall have to take in another basis S' for $\mu_{S'}$, the image of μ_S under the mapping $X_S \to X_{S'}$; that is,

$$\mu_{S'} = |\det K| \, \mathrm{d}X_{S'}^1 \ldots \mathrm{d}X_{S'}^n ,$$

where K is the matrix of the change of basis.

On the other hand, if E is a Euclidean space, there exists a canonical Lebesgue measure $\mathrm{d}X$ on E defined by

$$\int f(X) \, \mathrm{d}X = \int f_S(X_S) \sqrt{g(S)} \, \mathrm{d}X_S^1 \ldots \mathrm{d}X_S^n ,$$

where $g(S)$ is the Gram determinant of the basis of S (that is, $g(S) = \det G(S)$). For a change of basis K (implementing the transformation $X_{S'} \to X_S$), we have

$$g(S') = (\det K)^2 \, g(S) ,$$

which ensures invariance of the definition with respect to the basis chosen.

2. PROBABILITY DISTRIBUTIONS

Definitions. We say that a finite-dimensional vector space E is *probabilized* if a bounded positive measure with total mass equal to 1, called the *probability distribution*, is defined on E. The space E is called a *sample space.* Its elements are called *occurrences* or *outcomes.*

Every measurable subset A of E is called an *event*, and its measure is called the *probability* of A. We shall denote it by prob (A).

We say that an event A *occurs* in the outcome a (where $a \in E$) if $a \in A$. An event A is said to be *impossible* if $A = \varnothing$.

For two events A and B, the event $A \cap B$ will also be denoted by AB. This is the event that occurs if and only if A and B both occur. If AB is impossible (that is, if $AB = \varnothing$), then A and B are said to be *incompatible*. The union $A \cup B$ of two events will usually be called the *sum* of these two events. This is the event that occurs if and only if at least one of the two events A, B occurs. The sum of a family of arbitrary events is defined analogously.

The event $\complement A$ is called the *complementary* event (or the negation) of an event A. This is the event that occurs if and only if A fails to occur.

Examples. I. *Discrete distributions.* These are of the form

$$\sum_i p_i \delta_{a_i},$$

where $\{a_i\}$ is a finite or denumerable family of points of E and where the coefficients p_i satisfy the conditions

$$p_i \geqslant 0, \qquad \sum_i p_i = 1.$$

The coefficients p_i can be interpreted as the probabilities of the sequence of events $\{a_i\}$. We write

$$p_i = \text{prob}\,(a_i),$$

and identifying every occurrence a_i with the event $\{a_i\}$ consisting only of this one event (that is, the event that occurs if and only if the occurrence a_i occurs).

If A is an event (that is, in the present case, any subset of E), we have

$$\text{prob}\,(A) = \sum_{a_i \in A} p_i.$$

II. *Absolutely continuous distributions.* These are distributions defined by a density with respect to Lebesgue measure.

If μ is a distribution of probability density θ (that is, if $\mu^X = \theta(X)\mathrm{d}X$), the probability of an event A is

$$\text{prob}\,(A) = \mu(A) = \int_A \theta(X)\,\mathrm{d}X.$$

We remark that we should have

$$\begin{cases} \theta \geqslant 0 \\ \theta \in \mathbf{L}^1 \end{cases}.$$

We cite the following particular cases:

Uniform distribution over a segment $[a, b]$ *of* **R.** This is the measure of density

$$\theta(X) = \begin{cases} \dfrac{1}{b-a} & \text{if } \ X \in [a, b] \\ 0 & \text{if } \ X \in [a, b] \,. \end{cases}$$

Reduced Gaussian distribution on **R.** This is the distribution of density

$$\theta(X) = \frac{1}{\sqrt{2\pi}} \cdot e^{-X^2/2} .$$

We have $\theta \geqslant 0$. Let us verify that

$$\int_{-\infty}^{+\infty} \theta(X) \, dX = 1 .$$

We have

$$\int_{-\infty}^{+\infty} e^{-X^2/2} \, dX = 2\sqrt{2} \int_{0}^{+\infty} e^{-Y^2} \, dY = \sqrt{2} \, \Gamma(1/2) = \sqrt{2\pi} .$$

Gaussian distributions on a vector space. Let E denote an n-dimensional vector space. Suppose that there is a positive-definite quadratic form expressed in each basis S in the form $X \to \overline{X}_S \, \mathcal{A}(S) \, X_S$. For each basis S, consider the measure

$$\mu_S = \frac{\sqrt{\det \mathcal{A}(S)}}{(2\pi)^{n/2}} \exp(-\tfrac{1}{2} \overline{X}_S \, \mathcal{A}(S) \, X_S) \, dX_S^1 \dots dX_S^n .$$

For every change of basis defined by $X_S = K X_{S'}$, the image of μ_S is equal to

$$\frac{\sqrt{\det \mathcal{A}(S)}}{(2\pi)^{n/2}} \exp(-\tfrac{1}{2} \overline{X}_{S'} \, \mathcal{A}(S') \, X_{S'}) \, | \det(K) | \, dX_{S'}^1 \dots dX_{S'}^n .$$

Since $\mathcal{A}(S') = \overline{K} \mathcal{A}(S) \, K$ and hence

$$\det \mathcal{A}(S') = \det \mathcal{A}(S) \, (\det(K))^2 ,$$

this image is indeed equal to $\mu_{S'}$.

Therefore, there exists a measure μ whose representation in each basis S is equal to μ_S. Such a measure is called a *Gaussian distribution on E.* We have $\mu \geqslant 0$. To show that $\int m = 1$, we need only show that $\int \mu_S = 1$ for a particular basis S. Let us choose S such that $\mathcal{A}(S) = \mathbf{1}$. We then have (omitting the subscript S in the right-hand member)

$$\mu_S = \frac{1}{(2\pi)^{n/2}} \exp\left[-\tfrac{1}{2}\sum_i (X^i)^2\right] dX^1 \cdots dX^n ,$$

so that

$$\int \mu_S = \frac{1}{(2\pi)^{n/2}} \prod_i \int_{-\infty}^{+\infty} \exp\left[-\tfrac{1}{2}(X^i)^2\right] dX^i = 1 .$$

Theorem 1 (on total probabilities). *Let E denote a probabilized vector space and let A denote an event that is the sum of a finite or denumerable family of pairwise incompatible events A_i. Then*

$$\text{prob}\,(A) = \sum_i \text{prob}\,(A_i) .$$

Proof. Since μ is the probability distribution, we have

$$\text{prob}\,(A) = \mu(A) = \sum_i \mu(A_i) = \sum_i \text{prob}\,(A_i) .$$

Corollaries. I. $\text{prob}\,(\complement A) = 1 - \text{prob}\,(A)$.

II. *For arbitrary events A and B,*

$$\text{prob}\,(A + B) = \text{prob}\,(A) + \text{prob}\,(B) - \text{prob}\,(AB) .$$

Definitions. An event of probability 1 is said to be *almost certain.* An event of probability 0 is said to be *almost impossible.* Two events A and B such that AB is almost impossible are said to be *almost incompatible.*

Definition. Let A be an event of nonzero probability. We call the probability distribution

$$\mu_A = \frac{\varepsilon_A \, \mu}{\text{prob}\,(A)} ,$$

where ε_A is the characteristic function of the set A, the *probability distribution conditioned by A.*

We always have $\int \mu_A = 1$. For every event B, we denote by prob $(B \mid A)$ the probability of B with respect to the distribution μ_A and we call it the conditional probability of B, when A is given. We have

$$\text{prob}\,(B \mid A) = \frac{\text{prob}\,(BA)}{\text{prob}\,(A)} ,$$

that is

$$\text{prob}\,(BA) = \text{prob}\,A \,.\, \text{prob}\,(B \mid A)\,.$$

Theorem 2 (on composite probabilities). *Let $\{A_i\}$ denote a finite or denumerable family of almost incompatible events the sum of which is almost certain. Let B denote any event. Then,*

$$\text{prob}\,(B) = \sum_i \text{prob}\,(A_i)\,\text{prob}\,(B \mid A_i)\,.$$

Proof. The formula follows from the fact that

$$\text{prob}\,(B) = \sum_i \text{prob}\,(BA_i)\,.$$

3. RANDOM VARIABLES

Definitions. Let E denote a probabilized vector space. Then any scalar- or vector-valued measurable function on E is called a *random variable.*

The *mean* of a random variable $Y = f(X)$ is defined as the quantity

$$\mathbf{E}(Y) = \int f(X)\,.\,\mu^X,$$

where μ^X is the probability distribution on E. This mean is defined subject to the following conditions:

(a) if Y is a scalar variable, $\mathbf{E}(Y)$ is defined if $f \in \mathbf{L}^1$;

(b) if Y is a vector-valued function with range in F, the preceding formula means that, for every linear form u on F,

$$\mathbf{E}(u(Y)) = \int u(f(X))\,.\,\mu^X$$

or that, for a particular basis S of F,

$$\mathbf{E}(Y_S^i) = \int f_S^i(X)\,.\,\mu^X.$$

(This formula remains valid for any other basis for F.)

Thus, $\mathbf{E}(Y)$ is defined if $u \circ f \in \mathbf{L}^1$ for every linear form u on F or in an equivalent manner if $f_S^i \in \mathbf{L}^1$ for a particular basis S and for every i.

Properties. We have

$$\mathbf{E}(\lambda_1 Y_1 + \lambda_2 Y_2) = \lambda_1 \mathbf{E}(Y_1) + \lambda_2 \mathbf{E}(Y_2),$$

$$\mathbf{E}(a) = a \quad \text{if a is a constant.}$$

In particular,

$$\mathbf{E}(X) = \int X \,.\, \mu^X .$$

Two random variables $Y_1 = f_1(X)$ and $Y_2 = f_2(X)$ are said to be *almost certainly equal* if $f_1(X) = f_2(X)$ almost everywhere with respect to the measure μ^X, that is, if the event $f_1(X) = f_2(X)$ is almost certain.

Examples. I. If $\mu = \sum_i p_i \delta_{a_i}$, we have

$$\mathbf{E}(Y) = \sum_i p_i f(a_i) = \sum_i p_i Y^i,$$

by setting $Y^i = f(a_i)$.

If the events a_i are finite in number (let us say that there are m of them), then $\mathbf{E}(Y)$ always exists. We can identify the variable Y with the m-tuple

$$\begin{bmatrix} Y^1 \\ \cdot \\ \cdot \\ \cdot \\ Y^m \end{bmatrix},$$

the components of which are the values assumed by Y for the events a_i.* The mean $\mathbf{E}(Y)$ is the weighted mean of these values, the weights being the numbers p_i. If the events a_i constitute a denumerable family, $\mathbf{E}(Y)$ will be defined only if the series $\sum_i p_i f(a_i)$ is absolutely convergent.

II. If $\mu^X = \theta(X)\,\mathrm{d}X$, we have

$$\mathbf{E}(Y) = \int_{-\infty}^{+\infty} f(X)\,\theta(X)\,\mathrm{d}X .$$

III. For every event A, let us define a random variable, known as *Bernoulli's variable* of the event A by

$$\varepsilon_A(X) = \begin{cases} 1 & \text{if } X \in A, \\ 0 & \text{if } X \notin A. \end{cases}$$

Thus, ε_A is simply the characteristic function of the set A. We have

$$\mathbf{E}(\varepsilon_A) = \int_A \mu = \mu(A) = \mathrm{prob}\,(A) .$$

---

*This m-tuple actually defines the class of Y for the equivalence relation where $a \sim b$ if and only if a is almost certainly equal to b.

Definition. A variable of X is said to be *centered* if $\mathbf{E}(X) = 0$. Centered variables constitute a vector space (a vector subspace of the vector space $\mathbf{L}^1$ of random variables that have a mean).

Let E denote a probabilized vector space, let X denote a number of E, and let $Y = f(X)$ denote a random variable defined on E. Let F denote a vector space to which Y belongs. The probability distribution μ^X on E has an image ν^Y under f, known as the *probability distribution* of the random variable Y. (We do indeed have $\nu^Y \geqslant 0$ and $\int \nu^Y = 1$.)

This image ν^Y is such that

$$\int g(Y) \cdot \nu^Y = \int g(f(X)) \cdot \mu^X ,$$

for every scalar function g that is measurable with respect to ν^Y whenever one or the other of the members of this equation is defined.

Let us set $Z = g(Y) = g(f(X))$. The variable Z can be considered either as a random variable on F or as a random variable on E. The left-hand member of the above equation represents the mean $\mathbf{E}_F(Z)$ considered as defined on F; the right-hand member represents the mean $\mathbf{E}_E(Z)$ considered as defined on E.

Thus, by the definition of the image probability distribution ν^Y, the mean of a random variable $Z = g(Y)$ can be calculated either on F with the aid of the distribution ν^Y or on E with the aid of the distribution μ^X. The value of this mean will be denoted simply by $\mathbf{E}(Z)$.

For example,

$$\mathbf{E}(Y) = \int f(X) \cdot \mu^X = \int Y \cdot \nu^Y .$$

This property is easily generalized to the case in which Z is a vector-valued random variable. In particular, if $Y^1, \ldots, Y^m$ are real random variables defined on E, we can set

$$Y = \begin{bmatrix} Y^1 \\ \vdots \\ Y^m \end{bmatrix} ,$$

and consider in $\mathbf{R}^m$ the probability distribution ν^Y of the m-tuple Y. This is the image under the mapping $X \to Y$ of the probability distribution μ^X. This distribution will enable us to evaluate the mean $\mathbf{E}(Z)$ for every $Z = g(Y^1, \ldots, Y^m) = g(Y)$. We have

$$\mathbf{E}(Z) = \int g(Y) \cdot \nu^Y .$$

4. DESCRIPTIVE ELEMENTS OF A PROBABILITY DISTRIBUTION

We shall confine ourselves to the most frequently used concepts in the description or characterization of a probability distribution.

A. Study of real random variables. Let X denote a scalar random variable with probability distribution μ^X. The following definition is a particularization of the definition of the cumulative function of a measure.

Definition. The function

$$F(x) = \int_{]-\infty,x[} \mu^X = \text{prob}\{X < x\},$$

is called the *cumulative function* or the *distribution function.* We have $F(-\infty) = 0$ and $F(+\infty) = 1$. The function F is an increasing function. Also,

$$\begin{cases} F \text{ is continuous on the left;} \\ F(x+0) - F(x) = \mu(\{x\}) = \text{prob}\{X = x\}. \end{cases}$$

To prove this, let $\{x_n\}$ denote a sequence such that $\lim_{n\to\infty} x_n = x$ and $x_n \leqslant x$. Let ε_n denote the characteristic function of the interval $]-\infty, x_n[$ and let i denote the characteristic function of the interval $]-\infty, x[$. We have

$$\varepsilon(X) = \lim_{n\to\infty} \varepsilon_n(X),$$

so that

$$F(x) = \int \varepsilon(X) \,.\, \mu^X = \lim_{n\to\infty} \int \varepsilon_n(X) \,.\, \mu^X = \lim_{n\to\infty} F(x_n).$$

Now, let $\{x_n\}$ denote a sequence such that

$$\lim_{n\to\infty} x_n = x \quad \text{and} \quad x_n > x.$$

We have

$$\lim_{n\to\infty} \varepsilon_n(X) = \begin{cases} 1 & \text{if} \quad X \in]-\infty, x] \\ 0 & \text{if} \quad X \notin]-\infty, x], \end{cases}$$

so that

$$\begin{aligned} F(x+0) = \lim_{n\to\infty} F(x_n) = \lim_{n\to\infty} \int \varepsilon_n(X) \,.\, \mu^X &= \int_{]-\infty,x]} \mu^X \\ &= \mu(]-\infty, x[) + \mu(\{x\}) = F(x) + \mu(\{x\}), \end{aligned}$$

which completes the proof.

Remark. If μ contains no point masses (in particular, if μ is defined by a density), the F is continuous. If μ is a finite sum of point measures, the F is a step function.

We recall that

$$\int_{[a,b[} \mu = F(b) - F(a),$$

that is, prob $(a \leqslant X < b) = F(b) - F(a)$.

At the end of the chapter is a table of the distribution function of a normalized Gaussian variable.

Notation. We denote by $\mathring{X}$ the random variable $X - \mathbf{E}(X)$. We have $\mathbf{E}(\mathring{X}) = 0$; in other words, $\mathring{X}$ is centered.

Definitions. The quantity $m_k(X) = \mathbf{E}(X^k)$ is called the *moment* of order k. The quantity $m_k(\mathring{X})$ is called the *centered moment* of order k.

These moments are defined for $k \leqslant k_1$, where k_1 depends on the probability distribution. We have

$$m_1(X) = \mathbf{E}(X).$$

The centered moment of order 2 is called the *variance*. We denote it by $\sigma^2(X)$, so that

$$\sigma^2(X) = \mathbf{E}((X - \mathbf{E}(X))^2).$$

We have

$$\sigma^2(X) = \mathbf{E}[X^2 - 2\,\mathbf{E}(X)\,X + (\mathbf{E}(X))^2],$$

from which we get

$$\sigma^2(X) = \mathbf{E}(X^2) - (\mathbf{E}(X))^2.$$

We note that

$$\sigma^2(X + a) = \sigma^2(X) \qquad \text{for every } a \in \mathbf{R},$$

$$\sigma^2(kX) = k^2\,\sigma^2(X).$$

The nonnegative square root $\sigma(X)$ of the variance is called the *standard deviation* of a random variable. A variable X is said to be normalized if $\mathbf{E}(X) = 0$ and $\sigma(X) = 1$.* For every X,

* This definition is compatible with the definition already given for a normalized Gaussian variable.

$$\frac{X - \mathbf{E}(X)}{\sigma(X)},$$

is normalized. (It is called the normalization of X.)

Example. Let ε_A denote Bernoulli's variable for an event A of probability p. Suppose that $q = 1 - p$. We have $\varepsilon_A^2 = \varepsilon_A$, so that

$$\mathbf{E}(\varepsilon_A^2) = \mathbf{E}(\varepsilon_A) = p,$$

and, consequently,

$$\sigma^2(\varepsilon_A) = p - p^2 = p(1-p) = pq \quad \text{and} \quad \sigma(\varepsilon_A) = \sqrt{pq}.$$

The *characteristic function* of a random variable X is defined as the Fourier transform φ_X of the probability distribution μ^X:

$$\varphi_X(u) = \int e^{iuX} \cdot \mu^X = \mathbf{E}(e^{iuX}).$$

The function φ_X is continuous. We have $\varphi_X(0) = 1$.

If the first k_1 moments $m_k(X)$ exist, then φ_X is k_1 times differentiable; we have

$$\varphi_X^{(k)}(u) = \int (i\,X)^k\, e^{iuX} \cdot \mu^X, \qquad \text{for } k \leqslant k_1,$$

and, consequently,

$$\varphi_X^{(k)}(0) = i^k\, m_k(X),$$

from which we get the finite expansion of φ_X in a neighborhood of 0:

$$\varphi_X(u) = 1 + i\,m_1\,u - \frac{m_2}{2}u^2 + \cdots + \frac{i^k\,m_k}{k!}u^k + o(u^k) \qquad (k \leqslant k_1).$$

Examples. I. Let X denote a normalized Gaussian variable with probability density

$$\theta(X) = \frac{1}{\sqrt{2\pi}}\, e^{-X^2/2}.$$

The characteristic function is

$$\begin{aligned}\varphi_X(u) &= \frac{1}{\sqrt{2\pi}} \int_{-\infty}^{+\infty} e^{iuX}\, e^{-X^2/2}\, dX \\ &= \frac{1}{\sqrt{2\pi}} \int_{-\infty}^{+\infty} \sum_{k=0}^{\infty} \frac{(i\,uX)^k}{k!} \cdot e^{-X^2/2}\, dX \\ &= \sum_{k=0}^{\infty} \frac{i^k\, m_k}{k!} u^k,\end{aligned}$$

where

$$m_k = \frac{1}{\sqrt{2\pi}} \int_{-\infty}^{+\infty} X^k \, e^{-X^2/2} \, dX .$$

Let us set $X = Y\sqrt{2}$. We have

$$m_k = \frac{1}{\sqrt{2\pi}} \cdot (\sqrt{2})^{k+1} \int_{-\infty}^{+\infty} Y^k \, e^{-Y^2} \, dY .$$

If k is odd, we have $m_k = 0$. If $k = 2h$, we have

$$m_{2h} = \frac{1}{\sqrt{\pi}} \cdot 2^h \int_{-\infty}^{+\infty} Y^{2h} \, e^{-Y^2} \, dY = \frac{2^h}{\sqrt{\pi}} \Gamma\left(h + \frac{1}{2}\right)$$

$$= \frac{2^h}{\sqrt{\pi}} \cdot \frac{1}{2} \cdot \frac{3}{2} \cdots \frac{2h-1}{2} \cdot \sqrt{\pi} = 1 \,.\, 3 \cdots (2h-1) = \frac{(2h)!}{2^h (h!)} .$$

From this, we get

$$\varphi_X(u) = \sum_{h=0}^{\infty} \frac{1^{2h} \, m_{2h}(X)}{(2h)!} u^{2h}$$

$$= \sum_{h=0}^{\infty} \frac{1}{2^h \, h!} (-u^2)^h$$

$$= \sum_{h=0}^{\infty} \frac{1}{h!} \left(-\frac{u^2}{2}\right)^h = e^{-u^2/2} .$$

Thus, the characteristic function of a normalized Gaussian variable is $e^{-u^2/2}$.

II. The distribution of an (almost certain) entire positive variable X such that

$$\text{prob}\{X = k\} = e^{-c} \frac{c^k}{k!}, \qquad \forall k \in \mathbf{Z}_+ ,$$

is called a *Poisson distribution* of density c. This probability distribution can therefore be written

$$\mu^X = \sum_{k=0}^{+\infty} e^{-c} \frac{c^k}{k!} \delta_k .$$

We have

$$\mathbf{E}(X) = e^{-c} \sum_{k=0}^{+\infty} k \frac{c^k}{k!} = e^{-c} \sum_{k=1}^{+\infty} \frac{c^k}{(k-1)!} = c ,$$

$$\mathbf{E}(X^2) = e^{-c} \sum_{k=0}^{+\infty} k^2 \frac{c}{k!} = e^{-c} \sum_{k=0}^{+\infty} [k(k-1) + k] \frac{c^k}{k!}$$

$$= e^{-c} \sum_{k=2}^{+\infty} \frac{c^k}{(k-2)!} + e^{-c} \sum_{k=1}^{+\infty} \frac{c^k}{(k-1)!} = c^2 + c ,$$

and

$$\sigma^2(X) = \mathbf{E}(X^2) - (\mathbf{E}(X))^2 = c .$$

The characteristic function is equal to

$$\varphi_X(t) = e^{-c} \sum_{k=0}^{+\infty} e^{itk} \frac{c^k}{k!} = e^{-c} \sum_{k=0}^{+\infty} \frac{(c\, e^{it})^k}{k!} = e^{-c}\, e^{ce^{it}} = e^{c(e^{it}-1)} .$$

The following theorem shows the effect of a homothetic transformation or a translation on the characteristic function.

Theorem 3. *Let φ_X, $\varphi_{\lambda X}$, and φ_{X+a} denote the characteristic functions of variables X, λX, and $X + a$. We have*

$$\varphi_{\lambda X}(u) = \varphi_X(\lambda u)$$

$$\varphi_{X+a}(u) = e^{iau}\, \varphi_X(u) .$$

Proof.

$$\varphi_{\lambda X}(u) = \mathbf{E}(e^{iu\lambda X}) = \varphi_X(\lambda u)$$

$$\varphi_{X+a}(u) = \mathbf{E}(e^{iu(X+a)}) = e^{iau}\, \varphi_X(u) .$$

Remark. The reader should be careful not to confuse the first formula in this theorem with the formula indicating the effect of a homothetic transformation on the Fourier transform of a function: If $g(x) = f(\lambda x)$, we have

$$(\mathcal{F} g)(\omega) = \frac{1}{\lambda} (\mathcal{F} f) \left(\frac{\omega}{\lambda}\right) .$$

The difference stems from the fact that a homothetic transformation changes the probability density.

Definitions. A scalar random variable X is said to be Gaussian if its normalization

$$\frac{X - \mathbf{E}(X)}{\sigma(X)} ,$$

is a normalized Gaussian variable. Let us set $\mathbf{E}(X) = m$ and $\sigma(X) = \sigma$. Since the variable $Y = (X - m)/\sigma$ is a normalized Gaussian variable, its probability distribution is

$$\frac{1}{\sqrt{2\pi}} e^{-Y^2/2}\, dY .$$

Consequently, X has probability distribution

$$\frac{1}{\sigma\sqrt{2\pi}} e^{-(X-m)^2/2\sigma^2}\, dX .$$

The characteristic function of Y is

$$\varphi_Y(u) = e^{-u^2/2} .$$

Consequently, the characteristic function of $X = \sigma Y + m$ is

$$\varphi_X(u) = e^{imu} \varphi_Y(\sigma u) = e^{imu} e^{-\sigma^2 u^2/2} .$$

B. Study of probability distributions in $\mathbf{R}^n$. Let us consider $\mathbf{R}^2$, for example. Let

$$X = \begin{bmatrix} X^1 \\ X^2 \end{bmatrix} \in \mathbf{R}^2.$$

The probability distributions of the scalar variables X^1 and X^2 are called *marginal distributions.* By definition, they are the images of the probability distribution μ^X in $\mathbf{R}^2$ under the mappings

$$\begin{bmatrix} X^1 \\ X^2 \end{bmatrix} \to X^1 \quad \text{and} \quad \begin{bmatrix} X^1 \\ X^2 \end{bmatrix} \to X^2 ,$$

that is, under projection onto the axes.

Examples. A discrete distribution will often be represented in the form

$$\mu = \sum p_{i,j} \delta_{a_i,b_j} .$$

Then,

$$p_{i,j} = \text{prob}\,(X^1 = a_i, X^2 = b_j) .$$

The probability distribution of X^1 is then

$$\nu_1 = \sum_i q_i \delta_{a_i} ,$$

where

$$q_i = \text{prob}\,(X^1 = a_i) = \sum_j p_{i,j} ,$$

and the probability distribution of X^2 is

$$\nu_2 = \sum_j r_j \delta_{b_j} ,$$

where

$$r_j = \text{prob}\,(X^2 = b_j) = \sum_i p_{i,j} .$$

In the case of a probability distribution of density $\theta(X^1, X^2)$, the probability distributions of X^1 and X^2 have densities

$$\zeta_1(X^1) = \int_{-\infty}^{+\infty} \theta(X^1, X^2)\, dX^2 ,$$

$$\zeta_2(X^2) = \int_{-\infty}^{+\infty} \theta(X^1, X^2)\, dX^1 .$$

Among conditional distributions, we shall pay especial attention to conditional distributions, conditioned by an event $X^1 \in A$ or $X^2 \in B$. In particular, for a discrete distribution, we shall consider together (1) conditional distributions of X^1 for $X^2 = b_j$. They are equal to

$$\sum_i q_{i|j}\, \delta_{a_i} , \qquad \text{with} \qquad q_{i|j} = \frac{p_{i,j}}{r_j} .$$

(2) Conditional distributions of X^2 for $X^1 = a_i$. They are equal to

$$\sum_j r_{j|i}\, \delta_{b_j} , \qquad \text{with} \qquad r_{j|i} = \frac{p_{i,j}}{q_i} .$$

Let us return to the case of $\mathbf{R}^m$. We shall refer to the quantities

$$m_{k_1 \dots k_m} = \mathbf{E}\big[(X^1)^{k_1} \dots (X^m)^{k_m}\big] ,$$

as *moments*. We can also define a characteristic function

$$\varphi(u_1, \dots, u_m) = \mathbf{E}\big[e^{i(u_1 X^1 + \dots + u_m X^m)}\big] ,$$

which is simply the Fourier transform of the probability distribution μ^X of X.

C. Study of vector-valued random variables. One can define the characteristic function of a random variable $X \in E$ with probability distribution μ^X as follows:

$$\varphi_X(U) = \mathbf{E}(e^{i<X,U>}) = \int e^{i<X,U>}\, \mu^X = (\mathcal{F}\mu^X)(U),$$

where U is any element belonging to the dual E' of E or of any space $\hat{E}$ in duality with E (in particular, E itself if E is a Euclidean space). By using a basis for E, we again obtain the analytic form presented above for the case in which $E = \mathbf{R}^n$.

Theorem 4. *Let $X \in E$ denote a random variable, let a denote a member of E, and let K denote a member of $L(E, F)$. Let φ_X, φ_{X+a}, and φ_{KX} denote the characteristic functions of X, $X + a$, and KX. Then*

$$\varphi_{X+a}(U) = \mathrm{e}^{\mathrm{i}<a,U>} \varphi_X(U) \qquad (U \in E)$$

$$\varphi_{KX}(V) = \varphi_X(K^\star V) \qquad (V \in F') .$$

Proof.

$$\varphi_{X+a}(U) = \mathbf{E}(\mathrm{e}^{\mathrm{i}<X+a,U>}) = \mathbf{E}(\mathrm{e}^{\mathrm{i}<a,U>} \mathrm{e}^{\mathrm{i}<X,U>}) = \mathrm{e}^{\mathrm{i}<a,U>} \varphi_X(U) .$$

$$\varphi_{KX}(V) = \mathbf{E}(\mathrm{e}^{\mathrm{i}<KX,V>}) = \mathbf{E}(\mathrm{e}^{\mathrm{i}<X,K^\star V>}) = \varphi_X(K^\star V) .$$

5. GEOMETRIC STUDY OF SCALAR RANDOM VARIABLES

Let E denote a probabilized vector space. Let H denote the space of real random variables X such that

$$m_2(X) = \mathbf{E}(X^2) < \infty .$$

It will be convenient to identify two such variables if they are almost certainly equal. Then, the space H is identical to the space of (measurable) functions square-integrable with respect to the measure of the probability distribution on E, two such functions being identified if they are equal almost everywhere.

The space H is equipped with the usual scalar product

$$\langle X, Y \rangle = \int f(\xi)\, g(\xi) \,.\, \mu^\xi \quad \text{if} \quad X = f(\xi) \quad \text{and} \quad Y = g(\xi)$$

$$= \mathbf{E}(XY) .$$

Equipped with this scalar product, H is a Hilbert space.

In the space H, we shall consider

$$\begin{cases} \text{the line } \{1\} \text{ consisting of constants,} \\ \text{the subspace } \overset{\circ}{H} \text{ consisting of centered variables.} \end{cases}$$

We have

$$\mathbf{E}(X) = \langle 1, X \rangle .$$

Since the condition $\mathbf{E}(X) = 0$ is written $\langle 1, X \rangle = 0$, we see that $\overset{\circ}{H}$ is simply the hyperplane orthogonal to $\{1\}$.

For every $X \in H$, * we have

$$X = \overset{\circ}{X} + \mathbf{E}(X), \quad \text{where} \quad \begin{cases} \overset{\circ}{X} \in \overset{\circ}{H} , \\ \mathbf{E}(X) \in \{1\} . \end{cases}$$

The mapping $X \to \overset{\circ}{X}$ is the orthogonal projection onto $\overset{\circ}{H}$. We shall call it the *centralizer*.

---

* The notation $X \in H$ is contradictory with the notation $X \in \mathbf{R}$, which would say that X is a real variable. The risk of confusion seems sufficiently slight for us not to seek to eliminate this difficulty since, to do so, we would need to make our notation more cumbrous.

Definitions. The *covariance* of two random variables $X, Y \in H$ is defined as the quantity

$$\mathbf{v}(X, Y) = \mathbf{E}(\overset{\circ}{X}\overset{\circ}{Y}).$$

We have

$$\mathbf{v}(X, Y) = \langle \overset{\circ}{X}, \overset{\circ}{Y} \rangle.$$

Furthermore, since $\{1\}$ and $\overset{\circ}{H}$ are orthogonal, the formulas

$$X = \overset{\circ}{X} + \mathbf{E}(X)$$
$$Y = \overset{\circ}{Y} + \mathbf{E}(Y),$$

imply

$$\langle X, Y \rangle = \langle \overset{\circ}{X}, \overset{\circ}{Y} \rangle + \mathbf{E}(X)\,\mathbf{E}(Y),$$

so that

$$\mathbf{E}(XY) = \mathbf{v}(XY) + \mathbf{E}(X)\,\mathbf{E}(Y);$$

that is,

$$\boxed{\mathbf{v}(X, Y) = \mathbf{E}(XY) - \mathbf{E}(X)\,\mathbf{E}(Y)}\ .$$

Figure 1 shows diagrammatically the relations among the principal elements considered above.

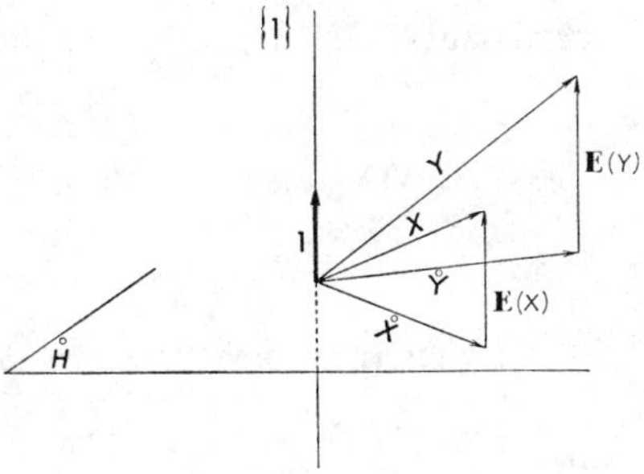

FIG. 1.

For $Y = X$, we find

$$\mathbf{v}(X, X) = \mathbf{E}(\overset{\circ}{X}{}^2) = \sigma^2(X).$$

Therefore, for the Hilbert structure of H, we have

$$\sigma^2(X) = \| \mathring{X} \|^2, \quad \text{that is,} \quad \sigma(X) = \| \mathring{X} \| .$$

Definition. The *correlational coefficient* of X and Y is defined as the cosine of the angle between $\mathring{X}$ and $\mathring{Y}$ in H, that is, the quantity

$$\rho(X, Y) = \frac{\langle \mathring{X}, \mathring{Y} \rangle}{\| \mathring{X} \| \, \| \mathring{Y} \|},$$

in other words,

$$\boxed{\rho(X, Y) = \frac{\mathbf{v}(X, Y)}{\sigma(X)\,\sigma(Y)}} .$$

We note that $\rho(X, Y)$ is indeterminant if $\sigma(X)$ or $\sigma(Y)$ is 0.

The correlational coefficient has the following properties:

$$\begin{cases} \rho(X, Y) \in [-1, +1] \\ \rho(X, Y) = 0 \Leftrightarrow \mathbf{v}(X, Y) = 0 \Leftrightarrow \mathbf{E}(XY) = \mathbf{E}(X)\,\mathbf{E}(Y) \\ \qquad\qquad\qquad\qquad\qquad\quad \Leftrightarrow \sigma^2(X + Y) = \sigma^2(X) + \sigma^2(Y) \\ \rho(X, Y) = \pm 1 \Leftrightarrow \exists \alpha \quad \text{such that } \mathring{Y} = \alpha \mathring{X} \\ \qquad\qquad\qquad\quad \Leftrightarrow \exists \alpha, \beta \quad \text{such that } Y = \alpha X + \beta . \end{cases}$$

Remark. The relation $\rho(X, Y) = 0$ is also written $\mathring{X} \perp \mathring{Y}$ (in H).

Definition. The *variance matrix* of p random variables $X_1, ..., X_p$ is defined as the Gram matrix of $\mathring{X}_1, ..., \mathring{X}_p$ for the Hilbert structure of H. We denote it by

$$\mathbf{V}(X_1, ..., X_p) .$$

We have, by definition,

$$\mathbf{V}(X_1, ..., X_p) = \begin{bmatrix} \sigma^2(X_1) & \mathbf{v}(X_1, X_2) \dots & \mathbf{v}(X_1, X_p) \\ \mathbf{v}(X_2, X_1) & \sigma^2(X_2) \quad \dots & \vdots \\ \vdots & & \vdots \\ \mathbf{v}(X_p, X_1) & \dots \quad \dots & \sigma^2(X_p) \end{bmatrix} .$$

Let E denote a probabilized vector space. Let S denote a basis for E. Then, the variance matrix $\mathbf{V}(X_S^1, ..., X_S^n)$ will be nonsingular unless there exist constants $\lambda_1, ..., \lambda_n$ such that

$$\sigma^2(\lambda_1 X_S^1 + \cdots + \lambda_n X_S^n) = 0 ,$$

that is, unless

$$\lambda_1 X_S^1 + \cdots + \lambda_n X_S^n = k,$$

almost certainly. Such a circumstance will obtain if and only if the support of the probability distribution is in an affine hyperplane of E.

Let $\hat{E}$ denote the space of linear random variables over E. Since $\hat{E}$ is isomorphic to the space of linear forms on E (because every linear random variable is the value of some linear form on E), it follows that $\hat{E}$ is in duality with E.

Let Y and Z denote members of $\hat{E}$. The mapping $Y, Z \to \mathbf{E}(\mathring{Y}\mathring{Z})$ is a bilinear form. For every basis S for E, the X_S^i constitute the dual basis $\hat{S}$ in $\hat{E}$. The bilinear form $\mathbf{E}(\mathring{Y}\mathring{Z})$ has components

$$\mathbf{E}(\mathring{X}_S^i \mathring{X}_S^j) = \mathbf{v}(X_S^i, X_S^j).$$

Consequently, the components of $\mathbf{V}(X_S^i, \ldots, X_S^n)$ are the components of a tensor two times covariant on $\hat{E}$ (that is, two times contravariant on E). In particular, if we make the change of basis defined in E by $X_S = KX_{S'}$, the matrix for the change of basis in $\hat{E}$ will be $K' = \overline{K}^{-1}$. Consequently,

$$\mathbf{V}(X_{S'}^1, \ldots, X_{S'}^n) = \overline{K'}\,\mathbf{V}(X_S^1, \ldots, X_S^n)\,K';$$

that is,

$$\mathbf{V}(X_S^1, \ldots, X_S^n) = K\mathbf{V}(X_{S'}^1, \ldots, X_{S'}^n)\,\overline{K}.$$

More generally, we have the following result (the proof of which below is independent of the preceding explanations).

Theorem 5. *Let X denote a random n-dimensional vector-valued variable and let Y denote MX where $M \in L(\mathbf{R}^n, \mathbf{R}^p)$. Then,*

$$\mathbf{V}(Y^1, \ldots, Y^p) = M\mathbf{V}(X^1, \ldots, X^n)\,\overline{M}.$$

Proof. We have

$$Y^h = \sum_i M_i^h X^i, \qquad Y^k = \sum_j M_j^k X^j,$$

so that

$$\mathbf{v}(Y^h, Y^k) = \sum_{ij} M_i^h\,\mathbf{v}(X^i, Y^j)\,M_j^k = \sum_{ij} M_i^h\,\mathbf{v}(X^i, Y^j)\,(\overline{M})_k^j,$$

which completes the proof.

The case of complex variables. Let E denote a probabilized *real* vector space. To study complex random variables, we shall find it convenient to introduce the space H consisting of variables X such that $\mathbf{E}(|X|^2) < \infty$. Just as in the real case, we shall consider the line $\{1\}$ of constants and the orthogonal hyperplane $\mathring{H}$ consisting of centered variables. Again we set $X = \mathring{X} + \mathbf{E}(X)$.

Let us denote by $\overline{X}$ the variable representing the complex conjugate of X.* We define the *covariance* of X and Y as the quantity

$$\mathbf{v}(X, Y) = \mathbf{E}(\overline{\mathring{X}}\mathring{Y}),$$

and the *variance* of X as the quantity

$$\sigma^2(X) = \mathbf{v}(X, X) = \mathbf{E}(|\mathring{X}|^2).$$

The definition of the matrix of variance remains unchanged.

Linear regression. Definition. Let $X_1, \ldots, X_p$ denote p real random variables such that $\mathbf{E}(X_i^2) < +\infty$ and let Y denote a random variable such that $\mathbf{E}(Y^2) < +\infty$. Then, the random variable

$$Y' = \alpha^1 X_1 + \cdots + \alpha^p X_p + \alpha^{p+1},$$

where the constants $\alpha^1, \ldots, \alpha^{p+1}$ are chosen so as to minimize $\mathbf{E}((Y - Y')^2)$ is called the *regression* of Y with respect to $X_1, \ldots, X_p$. The constants $\alpha^1, \ldots, \alpha^p$ are called the *coefficients of regression* of Y with respect to $X_1, \ldots, X_p$.

One may write

$$Y' = \alpha^1 X_1 + \cdots + \alpha^p X_p + \alpha^{p+1} 1.$$

Consequently, Y' is, in H, the projection of Y onto the subspace generated by $X_1, \ldots, X_p, 1$.

We can calculate $\alpha^1, \ldots, \alpha^p, \alpha^{p+1}$ by the usual methods: They necessitate finding the inverse of the Gram matrix of the variables $X_1, \ldots. X_p, 1$. One can simplify the numerical calculation by taking the following facts into consideration:

We necessarily have $Y - Y' \perp X^i$ and $Y - Y' \perp 1$. This last condition can be written

$$\mathbf{E}(Y) = \mathbf{E}(Y'),$$

that is,

$$\mathbf{E}(Y) = \alpha^1 \mathbf{E}(X_1) + \cdots + \alpha^p \mathbf{E}(X_p) + \alpha^{p+1}.$$

Thus, we have

$$\mathring{Y}' = \alpha^1 \mathring{X}_1 + \cdots + \alpha^p \mathring{X}_p,$$

---

*In certain cases, this notation might be confused with the notation for the transpose of a matrix. Then, we shall denote the transpose of a matrix by ${}^tA$ and we shall denote the complex conjugate of an n-tuple by ${}^cX$.

and

$$\mathbf{E}((Y - Y')^2) = \sigma^2(\mathring{Y} - \mathring{Y}').$$

Therefore, we need to choose $\alpha^1, \ldots, \alpha^p$ in such a way that $\sigma^2(\mathring{Y} - \mathring{Y}')$ is minimized. In other words, $\mathring{Y}'$ is the projection of $\mathring{Y}$ onto the subspace generated by $\mathring{X}_1, \ldots, \mathring{X}_p$. Consequently, if $\mathbf{V}(X_1 \ldots X_p)$ is nonsingular, we have

$$\begin{bmatrix} \alpha^1 \\ \vdots \\ \alpha^p \end{bmatrix} = \overset{-1}{\mathbf{V}}(X_1, \ldots, X_p) \cdot \begin{bmatrix} \mathbf{v}(X_1, Y) \\ \vdots \\ \mathbf{v}(X_p, Y) \end{bmatrix}.$$

A particular case. The regression Y' of Y with respect to a variable X is defined by

$$\mathring{Y}' = \frac{\mathbf{v}(X, Y)}{\sigma^2(X)} \mathring{X}, \quad \mathbf{E}(Y') = \mathbf{E}(Y),$$

that is,

$$Y' - \mathbf{E}(Y) = \frac{\mathbf{v}(X, Y)}{\sigma^2(X)} (X - \mathbf{E}(X)).$$

Definition. The line in the plane consisting of pairs $\begin{bmatrix} x \\ y \end{bmatrix}$ and corresponding to the equation

$$y - \mathbf{E}(Y) = \frac{\mathbf{v}(X, Y)}{\sigma^2(X)} (x - \mathbf{E}(X)),$$

is called the *line of regression* of Y with respect to X. Similarly, the equation of the line of regression of X with respect to Y is

$$x - \mathbf{E}(X) = \frac{\mathbf{v}(X, Y)}{\sigma^2(Y)} (y - \mathbf{E}(Y)).$$

These two lines intersect at the point

$$\begin{bmatrix} \mathbf{E}(X) \\ \mathbf{E}(Y) \end{bmatrix}.$$

A necessary and sufficient condition for them to coincide is that

$$\frac{\mathbf{v}(X, Y)}{\sigma^2(X)} = \frac{\sigma^2(Y)}{\mathbf{v}(X, Y)},$$

that is, that $\rho(X, Y) = \pm 1$.

Similarly, if the regression of Y with respect to X_1 and X_2 is

$$Y' - \mathbf{E}(Y) = \alpha^1(X_1 - \mathbf{E}(X_1)) + \alpha^2(X_2 - \mathbf{E}(X_2)),$$

the plane whose equation is

$$y - \mathbf{E}(Y) = \alpha^1(x_1 - \mathbf{E}(X_2)) + \alpha^2(x_2 - \mathbf{E}(X_2)),$$

is called the *plane of regression.*

Definition. The quantity

$$R(Y; X_1, \ldots, X_p) = \rho(Y, Y') = \frac{\sigma(Y')}{\sigma(Y)},$$

is called the *coefficient of total correlation* of Y with respect to $X_1, \ldots, X_p$. Note that

$$\sigma^2(Y') = \alpha^1 \, \mathbf{v}(X_1, Y) + \cdots + \alpha^p \, \mathbf{v}(X_p, Y).$$

6. STOCHASTIC INDEPENDENCE OF RANDOM VARIABLES

Definition. Let X^1 and X^2 denote any two real random variables, let μ denote the probability distribution of the pair

$$X = \begin{bmatrix} X^1 \\ X^2 \end{bmatrix},$$

and let ν_1 and ν_2 denote the probability distributions of X^1 and X^2. The random variables X^1 and X^2 are said to be *stochastically independent* if

$$\mu = \nu_1 \, \nu_2 \, .$$

Two *events* A and B are said to be independent if the associated Bernoulli variables are stochastically independent.

Let us recall that the measure $\nu_1 \, \nu_2$ is characterized by the fact that

$$(*) \qquad \int f(X^1) \, g(X^2) \, . \, \nu_1^{X_1} \, \nu_2^{X_2} = \int f(X^1) \, \nu_1^{X_1} . \int g(X^2) \, \nu_2^{X_2},$$

for every f and g in $\mathfrak{D}^0$ and that this relation remains valid for integrable f and g and *a fortiori* for square-integrable f and g. The relation (*) is written

$$\mathbf{v}(f(X^1), g(X^2)) = 0.$$

Then, we have the following theorem:

Theorem 6. *A necessary and sufficient condition for two random variables X^1 and X^2 to be independent is that*

$$\rho(f(X^1), g(X^2)) = 0,$$

for every f and g in $\mathfrak{D}^0$ or for every

$$f \in \mathbf{L}^2(\nu_1), \qquad g \in \mathbf{L}^2(\nu_2).$$

Corollaries. I. *If X^1 and X^2 are stochastically independent, so are $f(X^1)$ and $g(X^2)$.*

II. *If X^1 and X^2 are independent, the events $X^1 \in A_1$ and $X^2 \in A_2$ are independent regardless of the subsets A_1 and A_2 of $\mathbf{R}$ (which are assumed measurable with respect to ν_1 and ν_2, respectively).*

Theorem 7. *Each of the following conditions is necessary and sufficient for two events A and B to be independent:*

(1) $\text{prob}(AB) = \text{prob}(A)\,\text{prob}(B)$.
(2) $\text{prob}(A \mid B) = \text{prob}(A)$ (*provided* $\text{prob}(B) \neq 0$).
(3) $\text{prob}(A \mid B) = \text{prob}(A \mid \complement B)$ (— — — $\text{prob}(B) \neq 0$ et 1).
(4) $\mathbf{v}(\varepsilon_A, \varepsilon_B) = 0$.

Proof. This follows without difficulty from the fact that

$$(\alpha)\ \mathbf{v}(\varepsilon_A, \varepsilon_B) = \mathbf{E}(\varepsilon_A \varepsilon_B) - \mathbf{E}(\varepsilon_A)\,\mathbf{E}(\varepsilon_B) = \text{prob}(AB) - \text{prob}(A)\,\text{prob}(B),$$

and from the formulas

$$(\beta)\quad \text{prob}(AB) = \text{prob}(B)\,.\,\text{prob}(A \mid B)$$

$$(\gamma)\quad \text{prob}(A) = \text{prob}\,B\,.\,\text{prob}(A \mid B) + \text{prob}(\complement B)\,.\,\text{prob}(A \mid \complement B).$$

Specifically,

(a) Suppose that A and B are independent. We then have (4) (by definition), from which we get (1) (according to (α)) and (2) (according to (β)). Also,

$$\mathbf{v}(\varepsilon_A, \varepsilon_{\complement B}) = \mathbf{v}(\varepsilon_A, 1 - \varepsilon_B) = -\,\mathbf{v}(\varepsilon_A, \varepsilon_B) = 0,$$

so that $\text{prob}\,A = \text{prob}(A \mid \complement B)$ and, hence, (3) holds.

(b) The functions in ε_A (resp. ε_B) constitute a two-dimensional vector space a basis for which is ε_A and $1 - \varepsilon_A$ (resp. ε_B and $1 - \varepsilon_B$). Consequently,

$$(4) \Rightarrow \mathbf{v}(f(\varepsilon_A), g(\varepsilon_B)) = 0, \qquad \forall f, g.$$

Thus, (4) implies that the events of A and B are independent. Furthermore, (1) $\Rightarrow$ (4) (according to (α)); (2) $\Rightarrow$ (1) (according to (β)); (3) $\Rightarrow$ (2) (according to (γ) and the relation $\text{prob}(B) + \text{prob}(\complement B) = 1$).

In the case of p random variables, we have analogous definitions: Random variables $X^1, \ldots, X^p$ are said to be stochastically independent if the probability distribution μ^X of

$$X = \begin{bmatrix} X^1 \\ \vdots \\ X^p \end{bmatrix}$$

is equal to the product of the probability distributions $\nu_1, \dots, \nu_p$ of $X^1, \dots, X^p$. Events $A_1, \dots, A_p$ are said to be independent if the Bernoulli variables $\varepsilon_{A_1}, \dots, \varepsilon_{A_p}$ are independent.

Remarks. I. Let μ^X denote the distribution of

$$X = \begin{bmatrix} X^1 \\ \vdots \\ X^n \end{bmatrix}.$$

If μ^X is represented in the form

$$\mu^X = \nu^{X^1}, \dots, {}^{X^n}.$$

where $\nu^{X_1}, \dots, \nu^{X_n}$ are probability distributions, then $\nu^{X_1}, \dots, \nu^{X_n}$ are necessarily the probability distributions of $X^1, \dots, X^n$, and hence $X^1, \dots, X^n$ are independent. This is true because, for every $f \in \mathfrak{D}^0(\mathbf{R})$,

$$\int f(X^i) \cdot \mu^X = \int f(X^i) \cdot \nu^{X_i},$$

which shows that ν^{X^i} is the image of μ^X under the projection $X \to X^i$.

II. If $X^1, \dots, X^n$ are independent and if we set

$$X = \begin{bmatrix} X^\alpha \\ \hline X^\beta \end{bmatrix},$$

then X^α and X^β are independent.

III. If $X^1, \dots, X^n$ are independent scalar random variables, then

$$\mathbf{E}(X^1 \dots X^n) = \mathbf{E}(X^1) \dots \mathbf{E}(X^n),$$

as can be shown by induction.

Theorem 8. *Suppose that X and Y are two stochastically independent random variables with probability distributions μ^X and ν^Y. Then, the variable $X + Y$ has probability distribution $\mu * \nu$ and hence, its characteristic function is the product $\varphi_X \varphi_Y$ of the characteristic functions of X and Y.*

Proof. This follows from the definitions and the formula

$$\mathcal{F}(\mu * \nu) = \mathcal{F}(\mu) \mathcal{F}(\nu),$$

for two bounded measures.

Theorem 9. *Suppose that X and Y are independent scalar Gaussian variables and that $Z = X + Y$ is a Gaussian variable with mean* $\mathbf{E}(Z) = \mathbf{E}(X) + \mathbf{E}(Y)$ *and variance* $\sigma^2(Z) = \sigma^2(X) + \sigma^2(Y)$.

Proof. Let φ_X, φ_Y, and φ_Z denote the characteristic functions of X, Y, and Z. Then,

$$\varphi_X(u) = \exp(\mathrm{i}\, u\mathbf{E}(X)) \exp(-\tfrac{1}{2} u^2 \sigma^2(X)),$$

$$\varphi_Y(u) = \exp(\mathrm{i}\, u\mathbf{E}(Y)) \exp(-\tfrac{1}{2} u^2 \sigma^2(Y)),$$

from which we get

$$\varphi_Z(u) = \varphi_{X \dot{+} Y}(u) = \exp[\mathrm{i}\, u(\mathbf{E}(X) + \mathbf{E}(Y))]\exp[-\tfrac{1}{2} u^2\sigma^2((X) + \sigma^2(Y))],$$

which is indeed the characteristic function of a Gaussian variable of mean $\mathbf{E}(X) + \mathbf{E}(Y)$ and variance $\sigma^2(X) + \sigma^2(Y)$.

Application to Gaussian variables. Let us calculate the matrix of variance of n random variables $X^1, \ldots, X^n$ with centered Gaussian distribution. Suppose that

$$\frac{\sqrt{\det(\mathcal{A})}}{(2\pi)^{n/2}} \exp(-\tfrac{1}{2} \overline{X}\mathcal{A}X)\, \mathrm{d}X^1 \ldots \mathrm{d}X^n,$$

is their joint probability distribution.

We can consider this probability distribution as representing in a particular basis a distribution whose general representation is

$$\mu_S = \frac{\sqrt{\det \mathcal{A}(S)}}{(2\pi)^{n/2}} \exp(-\tfrac{1}{2} \overline{X}_S\, \mathcal{A}(S)\, X_S)\, \mathrm{d}X_S^1 \ldots \mathrm{d}X_S^n.$$

Let S' denote a basis such that $\mathcal{A}(S') = 1$. Then,

$$\mu_{S'} = \frac{1}{(2\pi)^{n/2}} \exp\Big(-\tfrac{1}{2} \sum_i (X_{S'}^i)^2\Big)\, \mathrm{d}X_{S'}^i \ldots \mathrm{d}X_{S'}^n$$

$$= \prod_i \frac{1}{\sqrt{2\pi}} \exp(-\tfrac{1}{2} (X_{S'}^i)^2)\, \mathrm{d}X_{S'}^i.$$

Thus, the variables $X_{S'}^i$ are independent normalized Gaussian variables, so that

$$\mathbf{V}(X_{S'}^1, \ldots, X_{S'}^n) = \mathbf{1}.$$

Let us return to an arbitrary basis by setting

$$X_{S'} = KX_S,\ X_S = \overset{-1}{K} X_{S'}.$$

We then have

$$\mathcal{A}(S) = \bar{K}\mathbf{1}K = \bar{K}K,$$

$$\mathbf{V}(X_S^1, \ldots, X_S^n) = K^{-1}\mathbf{1}\bar{K}^{-1} = K^{-1}.\bar{K}^{-1} = \overset{-1}{\mathcal{A}}(S).$$

Thus,

The variance matrix of n variables $X^1, \ldots, X^n$ of joint density

$$\frac{\sqrt{\det(\mathcal{A})}}{(2\pi)^{n/2}} \exp(-\tfrac{1}{2}\bar{X}\mathcal{A}X),$$

is equal to $\overset{-1}{\mathcal{A}}$.

Also, *if*

$$X = \begin{bmatrix} X^1 \\ \cdot \\ \cdot \\ \cdot \\ X^n \end{bmatrix},$$

has a Gaussian distribution and if $\rho(X^i, X^j) = 0$ for $i \neq j$, the variables $X^1, \ldots, X^n$ are independent.

In this case, we have, in fact,

$$\mathbf{V}(X^1, \ldots, X^n) = \begin{bmatrix} \sigma_1^2 & & 0 \\ & \ddots & \\ 0 & & \sigma_n^2 \end{bmatrix}.$$

Therefore, the probability distribution of X may be written

$$\mu^X = \frac{1}{(2\pi)^{n/2}\prod_i \sigma_i} \exp\left(-\frac{1}{2}\sum_i \frac{(X^i)^2}{\sigma_i^2}\right) dX^1 \ldots dX^n$$

$$= \prod_i \frac{1}{\sigma_i\sqrt{2\pi}} \cdot \exp\left(-\frac{1}{2}\frac{(X^i)^2}{\sigma_i^2}\right) dX^i.$$

Let us now evaluate the characteristic function of a Gaussian distribution of density

$$\frac{\sqrt{\det A}}{(2\pi)^{n/2}} \exp(-\tfrac{1}{2}\bar{X}\mathcal{A}X).$$

Let us consider this distribution as the representation of a probability distribution whose general representation is

$$\frac{\sqrt{\det \mathcal{A}(S)}}{(2\pi)^{n/2}} \exp(-\tfrac{1}{2}.\bar{X}_S\mathcal{A}(S)X_S).$$

Let S and S' denote two bases. Let us set $X_S = X$, $X_{S'} = X'$, and $\mathcal{A}(S) = \mathcal{A}$. Let us suppose that $\mathcal{A}(S') = \mathbf{1}$.

Let $\varphi_{X'}(U') = \mathbf{E}(e^{i\langle X', U'\rangle})$ denote the characteristic function of X'. We have

$$\langle X', U' \rangle = \sum_k U'^k X'^k$$

$$e^{i\langle X', U' \rangle} = \prod_k e^{iU'^k X'^k}.$$

Now, the variables X'^k are independent. The same is true of the variables

$$e^{iU'^k X'^k},$$

so that

$$\varphi_{X'}(U') = \mathbf{E}(e^{i\langle X', U' \rangle}) = \prod_k \mathbf{E}(e^{iU'^k X'^k}) = \prod_k \varphi_{X'^k}(U'^k)$$
$$= \exp(-\tfrac{1}{2}\sum_k (U'^k)^2) = \exp(-\tfrac{1}{2}\overline{U}' U').$$

Let us now set

$$X' = KX, \quad X = \overset{-1}{K} X'.$$

We have

$$\varphi_X(U) = \varphi_{X'}(\overline{\overset{-1}{K}} U') = \exp(-\tfrac{1}{2}\overline{U} \overset{-1}{K} \overline{\overset{-1}{K}} U).$$

Now, $\mathcal{A} = \overline{K}K$, from which we get, upon resuming the original notations,

$$\varphi_X(U) = \exp(-\tfrac{1}{2}\overline{U} \overset{-1}{\mathcal{A}} U).$$

Thus, we have the result:

The characteristic function of a Gaussian distribution of density

$$\frac{\sqrt{\det \mathcal{A}}}{(2\pi)^{n/2}} \exp(-\tfrac{1}{2}\overline{X}\mathcal{A}X),$$

is

$$\varphi_X(U) = \exp(-\tfrac{1}{2}\overline{U} \overset{-1}{\mathcal{A}} U).$$

From this result, we can derive the following one (the same proof as in the scalar case):

Let X and Y denote two independent random n-tuples of densities

$$\frac{\sqrt{\det \mathcal{A}}}{(2\pi)^{n/2}} \exp(-\tfrac{1}{2}\overline{X}\mathcal{A}X),$$

and

$$\frac{\sqrt{\det \mathcal{B}}}{(2\pi)^{n/2}} \exp(-\tfrac{1}{2}\overline{Y}\mathcal{B}Y).$$

Then, the n-tuple $X + Y$ has a Gaussian probability distribution of density

$$\frac{\sqrt{\det \mathcal{C}}}{(2\pi)^{n/2}} \exp(-\tfrac{1}{2}\overline{X}\mathcal{C}X),$$

where $\overset{-1}{\mathcal{C}} = \overset{-1}{\mathcal{A}} + \overset{-1}{\mathcal{B}}$.

Remark. The formula $\overset{-1}{\mathcal{C}} = \overset{-1}{\mathcal{A}} + \overset{-1}{\mathcal{B}}$ expresses the additivity of the matrices of X and Y. This additivity remains valid in a case more general than of the case of independence of X and Y. Specifically, we have the following proposition:

Proposition 1. *Let X and Y denote two random n-tuples such that $\rho(X^i, Y^j) = 0 \ \forall i, j$. Then,*

$$\mathbf{V}(X^1 + Y^1, \ldots, X^n + Y^n) = \mathbf{V}(X^1, \ldots, X^n) + \mathbf{V}(Y^1, \ldots, Y^n).$$

Proof. We have

$$\begin{aligned} \mathbf{v}(X^i + Y^i, X^j + Y^j) &= \mathbf{v}(X^i, X^j) + \mathbf{v}(Y^i, Y^j) + \mathbf{v}(Y^i, X^j) + \mathbf{v}(X^i, Y^j) \\ &= \mathbf{v}(X^i, X^j) + \mathbf{v}(Y^i, Y^j). \end{aligned}$$

Theorem 10. *Let X denote a centered Gaussian n-tuple of density*

$$\frac{\sqrt{\det \mathcal{A}}}{(2\pi)^{n/2}} \exp(-\tfrac{1}{2}\overline{X}\mathcal{A}X).$$

Suppose that $M \in L(\mathbf{R}^n, \mathbf{R}^p)$. Then $Y = MX$ is a centered Gaussian p-tuple with variance matrix $M\overset{-1}{\mathcal{A}}\overline{M}$ if M is of rank p.

Proof. The characteristic function of X is

$$\varphi_X(U) = \exp(-\tfrac{1}{2}\overline{U}\overset{-1}{\mathcal{A}}U),$$

so that

$$\varphi_{MX}(V) = \exp(-\tfrac{1}{2}\overline{V}M\overset{-1}{\mathcal{A}}\overline{M}V).$$

A particular case. If $w \in (\mathbf{R}^n)'$, the random variable $Y = wX$ is a centered Gaussian variable with variance $w \,.\, \overset{-1}{\mathcal{A}} \,.\, \overline{w}$. Consequently, if X is a centered Gaussian n-tuple, all its components are centered Gaussian random variables.

7. CONVERGENCE OF PROBABILITY DISTRIBUTIONS AND RANDOM VARIABLES

Let us equip the set of probability distributions with the narrow topology. We shall say that a sequence X_n of random variables converges in distribution to a random variable X if the sequence of probability distributions μ_n of X_n converges narrowly to the probability distribution μ of X, that is, if for every continuous bounded function f

$$\lim_{n\to\infty} \int f \cdot \mu_n = \int f \cdot \mu .$$

According to the results obtained in the theory of the Fourier transformation, we have the following theorem:

Theorem 11. *For $\{X_n\}$ to converge in distribution to X, it is necessary and sufficient that the sequence $\{\varphi_n\}$ of characteristic functions φ_n of X converge simply to the characteristic function φ of X.*

If the sequence $\{\varphi_n\}$ of the characteristic functions of the random variables X_n converges simply to a function φ, uniformly on a neighborhood of 0, then φ is a characteristic function of some random variable X and $\{X_n\}$ converges in distribution to X.

Theorem 12 (called the "limit theorem"). *Let $\{X_n\}$ denote a sequence of independent* scalar random variables with the same probability distribution, with mean m, and with standard deviation σ. Then the sequence $\{Y_n\}$ defined by*

$$Y_n = \frac{(X_1 + \cdots + X_n) - nm}{\sigma\sqrt{n}} ,$$

converges in distribution to a reduced Gaussian variable.

Proof. Note that Y_n is normalized. Let us set

$$X'_k = \frac{X_k - m}{\sigma} .$$

We have

$$Y_n = \frac{X'_1 + \cdots + X'_n}{\sqrt{n}} \quad \text{with } \mathbf{E}(X'_k) = 0 , \qquad \sigma(X'_k) = 1 .$$

Let φ denote the common characteristic function of the variables X'_k. We have

* That is, such that $X_1, \dots, X_n$ are independent for every n.

$$\varphi_{Y_n}(u) = \left(\varphi\left(\frac{u}{\sqrt{n}}\right)\right)^n.$$

The expression on the right takes the form 1^∞ as $n \to \infty$. We have

$$\ln \varphi_{Y_n}(u) = n \ln \varphi\left(\frac{u}{\sqrt{n}}\right) \sim n\left[\varphi\left(\frac{u}{\sqrt{n}}\right) - 1\right].$$

Now,

$$\varphi(u) = 1 - \frac{u^2}{2} + o(u^2) \qquad (\text{as } u \to 0),$$

since the X'_k are normalized variables, so that

$$\ln \varphi_{Y_n}(u) \sim n\left(-\frac{u^2}{2n}\right) = -\frac{u^2}{2},$$

and

$$\lim_{n\to\infty} \varphi_{Y_n}(u) = e^{-u^2/2}.$$

This implies that $\{Y_n\}$ converges in distribution to a variable with characteristic function $e^{-u^2/2}$, that is, a normalized Gaussian variable.

This theorem can be generalized as follows:

Theorem 13. *Let $X_1, \ldots, X_n, \ldots$ denote a sequence of independent centered random m-tuples with the same probability distribution and suppose that the common variance matrix is* **V**. *Then, the m-tuple*

$$\frac{X_1 + \cdots + X_n}{\sqrt{n}},$$

converges in law to a centered Gaussian random m-tuple of density

$$\frac{1}{(2\pi)^{n/2}\sqrt{\det \mathbf{V}}} \exp(-\tfrac{1}{2}\bar{Y}\,\mathbf{V}^{-1}\,Y).$$

Topology over random variables. Let E denote a probabilized vector space and let H denote the space of random variables such that $\mathbf{E}(|X|^2) < \infty$. In H, let us consider the usual Hilbert topology. We write

$$X_n \xrightarrow{\text{ms}} X,$$

(to be read "X_n converges to X in mean square") if $\{X_n\}$ converges to X in H, that is, if

$$\mathbf{E}(|X_n - X|^2) \to 0 .$$

We cite as a reminder two other convergences that are used in the space of random variables:

(1) Suppose that $\xi \in E, X_n = f_n(\xi), X = f(\xi)$. We can say that $\{X_n\}$ converges ***almost certainly*** to X if $\{f_n(\xi)\}$ converges to $f(\xi)$ almost everywhere with respect to the probability distribution over E.

(2) We say that X_n ***converges in probability*** to X if, $\forall \varepsilon > 0$,

$$\text{prob}\{|X_n - X| \geqslant \varepsilon\} \to 0 \quad \text{as} \quad n \to \infty .$$

TABLE OF THE NORMALIZED GAUSSIAN DISTRIBUTION

The following table gives the values of the probability $F(x)$ that a normalized Gaussian random variable X will assume a value less than x:

x	$F(x)$	x	$F(x)$
0.0	0.50000	2.0	0.97725
0.1	0.53983	2.1	0.98214
0.2	0.57926	2.2	0.98610
0.3	0.61791	2.3	0.98928
0.4	0.65542	2.4	0.99180
0.5	0.69146	2.5	0.99379
0.6	0.72575	2,6	0.99534
0.7	0.75804	2.7	0.99653
0.8	0.78814	2,8	0.99744
0.9	0.81594	2,9	0.99813
1.0	0.84134	3.0	0.99865
1.1	0.86433	3.1	0.99903
1.2	0.88493	3.2	0.99931
1.3	0.90320	3.3	0.99952
1.4	0.91924	3.4	0.99966
1.5	0.93319	3.5	0.99977
1.6	0.94520	3.6	0.99984
1.7	0.95543	3.7	0.99989
1.8	0.96407	3.8	0.99993
1.9	0.97128	3.9	0.99995
		4.0	0.99997

For $x < 0$, one should use the formula $F(x) = 1 - F(-x)$.

6

Markov Chains

In Chapters VI and VII, we shall study several examples of random processes. We shall not give a precise general definition of the concept of a random process but simply mention that a random process is a phenomenon that takes place over a period of time and that has random characteristics. We shall make this concept precise only in the case of particular examples.

1. DEFINITION OF MARKOV CHAINS. GRAPH-THEORY CONCEPTS

Markov chains. **Definition.** A *stochastic matrix* is defined as any square matrix P such that

$$P_i^j \geqslant 0,$$

and

$$\sum_j P_i^j = 1 .$$

Examples. The following matrices are stochastic:

$$\begin{bmatrix} 0.7 & 0.4 \\ & \\ 0.3 & 0.6 \end{bmatrix}, \quad \begin{bmatrix} 0.2 & 0.5 & 0 \\ 0.5 & 0 & 1 \\ 0.3 & 0.5 & 0 \end{bmatrix}, \quad \begin{bmatrix} 0 & 1 & 1 \\ 0 & 0 & 0 \\ 1 & 0 & 0 \end{bmatrix}.$$

Definition. We shall say that the evolution of a system obeys a *Markov chain* if

(1) the instants at which the system is observable or observed constitute a sequence denoted by 0, 1, 2, . . . , m, . . . ,

(2) the system can assume a finite number of states (numbered from 1 to n),

(3) there exists a probabilized set over which the following events are defined: "The system is in the state i at the instant m",

(4) to every instant m is associated a stochastic matrix $P(m)$, and the conditional probability that the system will be in the state j at the instant $m + 1$ when it is given that it is in the state i at the instant m is equal, if it is defined, to $P_i^j(m)$, and

(5) the conditional probability that the system will be in the state j at the instant $m + 1$, when it is given that it is in the state i at the instant m and that the states of the system at the instants $0, 1, \ldots, m - 1$ have specified values, is also, if it is defined, equal to $P_i^j(m)$.

The system is said to have *transition matrix* $P(m)$ at the instant m. Property (2) is a simplifying hypothesis. Property (5) expresses the Markov nature of the process.

Let

$$x(m) = \begin{bmatrix} x^1(m) \\ \vdots \\ x^n(m) \end{bmatrix}$$

denote the probabilities of the different states at a given instant m.* We note that

$$x^i(m) \geqslant 0 \quad \text{and} \quad \sum_i x^i(m) = 1 .$$

The probabilities

$$x(m+1) = \begin{bmatrix} x^1(m+1) \\ \vdots \\ x^n(m+1) \end{bmatrix}$$

of the states at the following instant $m + 1$ are given by the theorem on total probabilities:

$$x^j(m+1) = \sum_i P_i^j(m)\, x^i(m) ,$$

that is,

$$x(m+1) = P(m)\, x(m) .$$

Consequently, if we denote by $x(0)$ the initial probability distribution of the different states, the distribution at later instants will be

---

* The n-tuple $x(m)$ defines the probability distribution of the random variable $I(m)$ that is equal to the state of the system at the instant m.

$$x(1) = P(0)\,x(0), \quad x(2) = P(1)\,x(1) = P(1)\,P(0)\,x(0)\,,$$

and, in general,

$$x(m) = P(m-1)\,\ldots\,P(1)\,P(0)\,x(0)\,.$$

We shall say that a Markov chain is *stationary* if it has a transition matrix P independent of m. We then have

$$x(m+1) = P\,x(m)\,.$$

Consequently,*

$$x(m) = \overset{m}{P}\,x(0)\,.$$

In this chapter, we shall need to study systems evolving according to stationary Markov chains having the same transition matrix but differing by initial distribution of the states $x(0)$.

We shall be especially interested in the asymptotic properties of $x(m)$ and the way in which these properties depend on $x(0)$.

Let P denote a stochastic matrix. We shall say that a transition $i \to j$ is permissible if $P_i^j \neq 0$.

The theory shows that the asymptotic properties of $\overset{m}{P}$ as $m \to \infty$ are closely connected with the study of permissible transitions. We shall make this study by using the fundamental concepts from graph theory.

Graph theory. Definitions. An *oriented graph* on a set X is defined as any mapping Γ of X into the collection of subsets of X. In what follows, we shall always understand "graph" to mean "oriented graph".

The elements of X are called *vertices* or *points* of the graph.

An arc of the graph Γ is any ordered pair (x, y) of vertices such that $y \in \Gamma(x)$. The vertex x is called the *initial vertex* of the arc (x, y), and y is called the *terminal vertex*.

If X is a finite set, we can represent its elements by points in a plane. Then, Γ will be represented by the set of its arcs. Figure 1 gives an example of such a representation.

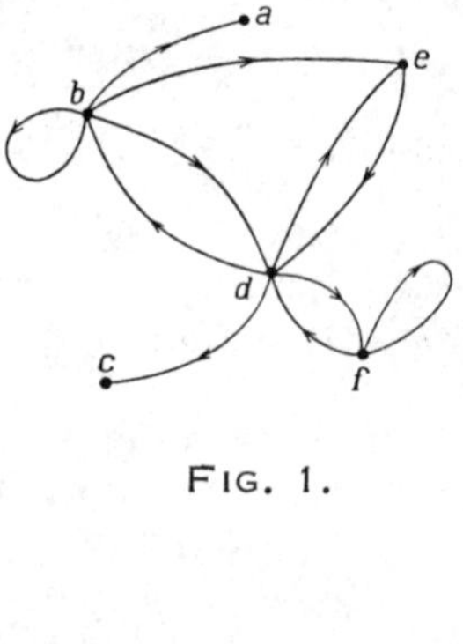

FIG. 1.

*Unless the contrary is stated, in this chapter we shall denote by $\overset{k}{A}$ the kth power of a matrix A.

A *path* is any sequence $x_0, x_1, ..., x_k$ of vertices such that $x_{i+1} \in \Gamma(x_i)$. The nonnegative number k is called the *length* of the path. A path of nonzero length can be defined by the sequence of its arcs $(x_0, x_1), (x_1, x_2), ..., (x_{k-1}, x_k)$. The vertex x_0 is called the *initial vertex* of the path, and the vertex x_k is called its *terminal vertex*. The path is said to be *elementary* if $x_i \neq x_j$ whenever $i \neq j$.

Any path whose initial vertex is also its terminal vertex is called a *circuit*. Every path of zero length is a circuit. A circuit of length 1 is called a *loop*.

Beginning with a graph Γ, one can define the following graphs Γ^k (for $k \geqslant 0$):

$y \in \Gamma^k(x)$ if the graph Γ has a path of length k of which x is the identified vertex and y the terminal vertex. In particular,

$$y \in \Gamma^0(x) \quad \text{if} \quad y = x \quad \text{(in other words, } \Gamma^0(x) = \{x\},\ \forall x \in X\text{);}$$
$$\Gamma^1 = \Gamma.$$

Furthermore, a graph $\hat{\Gamma}$ is defined as follows:

$y \in \hat{\Gamma}(x)$ if the graph Γ has a path (of any length) whose initial vertex is x and whose terminal vertex is y.

Obviously,

$$\hat{\Gamma}(x) = \bigcup_{k=0}^{\infty} \Gamma^k(x)\,.$$

We note that the relation $y \in \hat{\Gamma}(x)$ is a quasiordering (in particular, $x \in \hat{\Gamma}(x), \forall x \in X$). If $y \in \hat{\Gamma}(x)$, we shall say that y is *accessible* from x.

A graph Γ is said to be *strongly connected* if, for every $x, y \in X$, there exists a path in Γ with initial vertex at x and terminal vertex at y, in other words, if $\hat{\Gamma}(x) = X$ for every $x \in X$.

An equivalence relation, known as the *canonical equivalence relation* and defined by

$$x \sim y \quad \text{if} \quad y \in \hat{\Gamma}(x) \quad \text{and} \quad x \in \hat{\Gamma}(y)$$

is associated with every graph. This relation is simply the equivalence relation associated with the partial ordering $y \in \hat{\Gamma}(x)$. These equivalence classes will be called simply the *classes* of the graph.

The quasiordering relation $y \in \hat{\Gamma}(x)$ defines a quotient by the canonical equivalence relation and it defines an ordering relation over the classes of the graph. From time to time, we shall use the following terminology "a class C_2 is accessible from the class C_1 if every $x_2 \in C_2$ is accessible from every $x_1 \in C_1$ (as will be the case if this is true for *any* particular x_1 and x_2).

A subset A of X is called a *closed set* if

$$x \in A, \quad y = \hat{\Gamma}(x) \quad \Rightarrow \quad y \in A,$$

in other words, if

$$x \in A \quad \Rightarrow \quad \hat{\Gamma}(x) \subset A$$

or, if there fails to exist any arc with initial vertex in A and terminal vertex in the complement of A.

A class is said to be *final* if it is a closed set, in other words, if it is a maximal element for the ordering relation between the classes. If there exists only a finite number of classes (in particular, if X is finite), there exists at least one final class. A non-final class is said to be *transitory*.

Graphs associated with stochastic matrices. Let P denote an $n \times n$ stochastic matrix. Let us associate with it the graph, the set of vertices of which is $\{1, 2, \ldots, n\}$ and that is defined by

$$j \in \Gamma(i) \quad \Leftrightarrow \quad P_i^j > 0 .$$

If we interpret the matrix P as transition matrix of a Markov chain, the relation $j \in \Gamma(i)$ can be expressed by saying that the transition from the state i to the state j (in one stage) is permissible.

Note that corresponding to the matrix $\overset{k}{P}$ is the graph Γ^k.

2. SPECTRAL ANALYSIS OF STOCHASTIC MATRICES

In $\mathbf{C}^n$, we shall use the following norm:

$$x \rightarrow \| x \| = \sum_i | x^i |$$

and in $L(\mathbf{C}^n, \mathbf{C}^n)$, we shall use the associated norm

$$A \rightarrow \| A \| = \sup_{\|x\| \leqslant 1} \| Ax \| .$$

Proposition 1.

$$\| A \| = \sup_i \sum_j | A_i^j | = \sup \| A_i \| .$$

Proof. We have

$$Ax = \sum_i x^i A_i ,$$

where

$$(\alpha) \qquad \| Ax \| \leqslant \sum_i | x^i | \, \| A_i \| \leqslant \sup_i \| A_i \| \, . \, \sum_i | x^i | \, ;$$

that is,

$$\| Ax \| \leqslant \sup_i \| A_i \| \, . \, \| x \| \, ,$$

so that

$$\| A \| \leqslant \sup_i \| A_i \| \, .$$

Now, let i_0 be such that

$$\sup_i \| A_i \| = \| A_{i_0} \| \, .$$

We have

$$A_{i_0} = A\varepsilon_{i_0}, \quad \| \varepsilon_{i_0} \| = 1, \text{ so that } \quad \| A \| \geqslant \sup_i \| A_i \|$$

and, finally,

$$\| A \| = \sup_i \| A_i \| \, .$$

Proposition 2. *The equation* $\| Ax \| = \| A \| \, \| x \|$ *can hold only if the inequality* $x^i \neq 0$ *implies* $\| A_i \| = \| A \|$.

Proof. The second conditional inequality (α) in the preceding proof can become equality only if

$$| x^i | \, \| A_i \| = | x^i | \sup_i \| A_i \| \, .$$

Now, let us consider a stochastic matrix P. By definition,

$$P_i^j \geqslant 0 \, , \qquad \sum_j P_i^j = 1 \, .$$

Let us consider P as an element of $L(\mathbf{C}^n, \mathbf{C}^n)$. From the preceding relations, we get

Proposition 3. *The norm of every stochastic matrix is 1.*

Corollary. *The absolute value of no eigenvalue of a stochastic matrix exceeds 1.*

Proposition 4. *The value 1 is an eigenvalue of every stochastic matrix.*

Proof. Let $\eta = [1, ..., 1]$. Then $\eta P = \eta$, so that $\bar{P}\bar{\eta} = \bar{\eta}$. Consequently, 1 is an eigenvalue of $\bar{P}$. Hence, it is also an eigenvalue of P.

Proposition 5. *For every stochastic matrix P, the nilpotent parts associated with the eigenvalues of absolute value 1 are zero. In other words, if σ is an eigenvalue of absolute value 1 of multiplicity k, the vector subspace of the eigenvectors associated with the eigenvalue σ is dimension k.*

This proposition follows from the following three lemmas:*

Lemma 1. *If H is nilpotent, the kernel of H^{k-1} is properly contained in the kernel of H^k for all $k \geqslant 1$, so that the kernel of H^{k-1} is different from the entire space.*

Proof. We have trivially $\ker(H^k) \supset \ker(H^{k-1})$. Let us suppose that

$$\ker(H^k) = \ker(H^{k-1}).$$

Then,

$$\begin{aligned}\ker(H^{k+1}) &= \overset{-1}{H}(\ker(H^k)) = \overset{-1}{H}(\ker(H^{k-1}))\\ &= \ker(H^k)\,,\end{aligned}$$

from which we get $\ker(H^p) = \ker(H^{k-1})$ for $p \geqslant k-1$. This is absurd if the kernel of H^{k-1} is different from the entire space since $H^p = 0$ for sufficiently large p.

Lemma 2. *Let E denote a normed finite-dimensional vector space. Let us equip $L(E, E)$ with the norm associated with the norm on E:*

$$\| A \| = \sup_{\|x\| \leqslant 1} \| Ax \| .$$

Suppose that H is a nonzero nilpotent and that σ is an element of $\mathbf{C}$. Then,

$$\| \sigma + H \| > | \sigma | .$$

Proof. We know that $\| \sigma + H \| \geqslant | \sigma |$ since $\sigma + H$ has only the eigenvalue σ. Therefore, we can assume $\sigma = 1$. Now, let us consider $\mathbf{1} + H$. Let us show that $\| \mathbf{1} + H \| > 1$. Let x denote a vector such that $Hx \neq 0$ and $H^2 x = 0$. We have

---

*In these three lemmas, the kth power of a matrix A is denoted by A^k.

$$(\mathbf{1} + H)^k x = (1 + kH)\, x\,,$$

so that

$$\lim_{k \to +\infty} \| (\mathbf{1} + H)^k x \| = +\infty\,.$$

Now,

$$\| (\mathbf{1} + H)^k x \| \leqslant \| 1 + H \|^k \| x \|\,,$$

so that

$$\| \mathbf{1} + H \| > 1.$$

Lemma 3. *Let E denote a normed finite-dimensional vector space. Suppose that $L(E, E)$ is equipped with the norm associated with the norm on E. Let A denote a member of $L(E, E)$ and let σ denote an eigenvalue of E such that $|\sigma| = \| A \|$. Then, the nilpotent associated with σ is 0.*

Proof. Let M denote the spectral subspace associated with σ. The restriction of A to M is of the form $\sigma + H$. If $H \neq 0$, we have

$$\| A \| \geqslant \| \sigma + H \| > \sigma,$$

which contradicts the hypothesis.

Definition. We shall refer to the set of indices i such that $x^i \neq 0$ as the *support* of an element $x \in \mathbf{C}^n$.

Let α denote a set of indices. We shall denote by E_α the set of elements x such that $x^i = 0$ for $i \notin \alpha$, that is, the set of elements with support in α.

Suppose that $\alpha \neq \varnothing$. We see immediately that a necessary and sufficient condition for α to be closed is that E_α be stable with respect to P, in other words, that P admit a partition into submatrices of the following form:

$$P = \left[\begin{array}{c|c} P^\alpha_\alpha & P^\alpha_\beta \\ \hline 0 & P^\beta_\beta \end{array}\right].$$

Proposition 6. *If x is an eigenvector of a stochastic matrix P associated with an eigenvalue of absolute value 1, then the support of x is a closed set.*

Proof. Suppose that

$$x = \left[\begin{array}{c} x^\alpha \\ \hline 0 \end{array}\right], \quad \text{with} \quad \begin{cases} x^i \neq 0 \iff i \in \alpha \\ Px = \lambda x, \quad |\lambda| = 1 \end{cases}$$

and

$$P = \left[\begin{array}{c|c} P_\alpha^\alpha & P_\beta^\alpha \\ \hline P_\alpha^\beta & P_\beta^\beta \end{array}\right].$$

We have

$$P_\alpha^\alpha x^\alpha = \lambda x^\alpha ,$$

so that

$$\| P_\alpha^\alpha \| \, \| x^\alpha \| = \| x^\alpha \| ,$$

which implies $\| P_\alpha^\alpha \| = 1$ (we have trivially $\| P_\alpha^\alpha \| \leqslant 1$). Then, Proposition 2 implies that all the columns of P_α^α are of norm equal to 1 and hence that $P_\alpha^\beta = 0$. Consequently, α is a closed set.

Definition. A stochastic matrix is said to be *irreducible* if E_α is stable only when $\alpha = \varnothing$ and $\alpha = \{1, ..., n\}$, i.e., if P admits no closed set other than the empty set and the universal set, in other words, unless there fails to exist a partition of P of the form

$$P = \left[\begin{array}{c|c} P_\alpha^\alpha & P_\beta^\alpha \\ \hline 0 & P_\beta^\beta \end{array}\right].$$

A necessary and sufficient condition for P to be irreducible is that there exist only one class, in other words, that every state j be accessible from an arbitrary state i. This is true because if P is irreducible, every final class has to be the empty set. Therefore, the empty set is a class and there is only one class. The converse is immediate.

Proposition 7. *The eigenvalues of absolute value 1 of an irreducible stochastic matrix P are simple.*

Proof. Let P denote a stochastic matrix. Suppose that P has a multiple eigenvalue s of absolute value 1. Then, the subspace of the associated eigenvectors is of dimension 2 or greater. Therefore, it contains nonzero vectors having at least one zero component. In other words, there will exist eigenvectors whose support is neither the empty set nor the entire set. According to Proposition 6, this implies the existence of a nontrivial closed set, and P is not irreducible.

Let $\alpha, \beta, \ldots, \lambda$ denote the final classes of P. We have the following partition:*

$$P = \left[\begin{array}{c|c|c|c|c|c} P^{\alpha}_{\alpha} & 0 & & & & \\ \hline & P^{\beta}_{\beta} & 0 & & 0 & \\ \cline{2-3} 0 & & P^{\gamma}_{\gamma} & & & * \\ \cline{3-3} & 0 & & \cdots & & \\ & & 0 & & & \\ \cline{5-5} & & & & P^{\lambda}_{\lambda} & \\ \cline{5-6} & & & & 0 & Q \end{array}\right].$$

Proposition 8. *The eigenvalues of P of absolute value 1 are those of the stochastic matrices $P^{\alpha}_{\alpha}, \ldots, P^{\lambda}_{\lambda}$ that are induced on the final classes.*

Proof. The problem is to show that Q has no eigenvalue of absolute value 1. We have the following partition for $\overset{m}{P}$:

$$\overset{m}{P} = \left[\begin{array}{c|c|c|c|c} (\overset{m}{P})^{\alpha}_{\alpha} & 0 & & & \\ \hline & (\overset{m}{P})^{\beta}_{\beta} & & 0 & * \\ 0 & & \ddots & & \\ & 0 & & & \\ \cline{4-4} & & & (\overset{m}{P})^{\lambda}_{\lambda} & \\ \cline{4-5} & & & 0 & \overset{m}{Q} \end{array}\right] \quad \text{with} \quad \begin{cases} (\overset{m}{P})^{\alpha}_{\alpha} = \overbrace{P^{\alpha}_{\alpha}}^{m} \\ (\overset{m}{P})^{\beta}_{\beta} = \overbrace{P^{\beta}_{\beta}}^{m} \\ (\overset{m}{P})^{\lambda}_{\lambda} = \overbrace{P^{\lambda}_{\lambda}}^{m} \end{cases}.$$

Let $A = \alpha \cup \beta \cup \ldots \cup \lambda$ denote the union of the final classes and set $B = \complement A$.

For every $i \in B$, there exists in A a path with initial vertex i and terminal vertex j. Let l_i denote the length of such a path. It can be extended as the path "contained in A" of arbitrary length. Therefore, for every $l \geqslant l_i$, there exists a path beginning at i and ending in A.

Define $m = \sup_{i \in B} l_i$. Let us construct $\overset{m}{P}$. According to the construction, there exists a nonzero element over each column of $\overset{m}{Q}$. Consequently, $\| \overset{m}{Q} \| < 1$. Therefore, $\overset{m}{Q}$ cannot have an eigenvalue of absolute value 1. The same is true of Q. This completes the proof.

---

*In conformity with common usage, the asterisk is used to denote an arbitrary submatrix. Two asterisks appearing in the partitions of two different matrices do not necessarily represent the same submatrix.

Corollary. *The multiplicity of the eigenvalue 1 is equal to the number of final classes.*

The preceding proposition shows that the study of the eigenvalues of absolute value 1 leads to the case of irreducible stochastic matrices. It remains to study the case of eigenvalues different from 1 and of absolute value 1. To do this, we need to make a deeper combinatorial study of these irreducible stochastic matrices.

Definition. Let P denote an irreducible stochastic matrix. The greatest common divisor of the lengths of the circuits is called the *period* of P.

Proposition 9. *Let P denote an irreducible matrix of period d. Let l_1 and l_2 denote the lengths of two paths with the same initial vertex and the same terminal vertex. Then,*

$$l_1 \equiv l_2 \pmod{d}.$$

Proof. Suppose that l_1 and l_2 are the lengths of two paths extending from i to j and let l_3 denote the length of a path from j to i. Then,

$$l_1 + l_3 \equiv l_2 + l_3 \pmod{d},$$

so that

$$l_1 \equiv l_2 \pmod{d}.$$

Definition. The equivalence classes, where the equivalence relation is defined by

i is equivalent to j if and only if the length of every path from i to j is a multiple of d

are called *subclasses* of an irreducible matrix P.

Proposition 10. *There exist d subclasses, which we denote by $c_0, c_1, \ldots, c_{d-1}$, such that the length of every path from c_h to c_k is congruent to $k - h \pmod{d}$.*

Proof. Let us choose a arbitrarily. Then let l_1 and l_2 denote the lengths of two paths, both with initial vertex a, with terminal vertices i and j, respectively. Let l_3 denote the length of a path with initial vertex i and terminal vertex j. We have

$$l + l_3 \equiv l_2 \pmod{d}.$$

If i and j belong to the same subclass, we have $l_3 \equiv 0 \pmod{d}$, so that

$$l_1 \equiv l_2 \pmod{d}.$$

Conversely, if $l_1 \equiv l_2 \pmod{d}$, we have

$$l_3 \equiv 0 \pmod{d},$$

and i and j belong to the same subclass. Thus, every subclass is characterized by the length modulo d of the paths extending from a to a point in that subclass. Let c_h (for $h = 0, \ldots, d-1$) denote the subclass of the states i such that the length of a path from a to i is congruent to h. Suppose that $i \in c_h$ and $j \in c_k$. Let l_1 denote the length of a path from a to i, let l_2 denote the lengths of a path from a to j, and let l_3 denote the length of a path from i to j. Then,

$$\begin{cases} l_1 \equiv h & \pmod{d} \\ l_2 \equiv k & \pmod{d} \\ l_1 + l_3 \equiv l_2 & \pmod{d}, \end{cases} \quad \text{so that} \quad l_3 \equiv k - h \pmod{d},$$

which completes the proof.

Remark. It will be convenient to set $c_p = c_q$ if $p \equiv q \pmod{d}$. Using this notation, we have the

Corollary. *Let $i_0, i_1, \ldots, i_k$ denote successive states of a path. If $i_0 \in c_p$, we have $i_k \in c_{p+k}$.*

In particular:

If $j \in \Gamma(i)$ and $i \in c_p$, then $j \in c_{p+1}$;
If $j \in \Gamma^d(i)$, then i and j belong to the same subclass.

Corresponding to the partition of the set of states into subclasses is the following partition of the matrix P:

$$P = \left[\begin{array}{c|c|c|c|c} 0 & 0 & & & P^{o}_{d-1} \\ \hline P^1_0 & & & & \\ \hline & P^2_1 & \cdots & 0 & \\ \hline 0 & 0 & & & 0 \\ \hline & & & P^{d-1}_{d-2} & \end{array}\right],$$

where $P^k_h = P^{c_k}_{c_h}$.

Remark. The matrices P^k_h are not necessarily square since the subclasses do not necessarily have the same number of states.

Proposition 11 (Frobenius). *An irreducible stochastic matrix of period d has as eigenvalues of absolute value 1 the roots of the equation* $s^d = 1$, *and these are simple eigenvalues. All the components of the eigenvectors associated with the eigenvalue 1 have the same argument (there exists* $x \gg 0$ *such that* $Px = x$*).*

Proof. Let P denote a stochastic matrix of period d. Let s and $x \neq 0$ be such that

$$\begin{cases} Px = sx \\ |s| = 1 . \end{cases}$$

We recall that $x^i \neq 0$ for every i.

The relation $Px = sx$ may be written

$$(\beta) \qquad \sum_i P_i^j x^i = sx^j .$$

Since $|s| = 1$, it then follows that

$$\sum_j \left| \sum_i P_i^j x^i \right| = \sum_i |x^i| .$$

Since

$$\sum_j P_i^j = 1 \qquad (\forall i),$$

we may write

$$\sum_i |x^i| = \sum_i \left(\sum_j P_i^j\right) |x^i| ,$$

so that

$$\sum_j \left| \sum_i P_i^j x^i \right| = \sum_j \left(\sum_i P_i^j |x^i| \right) .$$

Since

$$\left| \sum_i P_i^j x^i \right| \leqslant \sum_i P_i^j |x^i| ,$$

we conclude that

$$\left| \sum_i P_i^j x^i \right| = \sum_i P_i^j |x^i|$$

$$\left(= \sum_i |P_i^j x^i|\right) \qquad (\forall j) .$$

Consequently, for every j, the quantity $\operatorname{Arg} P_i^j x^i$ is independent of i (for every value of i such that $P_i^j \neq 0$). According to (β), this can be written

$$\operatorname{Arg}(P_i^j x^i) = \operatorname{Arg}(sx^j)$$

(for every combination of j and i such that $P_i^j \neq 0$). Hence, if $j \in \Gamma(i)$, we have

$$\operatorname{Arg}(x^i) = \operatorname{Arg}(sx^j)\,.$$

From this we conclude that, for every closed curve of length l, we have $s^l = 1$. Since the set of those m such that $s^m = 1$ is an ideal and since this ideal contains all the lengths of the cycles, it contains the greatest common divisor of the lengths of the cycles. This implies

$$s^d = 1\,.$$

Furthermore, if $s = 1$ and $j \in \Gamma(i)$, we have

$$\operatorname{Arg}(x^i) = \operatorname{Arg}(x^j)\,.$$

The condition $j \in \Gamma(i)$ is superfluous since any two states can be connected by a path.

Now, we shall show that every root of $s^d = 1$ is an eigenvalue of P. Let us use the partitions of P, x, and Px associated with the subclasses. We have (the arrangement is such that every subcolumn of Px is the product by x of the subset of P that is at its left):

$$x = \begin{bmatrix} x^0 \\ x^1 \\ \vdots \\ x^{d-1} \end{bmatrix}$$

$$P = \begin{bmatrix} 0 & 0 & \cdots & 0 & P^0_{d-1} \\ P^1_0 & 0 & \cdots & 0 & 0 \\ 0 & P^2_1 & \cdots & 0 & 0 \\ \vdots & \vdots & & \vdots & \vdots \\ 0 & 0 & \cdots & P^{d-1}_{d-2} & 0 \end{bmatrix} \begin{bmatrix} P^0_{d-1}\,x^{d-1} \\ P^1_0\,x^0 \\ P^2_1\,x^1 \\ \vdots \\ P^{d-1}_{d-2}\,x^{d-2} \end{bmatrix} = Px\,.$$

On the other hand,

$$\overset{d}{P} = \left[\begin{array}{c|c|c|c} (\overset{d}{P})_0^0 & 0 & & \\ \hline & (\overset{d}{P})_1^1 & \cdots\cdots & 0 \\ 0 & & & \\ \cline{2-2}\cline{4-4} & 0 & & (\overset{d}{P})_{d-1}^{d-1} \end{array}\right],$$

where

$$(\overset{d}{P})_0^0 = P_{d-1}^0 \ldots P_1^2 \, P_0^1$$

and, in general,

$$(\overset{d}{P})_h^h = P_{h-1}^h \ldots P_0^1 \, P_{d-1}^0 \ldots P_{h+1}^{h+2} \, P_h^{h+1} \, .$$

The relation $Px = sx$ may be written

$$(\gamma) \qquad \begin{cases} P_{d-1}^0 \, x^{d-1} = sx^0 \\ P_0^1 \, x^0 = sx^1 \\ P_1^2 \, x^1 = sx^2 \\ \ldots\ldots\ldots\ldots\ldots\ldots \\ P_{d-2}^{d-1} \, x^{d-2} = sx^{d-1} \end{cases} .$$

Let x^0 denote a nonzero vector such that

$$(\overset{d}{P})_0^0 \, x^0 = x^0 \, .$$

Let us define $x^1, \ldots, x^{d-1}$ by the last $d-1$ of the equations (γ):

$$P_{d-2}^{d-1} \ldots P_1^2 \, P_0^1 \, x^0 = s^{d-1} \, x^{d-1} \, .$$

Multiplying by P_{d-1}^0, we have

$$(\overset{d}{P})_0^0 \, x^0 = s^{d-1} \, P_{d-1}^0 \, x^{d-1} \, ;$$

that is (since $s^d = 1$),

$$sx^0 = P_{d-1}^0 \, x^{d-1} \, .$$

Thus, x is the eigenvector associated with the eigenvalue s.

Remark. The matrix $\overset{d}{P}$ has d final classes, each of period 1.

Let us summarize the results that we have obtained.

Theorem 1. *Let P denote a stochastic matrix admitting p final classes with respective periods d_h. The eigenvalues of P of absolute value 1 are*

(1) *The eigenvalue 1 of multiplicity p,*
(2) *the solutions other than 1 of $s^{d_h} = 1$, the multiplicity of which is the number of equations of that type of which they are solutions.*

Example. If P admits five final classes with periods 1, 2, 3, 4, and 6, respectively, the eigenvalues are

1	of multiplicity 5,	
−1	of multiplicity 3	(solution of $s^2 = 1, s^4 = 1, s^6 = 1$),
$e^{2i\pi/3}$ and $e^{-2i\pi/3}$	of multiplicity 2	(solution of $s^3 = 1, s^6 = 1$),
i and −i	of multiplicity 1	(solution of $s^4 = 1$),
$e^{i\pi/3}$ and $e^{-i\pi/3}$	of multiplicity 1	(solution of $s^6 = 1$).

Definitions. A stochastic matrix is said to be *ergodic* if 1 is a simple eigenvalue of it, in other words, if there is only one final class. A stochastic matrix is said to be *regular* if 1 is a simple eigenvalue of it, if there is only one final class, and the period of that one is 1. A stochastic matrix P is said to be *primitive* if its only eigenvalue of absolute value 1 is 1, that is, if all the final classes are of period 1.

3. ASYMPTOTIC BEHAVIOR

Lemma 4. *If $|s| < 1$ and if H is nilpotent, then $\overset{m}{\widehat{s + H}}$ tends to zero as $m \to +\infty$.*

Proof. Suppose that $\overset{k+1}{H} = 0$. Then,

$$\overset{m}{\widehat{s + H}} = s^m + ms^{m-1} H + \cdots + m(m-1)\ldots(m-k+1)\, s^{m-k} \overset{k}{H},$$

from which the lemma follows, since all the terms approach 0 as $m \to \infty$.

Let

$$P = \sum_r \pi_r(s_r + H_r)$$

be the spectral decomposition of P with

$$\begin{cases} |s_r| \leqslant 1, \\ H_r = 0, \quad \text{if} \quad |s_r| = 1. \end{cases}$$

(The numbers s_r are the eigenvalues, the operators π_r are the spectral projections and the operators H_r are nilpotent.)

We have

$$\overset{m}{P} = \sum_r \pi_r \overbrace{(s_r + H_r)}^{m} = \sum_{|s_r|=1} \pi_r s_r^m + \varepsilon(m),$$

where $\varepsilon(m) \to 0$ as $m \to +\infty$.

Theorem 2. *The following three properties are equivalent:*

(i) *P is primitive;*

(ii) *$\overset{m}{P}$ has a limit as $m \to +\infty$;*

(iii) *no matter what the value of $x(0)$, the quantity $x(m) = \overset{m}{P}x(0)$ has a limit as $m \to +\infty$.*

If these properties are satisfied, we have

$$\lim_{m \to +\infty} \overset{m}{P} = \pi_1,$$

where π_1 is the spectral projection associated with the eigenvalue 1.

Proof. (i) $\Rightarrow$ (ii): If P is primitive, we have

$$\lim_{m \to +\infty} \overset{m}{P} = \pi_1,$$

where π_1 is the spectral projector associated with the eigenvalue 1.

(ii) $\Rightarrow$ (iii): Obvious.

(ii) $\Rightarrow$ (i): If $\overset{m}{P}$ has a limit, so does $\pi_r \overset{m}{P} = \pi_r s_r^m$ and consequently so does s_r^m, and therefore $s_r = 1$ if $|s_r| = 1$.

(iii) $\Rightarrow$ (ii): If $\overset{m}{P}x(0)$ has a limit for every $x(0)$, then $(\overset{m}{P})_i = \overset{m}{P}\varepsilon_i$ and hence $\overset{m}{P}$ also has a limit.

Theorem 3. *A necessary and sufficient condition for $x(m)$ to have a limit independent of $x(0)$ is that P be regular.*

Proof. A necessary and sufficient condition is in fact that P be primitive and the rank of π_1 be 1.

Now consider the following matrix:

$$Q(m) = \frac{1}{m}(1 + P + \cdots + \overset{m-1}{P}).$$

Since $Q(m)$ is a mean of stochastic matrices, it is a stochastic matrix itself.

Theorem 4.

$$\lim_{m \to +\infty} Q(m) = \pi_1 .$$

Proof: We have

$$Q(m) = \sum_{|s_r|=1} \pi_r \left(\frac{1}{m} \sum_{k=0}^{m-1} s_r^k \right) + \varepsilon_1(m) ,$$

where

$$\varepsilon_1(m) = \frac{1}{m} \sum_{k=0}^{m-1} \varepsilon(k) .$$

Therefore, $\varepsilon_1(m) \to 0$ as $m \to \infty$ since $\varepsilon(k) \to 0$ as $k \to +\infty$. Now,

$$\frac{1}{m} \sum_{k=0}^{m-1} s_r^k = \begin{cases} \dfrac{1}{m} \dfrac{s_r^m - 1}{s_r - 1} \to 0 & \text{if } s_r \neq 1 \\ 1 & \text{if } s_r = 1 \end{cases} ,$$

from which the conclusion of the theorem follows.

Corollary. *If we set* $x(m) = \overset{m}{P} x(0)$, *we have*

$$\lim_{m \to \infty} \frac{1}{m} [x(0) + \cdots + x(m-1)] = \pi_1 \, x(0) .$$

In other words, the sequence $x(0), \ldots, x(m), \ldots,$ *has a limit in the sense of Fejér.*

A necessary and sufficient condition for this limit to be independent of $x(0)$ is that the rank of π_1 be 1, that is, that P be ergodic.

Remark. If P is ergodic, then the rank of π_1 is 1 and all its columns are not only proportional but equal since π_1 is stochastic. The common value of these columns is the unique P-invariant vector ω that satisfies the relations $\omega \geqslant 0$ and $\sum_i \omega^i = 1$. The support of this vector is the unique final class. The vector ω is the limit in the sense of Fejér of $x(m)$ and is independent of $x(0)$.

If P is regular, ω is the limit of $x(m)$ and is independent of $x(0)$. In referring to the interpretation of P as transition matrix of a Markov chain, we say that ω is the limiting probability distribution of the states of the system.

The following table gives the essential results of the discussion:

Type	Combinatorial Properties	Spectral Properties	Asymptotic Properties
Primitive	All the final classes are of period 1	1 is the only eigenvalue of absolute value 1	$\overset{m}{P}$ has a limit
Ergodic	Only one final class	1 is a simple eigenvalue	The limit of $x(m)$ in the sense of Fejér is independent of $x(o)$
Regular (= primitive and ergodic)	Only one final class of period 1	1 is the only eigenvalue of absolute value 1, and it is a simple eigenvalue	$x(m)$ has a limit independent of $x(o)$
Irreducible	Only 1 class	1 is a simple eigenvalue, and the associated eigenvectors have only nonzero components.	The limit of $x(m)$ in the sense of Fejér is independent of $x(o)$ and all its components are nonzero.

7

Second-Order Stationary Random Processes

1. FUNCTIONS OF POSITIVE TYPE. BOCHNER'S THEOREM

We recall that a $k \times k$ complex matrix A is said to be *positive-semidefinite* if

$$\bar{X}AX \geqslant 0, \qquad \forall X \in \mathbf{C}^k ,$$

where $\bar{X}$ denotes the Hermitian transpose of H.

If a Hermitian complex matrix A (that is, one such that $A = \bar{A}$) satisfies the equation $\bar{X}AX = 0$ for every $X \in \mathbf{C}^k$, then $A = 0$. To see this, set $\varphi(X, Y) = \bar{X}AY$. Then, φ is a Hermitian sesquilinear form and we have

$$4\,\varphi(X, Y) = \varphi(X + Y, X + Y) - \varphi(X - Y, X - Y) - \\ - \mathrm{i}\,\varphi(X + \mathrm{i}\,Y, X + \mathrm{i}\,Y) + \mathrm{i}\,\varphi(X - \mathrm{i}\,Y, X - \mathrm{i}\,Y).$$

Consequently,

$$\varphi(X, X) = 0 \quad (\forall X) \quad \text{implies} \quad \varphi(X, Y) = 0 \quad (\forall X, Y);$$

that is,

$$\bar{X}AY = 0 \quad (\forall X, Y), \quad \text{so that} \quad AY = 0 \quad (\forall Y) \quad \text{and} \quad A = 0.$$

Now, let A denote any complex matrix. We may set

$$A = A_1 + \mathrm{i}\, A_2 \qquad \text{and} \qquad \begin{cases} A_1 = \dfrac{A + \overline{A}}{2}. \\ A_2 = \dfrac{A - \overline{A}}{2\,\mathrm{i}} \end{cases}$$
$$\overline{A} = A_1 - \mathrm{i}\, A_2$$

The matrices A_1 and A_2 are Hermitian and we have

$$\overline{X}AX = \overline{X}A_1\, X + \mathrm{i}\, \overline{X}A_2\, X\,.$$

Consequently, if $\overline{X}AX \geqslant 0$ $(\forall X)$, then $\overline{X}A_2\, X = 0$ $(\forall X)$, so that $A_2 = 0$. Thus,

Every positive-definite complex matrix is Hermitian.

We note that, over the field of real numbers, the relation $\overline{X}AX \geqslant 0$ does not imply that A is symmetric.

Let $A = \begin{bmatrix} a & c \\ \bar{c} & b \end{bmatrix}$ denote a positive-semidefinite matrix. Let us define $\varphi(X, Y) = \overline{X}AY$. We have

$$\varphi(X, X) = \overline{X}AX \geqslant 0$$
$$|\,\varphi(X, Y)\,|^2 \leqslant \varphi(X, X)\, \varphi(Y, Y)$$

(Schwarz' inequality).

Now,

$$\varphi(\varepsilon_1, \varepsilon_1) = a,\ \varphi(\varepsilon_2, \varepsilon_2) = b,\ \varphi(\varepsilon_1, \varepsilon_2) = c\,,$$

so that

$$a \geqslant 0,\ b \geqslant 0,\ |\, c\,| \leqslant \sqrt{ab}\,.$$

Example. In a Hermitian space, the Gram matrix of k vectors $e_1, \ldots, e_k$ (that is, the matrix of the coefficients $\langle\, e_i, e_j \,\rangle$) is positive-semidefinite.

Definition. A complex function f defined on a real vector space E is said to be *of positive type* if, for every $x_1, \ldots, x_k \in E$ (where k is arbitrary), the matrix of the coefficients $f(x_i - x_j)$ is positive-semidefinite.

If we take $k = 2$, $x_1 = 0$, and $x_2 = x$, we obtain the result that the following matrix is positive-semidefinite:

$$A = \begin{bmatrix} f(0) & f(x) \\ f(-\,x) & f(0) \end{bmatrix},$$

and from this we conclude that every function of positive type has the following properties:

$$f = \tilde{f}, \quad f(0) \geqslant 0, \quad |f(x)| \leqslant f(0).$$

Consequently, every function of positive type is bounded.

Proposition 1. *The Fourier transform of a bounded positive measure is a (continuous) function of positive type.*

Proof. To simplify the notation, let us suppose that μ is a bounded positive measure on $\mathbf{R}$. Define $f = \mathcal{F}\mu$:

$$f(x) = \int e^{i\omega x} \cdot \mu^{\omega}.$$

Suppose that $x_1, \ldots, x_k \in \mathbf{R}$ and $z_1, \ldots, z_k \in \mathbf{C}$. We have

$$f(x_i - x_j) = \int e^{i\omega(x_i - x_j)} \cdot \mu^{\omega},$$

so that

$$\sum_{i,j} f(x_i - x_j)\, z_i \bar{z}_j = \int \sum_{i,j} z_i \bar{z}_j\, e^{i\omega(x_i - x_j)} \cdot \mu^{\omega}$$

$$= \int \sum_{i,j} z_i e^{i\omega x_i}\, \overline{z_j e^{i\omega x_j}} \cdot \mu^{\omega}$$

$$= \int \Big| \sum_{i} z_i e^{i\omega x_i} \Big|^2 \cdot \mu^{\omega} \geqslant 0.$$

This proposition has a converse, known as Bochner's theorem.

Theorem 1 (Bochner). *Every continuous function of positive type is the Fourier transform of a bounded positive measure.*

Proof. We shall confine ourselves to the case of a continuous function of positive type defined on $\mathbf{R}$. We set $f = \mathcal{F}T$, where T is a tempered distribution:

(a) For every $\varphi \in \mathfrak{D}$, we have

$$\int f \cdot (\tilde{\varphi} * \varphi) \geqslant 0.$$

This is true because

$$\int f \cdot (\varphi * \tilde{\varphi}) = \iint f(x + y)\, \tilde{\varphi}(x)\, \varphi(y)\, dx\, dy$$

$$= \iint f(y - x)\, \overline{\varphi(x)}\, \varphi(y)\, dx\, dy$$

$$= \lim_{\varepsilon \to 0} \varepsilon^2 \sum_{m,n=-\infty}^{+\infty} f(m\varepsilon - n\varepsilon)\, \overline{\varphi(n\varepsilon)}\, \varphi(m\varepsilon) \geqslant 0.$$

(We note that the summation is actually a finite one since φ has compact support.)

(b) For every $\varphi \in \mathbf{S}$,

$$\int f \cdot \tilde{\varphi} * \varphi \geqslant 0.$$

Let $\{\varphi_n\}$ be a sequence such that $\varphi_n \in \mathfrak{D}$ and $\varphi_n \xrightarrow{\mathbf{S}} \varphi$. Then, $\tilde{\varphi}_n * \varphi_n \xrightarrow{\mathbf{S}} \tilde{\varphi} * \varphi$ since the mapping $\varphi, \psi \to \varphi * \psi$ is a continuous mapping of $\mathbf{S} \times \mathbf{S}$ into $\mathbf{S}$. This property can easily be proven by using the relation

$$\| \varphi * \psi \|_u \leqslant \| \varphi \|_{L^2} . \| \psi \|_{L^2} .) .$$

Thus, if we apply property (a) to each φ_n, we obtain property (b) by passing to the limit.

(c) For every $\psi \in \mathbf{S}$,

$$\int \overline{\psi}\psi \, . \, T \geqslant 0 .$$

To see this, let us set

$$\varphi = \mathcal{F}\psi, \quad \tilde{\varphi} = \mathcal{F}\overline{\psi} .$$

Then,

$$\int \overline{\psi}\psi \, . \, T = \frac{1}{2\pi} \int \mathcal{F}(\overline{\psi}\psi) \, . \, \mathcal{F}T = \frac{1}{4\pi^2} \int \tilde{\varphi} * \varphi \, . \, f \geqslant 0 .$$

(d) For every nonnegative $u \in \mathfrak{D}$, we have $\int u \cdot T \geqslant 0$.

To show this, it will be sufficient to show that u can be approached in $\mathfrak{D}$ by functions of the form $\overline{\psi}\psi$, where $\psi \in \mathfrak{D}$. Let α denote a positive function in $\mathfrak{D}$ that is equal to 1 in a neighborhood of the support of u. Then,

$$u = \lim_{\left\{ \substack{\varepsilon > 0 \\ \varepsilon \to 0} \right.} \alpha^2(u + \varepsilon) ,$$

where

$$\alpha^2(u + \varepsilon) = \psi^2, \quad \psi = \alpha\sqrt{u + \varepsilon} \in \mathfrak{D} .$$

Assertion (d) can also be worded

(d′) T is a positive measure.

(e) T is a bounded measure.

To see this, let us set $\theta_n(x) = e^{-x^2/2n^2}$. Then,

$$\int T = \lim_{n \to +\infty} \int \theta_n \, . \, T = \frac{1}{2\pi} \lim_{n \to +\infty} \int \mathcal{F}\theta_n \, . \, f .$$

Furthermore,

$$(\mathcal{F}\theta_n)(y) = n\sqrt{2\pi}\, e^{-n^2y^2/2} .$$

The sequence $\{\theta_n\}$ converges simply to 1 as $n \to +\infty$. We have

$$\int \mathcal{F}\theta_n = 2\pi\theta_n(0) = 2\pi .$$

Consequently (cf. Chapter III, Theorem 23), $\{\mathcal{F}\theta_n\}$ converges narrowly to $2\pi\delta$ and

$$\lim_{n\to+\infty} \int \mathcal{F}\theta_n \,.\, f = 2\pi f(0) < +\infty ,$$

which completes the proof.

2. UNITARY REPRESENTATIONS OF R IN A HILBERT SPACE

Definition. A mapping $t \to U_t$ of $\mathbf{R}$ into the set of unitary operators in a Hilbert space H is called a (continuous) *unitary representation* of $\mathbf{R}$ in H if it enjoys the following properties:

(i) $U_{s+t} = U_s U_t$,

(ii) $U_{-s} = (U_s)^{-1} (= U_s^*)$,

(iii) $\forall a \in H$, the mapping $t \to U_t\, a$ is continuous on $\mathbf{R}$ into H (equipped with the "strong" topology defined by the norm).

From (i) and (ii), we conclude that $U_0 = U_{s-s} = U_s U_{-s} = \mathbf{1}$.

Examples. I. In a two-dimensional oriented Euclidean space, the mapping that assigns to $t \in \mathbf{R}$ the rotation through an angle t about the origin O is a unitary representation of $\mathbf{R}$.

II. Let $\mathbf{L}^2$ denote the Hilbert space of functions measurable and square-integrable with respect to Lebesgue measure on $\mathbf{R}$. A mapping $U_t : f \to f * \delta_t$ is a unitary representation of $\mathbf{R}$.

Theorem 2. *If $t \to U_t$ is a unitary representation of $\mathbf{R}$ into H, the function $t \to \langle U_t\, a, a \rangle$ is of positive type for every $a \in H$.*

Proof. Suppose that $t_1, \ldots, t_k \in \mathbf{R}$. By setting $\varphi(t) = \langle U_t\, a, a \rangle$, we have

$$\varphi(t_i - t_j) = \langle U_{t_i - t_j}\, a, a \rangle = \langle U_{t_j}^* U_{t_i}\, a, a \rangle$$
$$= \langle U_{t_i}\, a, U_{t_j}\, a \rangle .$$

Consequently, the matrix of the coefficients $\varphi(t_i - t_j)$ is the Gram matrix of the vectors $U_{t_i}\, a$. Therefore, it is positive-semidefinite.

A consequence. Let us set

$$\varphi_{a,b}(t) = \langle U_t\, a, b \rangle$$
$$\varphi_{a,a}(t) = \langle U_t\, a, a \rangle .$$

The function $\varphi_{a,a}$ is the Fourier transform of a positive bounded measure $\varpi_{a,a}$ on **R**.

Furthermore, the mapping $a, b \to \varphi_{a,b}(t)$ is sesquilinear. From this, we conclude that

$$4\,\varphi_{a,b} = \varphi_{a+b,a+b} - \varphi_{a-b,a-b} - \mathrm{i}\,\varphi_{a+\mathrm{i}b,a+\mathrm{i}b} + \mathrm{i}\,\varphi_{a-\mathrm{i}b,a-\mathrm{i}b}\,.$$

Consequently, $\varphi_{a,b}$ is a linear combination of the Fourier transforms of bounded positive measures and hence is the Fourier transform of a bounded measure. Let us set

$$\varphi_{a,b} = \mathcal{F}\varpi_{a,b}\,.$$

The mapping $a, b \to \varpi_{a,b}$ is sesquilinear.

We have

$$\boxed{\varphi_{b,a} = \tilde{\varphi}_{a,b}}\,.$$

This is true because

$$\varphi_{b,a}(t) = \langle U_t\, b, a \rangle = \overline{\langle a, U_t\, b \rangle} = \overline{\langle U_{-t}\, a, b \rangle}$$
$$= \overline{\varphi_{a,b}(-t)} = \tilde{\varphi}_{a,b}(t)\,.$$

From this we conclude

$$\boxed{\varpi_{b,a} = \overline{\varpi}_{a,b}}\,.$$

The measures $\varpi_{a,b}$ are called *spectral measures.* We note that

$$\boxed{\int \varpi_{a,b} = \varphi_{a,b}(0) = \langle a, b \rangle}\,.$$

Theorem 3. *Suppose that $t \to U_t$ is a unitary representation of* **R** *in H. Then, for every bounded measure μ, there exists a continuous operator $U(\mu)$ in H such that*

$$\langle x, U(\mu)\, y \rangle = \int \langle x, U_t\, y \rangle \,.\, \mu^t\,.$$

Such an operator is called *Radon's operator.* We set

$$U(\mu) = \int U_t \,.\, \mu^t\,,$$

$$U(\mu)\, x = \int U_t\, x \,.\, \mu^t\,.$$

Proof. We have

$$\left|\int \langle x, U_t y \rangle \cdot \mu^t \right| \leqslant \sup_t |\langle x, U_t y \rangle| \cdot \|\mu\|$$
$$\leqslant \|x\| \, \|y\| \, \|\mu\| .$$

Therefore, for fixed y, the mapping

$$x \to \int \langle x, U_t y \rangle \cdot \mu^t$$

is a continuous semilinear mapping. Therefore, there exists an element y_1 such that

$$\int \langle x, U_t y \rangle \cdot \mu^t = \langle x, y_1 \rangle .$$

Obviously, y_1 depends linearly on y and we have

$$\|y_1\| \leqslant \|y\| \, \|\mu\| .$$

Therefore, the mapping $y \to y_1$ is a linear operator of norm not exceeding $\|\mu\|$. We set $y_1 = U(\mu)\, y$. Then,

$$\|U(\mu)\| \leqslant \|\mu\| .$$

Theorem 4. *The mapping* $\mu \to U(\mu)$ *is a linear mapping of* $\mathbf{M}$ *into* $L(H, H)$, *and we have*

(1) $$\|U(\mu)\| \leqslant \|\mu\|$$

(2) $$U(\tilde{\mu}) = U(\mu)^*$$

(3) $$U(\mu * \nu) = U(\mu)\, U(\nu) = U(\nu)\, U(\mu)$$

(4) $$U(\delta_t) = U_t .$$

Proof. (1) See the proof of the preceding theorem.

(2)
$$\begin{aligned}\langle x, U(\tilde{\mu})\, y \rangle &= \int \langle x, U_t y \rangle \cdot \tilde{\mu}^t \\ &= \overline{\int \overline{\langle x, U_{-t} y \rangle} \cdot \mu^t} \\ &= \overline{\int \overline{\langle U_t x, y \rangle} \cdot \mu^t} \\ &= \overline{\int \langle y, U_t x \rangle \cdot \mu^t} \\ &= \overline{\langle y, U(\mu)\, x \rangle} = \langle U(\mu)\, x, y \rangle .\end{aligned}$$

(3)
$$\begin{aligned}\langle x, U(\mu)\, U(\nu)\, y \rangle &= \int \langle x, U_t\, U(\nu)\, y \rangle \,.\, \mu^t \\ &= \int \langle U_{-t}\, x, U(\nu)\, y \rangle .\, \mu^t \\ &= \int \langle U_{-t}\, x, U_\theta\, y \rangle \,.\, \mu^t\, \nu^\theta \\ &= \int \langle x, U_{t+\theta}\, y \rangle \,.\, \mu^t\, \nu^\theta \\ &= \int \langle x, U_\tau y \rangle \,.\, (\mu * \nu)^\tau \\ &= \langle x, U(\mu * \nu)\, y \rangle .\end{aligned}$$

Therefore, $U(\mu)\, U(\nu) = U(\mu * \nu)$. Similarly,

$$U(\nu)\, U(\mu) = U(\nu * \mu) = U(\mu * \nu) .$$

Thus, the Radon operators commute.

(4) Follows immediately from the definition.

Theorem 5.

$$\begin{cases} \varphi_{x,U(\nu)y} = \varphi_{x,y} * \nu & (1) \\ \varphi_{U(\mu)x,y} = \varphi_{x,y} * \tilde{\mu} & (2) \\ \varpi_{x,U(\nu)y} = \mathcal{F}^\star \nu \,.\, \varpi_{x,y} & (3) \\ \varpi_{U(\mu)x,y} = \overline{\mathcal{F}^\star \mu} \,.\, \varpi_{x,y} . & (4) \end{cases}$$

Proof. (1) We have

$$\begin{aligned}\varphi_{x,U(\nu)y}(t) &= \langle U_t\, x, U(\nu)\, y \rangle \\ &= \int \langle U_t\, x, U_\theta\, y \rangle \,.\, \nu^\theta \\ &= \int \langle U_{t-\theta}\, x, y \rangle \,.\, \nu^\theta \\ &= \int \varphi_{x,y}(t-\theta) \,.\, \nu^\theta \\ &= (\varphi_{x,y} * \nu)\,(t) .\end{aligned}$$

(2) $\varphi_{U(\mu)x,y} = \tilde{\varphi}_{y,U(\mu)x} = (\varphi_{y,x} * \mu)^\sim = \tilde{\varphi}_{y,x} * \tilde{\mu} = \varphi_{x,y} * \tilde{\mu}$.

(3) The relation

$$\varphi_{x,U(\nu)y} = \nu * \varphi_{x,y}$$

may be written

$$\mathcal{F}\varpi_{x,U(\nu)y} = \nu * \mathcal{F}\varpi_{x,y} .$$

Let us apply $\mathcal{F}^\star$ to both sides:

(4)
$$\varpi_{x,U(\nu)y} = \mathcal{F}^\star \nu \,.\, \varpi_{x,y}\,.$$
$$\varpi_{U(\mu)x,y} = \overline{\varpi_{y,U(\mu)x}} = \overline{\mathcal{F}^\star \mu \,.\, \varpi_{y,x}}\,.$$
$$= \overline{\mathcal{F}^\star \mu} \,.\, \varpi_{x,y}$$

Corollary.

$$\langle U(\mu)\, x, y \rangle = \varphi_{U(\mu)x,y}(0) = \int \overline{\mathcal{F}^\star \mu} \,.\, \varpi_{x,y}$$

$$\langle x, U(\nu)\, y \rangle = \varphi_{x,U(\nu)y}(0) = \int \mathcal{F}^\star \nu \,.\, \varpi_{x,y}$$

$$\| U(\mu)\, x \|^2 = \int |\, \mathcal{F}^\star \mu \,|^2 \,.\, \varpi_{x,x}\,.$$

Definition. The topology of simple convergence on a Hilbert space H equipped with its Hilbert topology is called the *strong topology* on the space of operators in H.

Theorem 6. *Suppose that a sequence $\{\mu_n\}$ of bounded positive measures μ_n converges narrowly to μ. Then, the sequence $\{U(\mu_n)\}$ of the operators $U(\mu_n)$ converges strongly to the operator $U(\mu)$ (that is, $\{U(\mu_n)\, x\}$ converges to $U(\mu)\, x$ for all $x \in H$).*

Proof. Let us first prove the following

Lemma. *If a sequence $\{x_n\}$ of elements x_n of a Hilbert space H is such that*

$$\lim_{n\to\infty} \langle x_n, y \rangle = \langle x, y \rangle$$

for every $y \in H$ (that is, if $\{x_n\}$ converges "weakly" to x) and if, in addition,

$$\lim_{n\to\infty} \| x_n \| = \| x \|\,,$$

then $\lim_{n\to\infty} x_n = x$ (for the Hilbert topology, that is, "strongly").

Proof. We have

$$\| x_n - x \|^2 = \| x_n \|^2 - 2\,\mathcal{R}\,\langle x_n, x \rangle + \| x \|^2$$

and

$$\lim_{n\to\infty} \langle x_n, x \rangle = \| x \|^2, \quad \text{so that} \quad \| x_n - x \|^2 \to 0\,.$$

Let y denote a member of H. We have

$$\langle y, U(\mu_n)\, x \rangle = \int \langle y, U_t\, x \rangle \,.\, \mu_n^t\,.$$

Since the function $t \to \langle y, U_t x \rangle$ is continuous and bounded and since $\{\mu_n\}$ converges narrowly to μ, we have

$$\lim_{n\to\infty} \langle y, U(\mu_n)\, x \rangle = \int \langle y, U_t\, x \rangle \,.\, \mu^t = \langle y, U(\mu)\, x \rangle\,.$$

Furthermore, from the corollary to Theorem 5, we have

$$\| U(\mu_n)\, x\|^2 = \int | \mathcal{F}^\star \mu_n |^2 \,.\, \varpi_{x,x}\,.$$

There exists an M such that $\int \mu_n \leqslant M$, which implies that $| \mathcal{F}^\star \mu_n | \leqslant M$. Since, in addition $\mathcal{F}^\star \mu_n \to \mathcal{F}^\star \mu$ simply, we have

$$\lim_{n\to\infty} \| U(\mu_n)\, x \|^2 = \int | \mathcal{F}^\star \mu |^2 \,.\, \varpi_{x,x} = \| U(\mu) x \|^2\,.$$

Consequently,

$$\lim_{n\to\infty} U(\mu_n)\, x = U(\mu)\, x\,.$$

Remark. The preceding theorem remains valid for a sequence $\{\mu_n\}$ of arbitrary bounded measures μ_n that converges narrowly to μ. The proof of this, which is more delicate than the proof of the theorem given, will be omitted here.

We shall now establish certain additional properties of spectral measures.

For every $x \in H$, we denote by H_x the closed vector subspace generated by the elements $U_t\, x$ for $t \in \mathbf{R}$ and we denote by $\mathcal{M}_x$ the vector subspace (in general, not closed) consisting of the elements $U(f)\, x$ for $f \in \mathbf{S}$.

We have $\mathcal{M}_x \subset H_x$. Also, the relation $U_t\, U(f) x = U(f * \delta_t)\, x$ shows that $\mathcal{M}_x$ is stable with respect to the operators U_t. We note also that, if $f_n \in \mathbf{S}$ is positive and if the sequence $\{f_n\}$ converges narrowly to δ in $\mathbf{M}$, then $\{U(f_n)\}$ converges strongly to $\mathbf{1}$ and hence $\{U(f_n) x\}$ converges to x. Consequently, x is a limit point of the set $\mathcal{M}_x$. Thus, the closure of $\mathcal{M}_x$ is a closed vector space, containing x and stable with respect to all the operators U_t. Therefore, this closure is identical to H_x. In other words, $\mathcal{M}_x$ is a dense subspace of H_x.

Now, consider the formula

$$\langle U(f)\, x, U(g)\, x \rangle = \int \overline{\mathcal{F}^\star f}\, \mathcal{F}^\star g \,.\, \varpi_{x,x}\,.$$

In particular, this formula yields

$$\|U(f)\,x\|^2 = \int |\mathcal{F}^\star f|^2 \,.\, \varpi_{x,x}$$

and shows that the relation $U(f)\,x = 0$ implies $\mathcal{F}^\star f = 0$ almost everywhere with respect to $\varpi_{x,x}$. Consequently, the relation $U(f_1)\,x = U(f_2)\,x$ implies $\mathcal{F}^\star f_1 = \mathcal{F}^\star f_2$ almost everywhere with respect to $\varpi_{x,x}$.

If to every $y = U(f)\,x$ for $f \in \mathbf{S}$, we assign that element of $\mathbf{L}^2(\varpi_{x,x})$ defined by $\mathcal{F}^\star f$, we define a mapping of $\mathcal{M}_x$ into $\mathbf{L}^2(\varpi_{x,x})$ that is an isomorphism of the pre-Hilbert space $\mathcal{M}_x$ onto the dense subspace of $\mathbf{L}^2(\varpi_{x,x})$ consisting of the elements of $\mathbf{L}^2(\varpi_{x,x})$ that can be represented by functions belonging to $\mathbf{S}$. Let z denote an element of H_x. There exists a sequence $\{f_n\}$ of elements $f_n \in \mathbf{S}$ such that

$$\lim_{n \to +\infty} U(f_n)\,x = z.$$

We have

$$\|\,z_n - z_m\,\|^2 = \int |\,\mathcal{F}^\star f_n - \mathcal{F}^\star f_m\,|^2 \,.\, \varpi_{x,x}\,.$$

Consequently, the functions $\mathcal{F}^\star f_n$ constitute a Cauchy sequence in $\mathbf{L}^2(\varpi_{x,x})$. Therefore, there exists a function $\psi_{z|x} \in \mathbf{L}^2(\varpi_{x,x})$ such that

$$\psi_{z|x} = \lim_{n \to \infty} \mathcal{F}^\star f_n\,, \qquad \text{in} \qquad \mathbf{L}^2(\varpi_{x,x})\,.$$

If $z = U(f)\,x$, we have, by definition, $\psi_{z|x} = \mathcal{F}^\star f$, almost everywhere with respect to $\varpi_{x,x}$.

Proposition 2. *For every* $z \in H_x$, *we have*

$$\varpi_{x,z} = \psi_{z|x}\,\varpi_{x,x}.$$

Proof. Let h denote a member of $\mathbf{S}$. We have

$$\begin{aligned}\int \overline{\mathcal{F}^\star h} \,.\, \varpi_{x,z} = \langle\, U(h)\,x, z\,\rangle &= \lim_{n \to \infty} \langle\, U(h)\,x, z_n\,\rangle \\ &= \lim_{n \to \infty} \langle\, U(h)\,x, U(f_n)\,x\,\rangle \\ &= \lim_{n \to \infty} \int \overline{\mathcal{F}^\star h} \,.\, \mathcal{F}^\star f_n \,.\, \varpi_{x,x} \\ &= \int \overline{\mathcal{F}^\star h} \,.\, \psi_{z|x} \,.\, \varpi_{x,x}\,,\end{aligned}$$

from which the assertion of the proposition follows.

Proposition 3. *For every* $z \in H_x$, *we have* $\varpi_{z,z} = |\,\psi_{z|x}\,|^2\,\varpi_{x,x}$.

Proof. Let h denote a member of $\mathbf{S}$. Then,

$$\int \overline{\mathcal{F}^\star h} \,.\, \varpi_{z,z} = \langle U(h) z, z \rangle = \lim_{n\to\infty} \langle U(h) z_n, z_n \rangle$$

$$= \lim_{n\to\infty} \langle U(h) U(f_n) x, U(f_n) x \rangle$$

$$= \lim_{n\to\infty} \int \overline{\mathcal{F}^\star h} \,.\, \overline{\mathcal{F}^\star f_n} \,.\, \mathcal{F}^\star f_n \,.\, \varpi_{x,x} \,.$$

In $\mathbf{L}^2(\varpi_{x,x})$, we have

$$\lim_{n\to\infty} \mathcal{F}^\star f_n = \psi_{z|x}$$

and, since $\mathcal{F}^\star h$ is bounded,

$$\lim_{n\to\infty} \overline{\mathcal{F}^\star h} \,.\, \mathcal{F}^\star f_n = \overline{\mathcal{F}^\star h} \,.\, \psi_{z|x} \,.$$

From this, we conclude that

$$\int \overline{\mathcal{F}^\star h} \,.\, \varpi_{z,z} = \int \overline{\mathcal{F}^\star h} \,.\, |\psi_{z,x}|^2 \,.\, \varpi_{x,x} \,.$$

which completes the proof of the proposition.

Remark. For every $z \in H$, we have $\varpi_{x,z} = \varpi_{x,z'}$, where z' is the projection of z onto H_x. This is true because, for every $f \in \mathbf{S}$,

$$\int \overline{\mathcal{F}^\star f} \,.\, (\varpi_{x,z} - \varpi_{x,z'}) = \langle U(f) x, z - z' \rangle = 0 \,.$$

Consequently,

$$\varpi_{x,z} = \psi_{z'|x} \, \varpi_{x,x} \,.$$

Thus, every spectral measure $\varpi_{x,z}$ has a density with respect to $\varpi_{x,x}$, and this density is square-integrable with respect to $\varpi_{x,x}$. We have

$$\int |\psi_{z'|x}|^2 \,.\, \varpi_{x,x} = \int \varpi_{z',z'} \,.$$

3. SECOND-ORDER STATIONARY RANDOM PROCESSES

In this section, we shall study random processes represented by a physical variable X of which the value at each instant t is a random variable $X(t)$. We shall constantly be using the geometric study made in Chapter 5, section 5.

Definitions. Suppose that we have a probabilized space. Let H denote the space of random variables Z such that $\mathbf{E}(|Z|^2) < \infty$. Suppose that we have a random process consisting of a mapping $X : t \to X(t)$ that assigns to every instant $t \in \mathbf{R}$ a scalar- or vector-valued random variable $X(t)$. This process is said to be *centered* if $\mathbf{E}(X(t)) = 0$.

Such a process will be called ***stationary*** if the probability distribution of the variables $X(t_1 - \theta), \dots, X(t_k - \theta)$ is identical to that of the variables $X(t_1), \dots, X(t_k)$ for all the times $t_1, \dots, t_k$ (where k is arbitrary) and all $\theta \in \mathbf{R}$.

In the case in which $X(t)$ is scalar, such a process is said to be stationary of the second order if

$$\begin{cases} X(t) \in H\,, & \forall t \in \mathbf{R}\,, \\ \mathbf{E}(X(t)) \quad \text{is independent of } t, & \\ \mathbf{v}(X(t_1 - \theta), X(t_2 - \theta)) = \mathbf{v}(X(t_1), X(t_2)) & (\forall t_1, t_2, \theta \in \mathbf{R})\,. \end{cases}$$

We note that if $X(t)$ is a stationary process, it is stationary of the second order if $X(t) \in H$.

If $X(t)$ is a process that is stationary of the second order, the process $\mathring{X}$ defined by

$$t \to \mathring{X}(t) = X(t) - m_X, \quad \text{where} \quad m_X = \mathbf{E}(X(t))$$

is a centered process stationary of the second order. The constant m_X is called the mean of the process X.

A process $t \to X(t)$ stationary of the second-order is said to be *continuous* (in mean square) if $t \to X(t)$ is a continuous mapping of $\mathbf{R}$ onto H.

A process

$$t \to X(t) = \begin{bmatrix} X^1(t) \\ \vdots \\ X^m(t) \end{bmatrix}$$

will be called ***stationary of the second order*** if

$$\begin{cases} X^i(t) \in H\,, & (\forall t \in \mathbf{R}, i = 1, \dots, m)\,, \\ \mathbf{E}(X(t)) \quad \text{is independent of } t, & \\ \mathbf{v}(X^i(t_1 - \theta), X^j(t_2 - \theta)) = \mathbf{v}(X^i(t_1), X^j(t_2)) & (\forall i, j, \theta, t_1, t_2)\,. \end{cases}$$

We shall say that such a process is *centered* if $\mathbf{E}(X(t)) = 0$. The process

$$t \to \mathring{X}(t) = X(t) - m_X, \quad \text{where} \quad m_X = \mathbf{E}(X(t))$$

is centered. The constant m-tuple m_X is called the ***mean*** of the process X.

We shall say that such a process is *continuous* (in mean square) if the mappings $t \to X^i(t)$ are continuous from $\mathbf{R}$ into H.

If the process

$$t \to X(t) = \begin{bmatrix} X^1(t) \\ \cdot \\ \cdot \\ X^m(t) \end{bmatrix}$$

is stationary of the second order, so is every process $X^\alpha : t \to X^\alpha(t)$, where α is a family of indices contained in $\{1, ..., m\}$ and, in particular, so is every process $X^i : t \to X^i(t)$. On the other hand, the fact that each of the processes X^i is stationary of the second order does not imply that the process X is stationary of the second order.

Notation. We denote by H_X the closed subspace of H generated by the elements $X^i(t)$ for $t \in \mathbf{R}$, where $i = 1, ..., m$.

Theorem 7. *A necessary and sufficient condition for the centered process*

$$X : t \to X(t) = \begin{bmatrix} X^1(t) \\ \cdot \\ \cdot \\ X^m(t) \end{bmatrix}$$

to be a continuous process stationary of the second order is that there exist a continuous unitary representation $\theta \to U_\theta$ of $\mathbf{R}$ into H such that

$$X^i(t - \theta) = U_\theta X^i(t) .$$

Proof of the sufficiency: If the representation $\theta \to U_\theta$ exists, we have

$$\begin{aligned} \mathbf{v}(X^i(t_1 - \theta), X^j(t_2 - \theta)) &= \langle X^i(t_1 - \theta), X^j(t_2 - \theta) \rangle \\ &= \langle U_\theta X^i(t_1), U_\theta X^j(t_2) \rangle = \langle X^i(t_1), X^j(t_2) \rangle \\ &= \mathbf{v}(X^i(t_1), X^j(t_2)) . \end{aligned}$$

Furthermore, the mapping $t \to X^i(t) = U_{-t} X^i(0)$ is continuous.

Proof of the necessity: Let $\mathcal{N}_X$ denote the subspace of H consisting of elements of the form

$$\xi = \sum_{ih} a_i^h X^i(t_h) ,$$

where the times t_h are arbitrary. The subspace H_X of H is the closure of $\mathcal{N}_X$.

Consider the set Γ of pairs $\begin{bmatrix} \xi \\ \eta \end{bmatrix}$ such that ξ and η have representations of the form

$$\xi = \sum_{ih} a_i^h X^i(t_h),$$
$$\eta = \sum_{ih} a_i^h X^i(t_h - \theta).$$

Obviously, the set Γ is a vector subspace of $H_X \times H_X$ (in general, Γ is not closed).

Let $\begin{bmatrix} \xi' \\ \eta' \end{bmatrix}$ denote another pair belonging to Γ. We can always assume that the times t_h appearing in the representation ξ' are the same as those appearing in the representation of ξ. Therefore,

$$\begin{cases} \xi' = \sum_{jk} a_j'^k X^j(t_k) \\ \eta' = \sum_{jk} a_j'^k X^j(t_k - \theta). \end{cases}$$

We then have

$$\begin{aligned} \langle \eta, \eta' \rangle &= \sum_{ihjk} \overline{a_i^h}\, a_j'^k \langle X^i(t_h - \theta), X^j(t_k - \theta) \rangle \\ &= \sum_{ihjk} \overline{a_i^h}\, a_j'^k \langle X^i(t_h), X^j(t_k) \rangle \\ &= \langle \xi, \xi' \rangle \cdot \end{aligned}$$

From this, it follows, in particular, that $\| \xi \| = \| \eta \|$ and, since Γ is a vector subspace,

$$\| \xi - \xi' \| = \| \eta - \eta' \|.$$

Consequently, the equation $\xi = \xi'$ implies $\eta = \eta'$. In other words, for every $\xi \in \mathcal{N}_X$, there exists one and only one $\eta \in \mathcal{N}_X$ such that $\begin{bmatrix} \xi \\ \eta \end{bmatrix} \in \Gamma$. Let us set

$$\eta = U_\theta \xi.$$

U_θ is a linear operator defined on the dense subspace $\mathcal{N}_X$ of H_X that satisfies the equation

$$\langle U_\theta \xi, U_\theta \xi' \rangle = \langle \xi, \xi' \rangle;$$

It can be extended as an isometric continuous operator over H_X such that

$$U_\theta X^i(t) = X^i(t - \theta) \qquad (\forall i)$$

(where we still use the notation U_θ for the extended operator), and this condition defines it unambiguously on H_X. From this we conclude that

$$\begin{aligned} U_{\theta_1+\theta_2} &= U_{\theta_1} U_{\theta_2} \\ U_{-\theta} &= (U_\theta)^{-1} . \end{aligned}$$

Therefore, the mapping $\theta \to U_\theta$ is a unitary representation of $\mathbf{R}$ in H_X. This representation is continuous. This is true because $U_{\theta_n} \xi \to U_{\theta_0} \xi$ as $\theta_n \to \theta_0$ for every $\xi \in \mathcal{N}_X$. Since $\| U_{\theta_n} \| = 1$, the topology of simple convergence on $\mathcal{N}_X$ coincides with the topology of simple convergence on H_X. Therefore, $U_{\theta_n} x \to U_\theta x$ for every $x \in H_X$. We can extend U_θ onto all H by setting $U_\theta \xi = \xi$ for $\xi \in H_X^\perp$.

Theorem 7a. *A necessary and sufficient condition for the process*

$$t \to X(t) = \begin{bmatrix} X^1(t) \\ \vdots \\ X^m(t) \end{bmatrix}$$

to be a continuous process stationary of the second order is that there exist a continuous unitary representation $\theta \to U_\theta$ *of* $\mathbf{R}$ *into* H *such that*

$$\begin{aligned} X^i(t-\theta) &= U_\theta X^i(t) \\ U_\theta 1 &= 1 . \end{aligned}$$

Proof of the sufficiency. We can confine ourselves to the case in which $H = H_X$. The operator U_θ commutes with the orthogonal projection onto $\mathring{H}$ since $\{1\}$ and $\mathring{H}$ are stable under U_θ. Therefore,

$$\mathring{X}^i(t-\theta) = U_\theta \mathring{X}^i(t) .$$

Hence, the process $\mathring{H}$ is a continuous process stationary of the second order.

Furthermore,

$$\begin{aligned} \mathbf{E}(X^i(t)) = \langle 1, X^i(t) \rangle = \langle 1, U_t^\star X^i(0) \rangle &= \langle U_t 1, X^i(0) \rangle \\ &= \langle 1, X^i(0) \rangle = \mathbf{E}(X^i(0)) . \end{aligned}$$

Proof of the necessity. Let us choose U_θ such that

$$U_\theta \mathring{X}^i(t-\theta) = U_\theta \mathring{X}^i(t) \quad \text{and} \quad U_\theta \xi = \xi \quad \text{for} \quad \xi \in H_X^\perp .$$

In what follows, all the stationary random processes that we shall be considering will be assumed continuous.

Suppose that $t \to \begin{bmatrix} X(t) \\ Y(t) \end{bmatrix}$ is a centered random process stationary of the second order, where $X(t)$ and $Y(t)$ are scalar random variables. The function

$$\theta \to \mathbf{v}(X(t-\theta), Y(t)) = \langle X(t-\theta), Y(t) \rangle = \langle U_\theta X(t), Y(t) \rangle$$

is called the ***intercorrelation function*** of X and Y. By hypothesis, it is independent of t. Following the notations of paragraph 2, it should be denoted by $\varphi_{X(t),\,Y(t)}$. Since it is independent of t, we shall simply denote it by $\varphi_{X,Y}$. If $Y = X$, the function $\varphi_{X,X}$ is called the ***autocorrelation function***. By definition,

$$\begin{aligned} \varphi_{X,Y}(\theta) = \varphi_{X(t),Y(t)}(\theta) &= \langle U_\theta X(t), Y(t) \rangle \\ &= \langle X(t-\theta), Y(t) \rangle = \mathbf{v}(X(t-\theta), Y(t)). \end{aligned}$$

We have

$$\varphi_{X,Y} = \varphi_{X(t),Y(t)} = \mathcal{F}\varpi_{X(t),Y(t)},$$

where $\varpi_{X(t),Y(t)}$ is a bounded measure independent of t, which we shall denote by $\varpi_{X,Y}$. Therefore, we have

$$\boxed{\varphi_{X,Y} = \mathcal{F}\varpi_{X,Y}}$$

and, in particular,

$$\varphi_{X,X} = \mathcal{F}\varpi_{X,X},$$

where $\varpi_{X,X}$ is a positive bounded measure. The measures $\varpi_{X,Y}$ and $\varpi_{Y,X}$ are called ***joint power spectral distributions*** of X and Y. The measure $\varpi_{X,X}$ is called the ***power spectral distribution*** of X. We have

$$\boxed{\varphi_{Y,X} = \tilde{\varphi}_{X,Y} \qquad \varpi_{Y,X} = \overline{\varpi_{X,Y}}}\,.$$

For noncentered X and Y, we set

$$\varphi_{X,Y} = \varphi_{\overset{\circ}{X},\overset{\circ}{Y}}, \qquad \varpi_{X,Y} = \varpi_{\overset{\circ}{X},\overset{\circ}{Y}}.$$

Note that

$$\int \varpi_{X,Y} = \mathbf{v}(X(t), Y(t)) \qquad \text{and} \qquad \int \varpi_{X,X} = \boldsymbol{\sigma}^2(X(t)).$$

If $\varpi_{X,X}$ (resp. $\varpi_{X,Y}$) has a density $\Phi_{X,X}$ (resp. $\Phi_{X,Y}$) with respect to Lebesgue measure, the quantity $\Phi_{X,X}$ (resp. $\Phi_{X,Y}$) will be called the ***power spectral density of X*** (***resp. the joint power spectral density of X and Y***).

4. CONVOLUTIONS ON A STATIONARY PROCESS OF THE SECOND ORDER

In what follows, we shall simply say "stationary" to mean "stationary of the second order".

Theorem 8. *Let* X: $t \to X(t)$ *denote a scalar stationary random process and let* $\theta \to U_\theta$ *denote an associated unitary representation (satisfying the hypotheses of Theorem 7 or Theorem 7a). Then, for every bounded measure* μ, *the process* Z *defined by*

$$Z(t) = \int X(t - \theta) \,.\, \mu^\theta = U(\mu)\, X(t)\ ,$$

and the process

$$t \to \begin{bmatrix} X(t) \\ Z(t) \end{bmatrix} .$$

are stationary.

We shall write $Z = \mu * X$.

Proof. We have

$$U(\mu)\, X(t) = \int U_\theta\, X(t) \,.\, \mu^\theta = \int X(t - \theta) \,.\, \mu^\theta\ ,$$

and

$$Z(t - \tau) = U(\mu)\, X\,(t - \tau) = U(\mu)\, U_\tau\, X(t)$$
$$= U_\tau\, U(\mu)\, X(t) = U_\tau\, Z(t)\,.$$

Consequently, the process Z is stationary and so is the process $t \to \begin{bmatrix} X(t) \\ Z(t) \end{bmatrix}$.

Remark. $Z(t) \in H_X \quad (\forall t \in \mathbf{R})$.

Theorem 9 (on properties of a convolution). *If* $t \to \begin{bmatrix} X(t) \\ Y(t) \end{bmatrix}$ *is a random process and* μ *and* ν *are bounded measures, then*

(1) $$(\mu * \nu) * X = \mu * (\nu * X)$$

(2) $$(\delta_\tau * X)\,(t) = X(t - \tau)$$

(3) $$\varphi_{\mu * X, Y} = \tilde{\mu} * \varphi_{X,Y}$$

(4) $$\varphi_{X, \nu * Y} = \nu * \varphi_{X,Y}$$

(5) $$\varpi_{\mu * X, Y} = \overline{\mathcal{F}^\star \mu} \,.\, \varpi_{X,Y}$$

(6) $$\varpi_{X, \nu * Y} = \mathcal{F}^\star \nu \,.\, \varpi_{X,Y}$$

in particular,

$$(7)\quad \begin{aligned} \varpi_{\mu*X,\mu*X} &= |\mathcal{F}^\star \mu|^2 \,.\, \varpi_{X,X} \,, \\ m_{\mu*X} &= m_X \int \mu \,. \end{aligned}$$

Proof. For properties (1)–(6), we may assume that X and Y are centered. Then,

$$\begin{aligned}
&(1)\ \ [(\mu * \nu) * X](t) = U(\mu * \nu) X(t) = U(\mu) U(\nu) X(t) = [\mu * (\nu * X)](t) \\
&(2)\ \ (\delta_\tau * X)(t) = U(\delta_\tau) X(t) = U_\tau X(t) = X(t-\tau) \\
&(3)\ \ \varphi_{\mu*X,Y} = \varphi_{U(\mu)X(t),Y(t)} = \tilde{\mu} * \varphi_{X(t),Y(t)} = \tilde{\mu} * \varphi_{X,Y} \\
&(4)\ \ \varphi_{X,\nu*Y} = \varphi_{X(t),U(\nu)Y(t)} = \nu * \varphi_{X(t),Y(t)} = \nu * \varphi_{X,Y} \\
&(5)\ \ \varpi_{\mu*X,Y} = \varpi_{U(\mu)X(t),Y(t)} = \overline{\mathcal{F}^\star \mu} \,.\, \varpi_{X(t),Y(t)} = \overline{\mathcal{F}^\star \mu} \,.\, \varpi_{X,Y} \\
&(6)\ \ \varpi_{X,\nu*Y} = \varpi_{X(t),U(\nu)Y(t)} = \mathcal{F}^\star \nu \,.\, \varpi_{X(t),Y(t)} = \mathcal{F}^\star \nu \,.\, \varpi_{X,Y} \\
&(7)\ \ m_{\mu*X} = \mathbf{E}((\mu * X)(t)) = \langle 1, (\mu * X)(t) \rangle \\
&\qquad = \int \langle 1, X(t) \rangle \,.\, \mu^t = \int m_X \,.\, \mu = m_X \int \mu \,.
\end{aligned}$$

Remark. Property (7) may also be written

$$m_{\mu*X} = m_X \,.\, (\mathcal{F}^\star \mu)(0) \,.$$

Thus, the bounded measures are the universal convolvers on stationary processes of the second order.

5. EXTENSION OF A CONVOLUTION

First of all, let us again take a unitary representation of $\mathbf{R}$ in the Hilbert space H. We continue to denote by $\mathcal{M}_x$ the vector subspace consisting of elements of the form $U(f)x$ for $f \in \mathbf{S}$. For every $T \in \mathbf{O}'_c$, we propose to define an operator $U(T)$ whose domain of definition is a dense vector subspace of H by

$$\begin{cases} \text{If } x \in \operatorname{def}(U(T)), \quad \text{then} \quad U(T)\,x \in H_x \,, \\ \langle U(T)\,x, U(f)\,x \rangle = \langle x, U(\tilde{T} * f)\,x \rangle, \qquad (\forall f \in \mathbf{S}) \,. \end{cases}$$

We note first of all that the equation $U(f)x = 0$ implies $\langle x, U(\tilde{T} * f)\, x \rangle = 0$. This is true because the equation $U(f)\,x$ implies $\mathcal{F}^\star f = 0$ almost everywhere with respect to $\varpi_{x,x}$ and hence

$$\langle x, U(\tilde{T} * f)\, x \rangle = \int \mathcal{F}^\star(\tilde{T} * f) \,.\, \varpi_{x,x} = \int \overline{\mathcal{F}^\star T} \,.\, \mathcal{F}^\star f \,.\, \varpi_{x,x} = 0 \,.$$

If $\mathcal{F}^\star T \in \mathbf{L}^2(\varpi_{x,x})$, we have

$$|\langle x, U(\tilde{T}*f)\,x\rangle|^2 \leqslant \int |\mathcal{F}^\star T|^2 \,.\, \varpi_{x,x} \,.\, \int |\mathcal{F}^\star f|^2 \,.\, \varpi_{x,x}\,,$$

that is,

$$|\langle x, U(\tilde{T}*f)\,x\,| \leqslant \left(\int |\mathcal{F}^\star T|^2 \,.\, \varpi_{x,x}\right)^{1/2} \|U(f)\,x\| \,.$$

Consequently, the mapping

$$U(f)\,x \to \langle x, U(\tilde{T}*f)\,x\rangle$$

is a continuous linear form. Therefore, there exists an element $y \in \mathcal{M}_x$ such that

$$\langle y, U(f)\,x\rangle = \langle x, U(\tilde{T}*f)\,x\rangle \,.$$

Let us set $y = U(T)\,x$. Then, we have

Theorem 10. *If* $T \in \mathbf{O}'_c$ *and* $\int |\mathcal{F}^\star T|^2 \,.\, \varpi_{x,x} < +\infty$, *there exists exactly one element, which we denote by* $U(T)\,x$, *such that*

$$\begin{cases} U(T)\,x \in H_x\,, \\ \langle U(T)\,x, U(f)\,x\rangle = \langle x, U(\tilde{T}*f)\,x\rangle\,, \qquad (\forall f \in \mathbf{S})\,. \end{cases}$$

Thus, the conditions $x \in \operatorname{def}(U(T))$ and

$$\int |\mathcal{F}^\star T|^2 \,.\, \varpi_{x,x} < +\infty\,,$$

are equivalent.

Proposition 4. *The set* $\operatorname{def}(U(T))$ *is a vector subspace of* H.

The proof is a consequence of the following lemma:

Lemma 1. *For every* x *and* y *in* H, *we have*

$$\varpi_{x+y,x+y} \leqslant 2(\varpi_{x,x} + \varpi_{y,y})\,.$$

Proof. For every $f \in \mathfrak{D}^0_+$, the mapping $x \to \int f \,.\, \varpi_{x,x}$ is a positive-semidefinite quadratic form associated with the Hermitian sesquilinear form

$$x, y \to \int f \,.\, \varpi_{x,y}\,.$$

Now, for every Hermitian sesquilinear form $\mathcal{A}$, such that $\mathcal{A}(x, x) \geqslant 0$,

$$\mathcal{A}(x + y, x + y) \leqslant 2(\mathcal{A}(x, x) + \mathcal{A}(y, y)) ,$$

so that

$$\int f \cdot \varpi_{x+y,x+y} \leqslant 2 \int f \cdot (\varpi_{x,x} + \varpi_{y,y}) ,$$

which completes the proof.

Proposition 5. *For every* $x \in \operatorname{def}(U(T))$ *and* $z \in H$, *we have*

$$\langle U(T)\, x, z \rangle = \int \overline{\mathcal{F}^\star T} \cdot \varpi_{x,z} .$$

Proof. (a) The case $z = U(f)\, x$ for $f \in \mathbf{S}$. We have

$$\langle U(T)\, x, U(f)\, x \rangle = \int \overline{\mathcal{F}^\star T} \cdot \mathcal{F}^\star f \cdot \varpi_{x,x} = \int \overline{\mathcal{F}^\star T} \cdot \varpi_{x,U(f)x} .$$

(b) The case $z \in H_x$. Let us set $z = \lim_{n\to\infty} z_n$, $z_n = U(f_n)\, x$, $f_n \in \mathbf{S}$. Then,

$$\langle U(T)\, x, z \rangle = \lim_{n\to\infty} \langle U(T)\, x, U(f_n)\, x \rangle = \lim_{n\to\infty} \int \overline{\mathcal{F}^\star T} \cdot \mathcal{F}^\star f_n \cdot \varpi_{x,x}$$

$$= \int \overline{\mathcal{F}^\star T} \cdot \psi_{z|x} \cdot \varpi_{x,x} = \int \overline{\mathcal{F}^\star T} \cdot \varpi_{x,z} .$$

(c) The general case $z \in H$. Let z' denote the projection of z onto H_x. We have

$$\langle U(T)\, x, z \rangle = \langle U(T)\, x, z' \rangle = \int \overline{\mathcal{F}^\star T} \cdot \varpi_{x,z'} = \int \overline{\mathcal{F}^\star T} \cdot \varpi_{x,z} .$$

Proposition 6. *If* x *and* y *belong to* $\operatorname{def}(U(T))$, *we have*

$$\begin{cases} U(T)\,(x + y) = U(T)\, x + U(T)\, y \\ U(T)\,(\lambda x) \quad = \lambda U(T)\, x . \end{cases}$$

Proof. The second equation is immediate. To prove the first, let us write

$$\langle U(T)\,(x + y), z \rangle = \int \overline{\mathcal{F}^\star T} \cdot \varpi_{x+y,z} = \int \overline{\mathcal{F}^\star T} \cdot (\varpi_{x,z} + \varpi_{y,z})$$

$$= \langle U(T)\, x, z \rangle + \langle U(T)\, y, z \rangle .$$

Also

$$\operatorname{def}(U(T)) \supset \mathcal{M}_x \quad (\forall x \in H) .$$

This is true because, if $y = U(f)\,x$ for $f \in \mathbf{S}$, we have

$$\varpi_{y,y} = |\mathcal{F}^\star f|^2 \,.\, \varpi_{x,x}\,.$$

Since $\mathcal{F}^\star T \in \mathbf{O}_m$, we have $\mathcal{F}^\star T \,.\, \mathcal{F}^\star f \in \mathbf{S}$ and, consequently,

$$\int |\mathcal{F}^\star T|^2 \varpi_{y,y} = \int |\mathcal{F}^\star T|^2 \,.\, |\mathcal{F}^\star f|^2 \varpi_{x,x} < \infty\,.$$

Thus, we finally have

Theorem 11. *For $T \in \mathbf{O}'_c$, the operator $U(T)$ is a linear operator whose domain of definition is a dense vector subspace of H.*

We remark that the definitions of $U(T)$ given for $T \in \mathbf{M}$ and $T \in \mathbf{O}'_c$ coincide on $\mathbf{M} \cap \mathbf{O}'_c$.

Proposition 7. *We have*

$$\left.\begin{aligned} \varpi_{U(T)x,y} &= \overline{\mathcal{F}^\star T} \,.\, \varpi_{x,y} \\ \varphi_{U(T)x,y} &= \tilde{T} * \varphi_{x,y} \end{aligned}\right\} \text{if } x \in \operatorname{def}\big(U(T)\big)$$

$$\left.\begin{aligned} \varpi_{x,U(T)y} &= \mathcal{F}^\star T \,.\, \varpi_{x,y} \\ \varphi_{x,U(T)y} &= T * \varphi_{x,y} \end{aligned}\right\} \text{if } y \in \operatorname{def}\big(U(T)\big)$$

$$\left.\begin{aligned} \varpi_{U(T)x,U(T)x} &= |\mathcal{F}^\star T|^2 \,.\, \varpi_{x,x} \\ \varphi_{U(T)x,U(T)x} &= \tilde{T} * T * \varphi_{x,x} \end{aligned}\right\} \text{if } x \in \operatorname{def}\big(U(T)\big)\,.$$

Proof.

$$\langle U(T)\,x, z \rangle = \int \overline{\mathcal{F}^\star T} \,.\, \varpi_{x,z}\,,$$

so that, for all $f \in \mathbf{S}$,

$$\int \mathcal{F}^\star f \,.\, \varpi_{U(T)x,y} = \langle U(T)\,x, U(f)\,y \rangle = \int \overline{\mathcal{F}^\star T} \,.\, \varpi_{x,U(f)y}$$

$$= \int \overline{\mathcal{F}^\star T} \,.\, \mathcal{F}^\star f \,.\, \varpi_{x,y}\,,$$

and, consequently,

$$\varpi_{U(T)x,y} = \overline{\mathcal{F}^\star T} \,.\, \varpi_{x,y}\,.$$

Therefore,

$$\varpi_{x,U(T)y} = \overline{\varpi_{U(T)y,x}} = \overline{\overline{\mathcal{F}^\star T} \,.\, \varpi_{y,x}} = \mathcal{F}^\star T \,.\, \varpi_{x,y}\,.$$

The fifth equation follows immediately from the first and third. The other formulas are derived by use of the Fourier transformation.

Theorem 12. *Suppose that* $T_1, T_2 \in \mathbf{O}'_c$. *If* $x \in \operatorname{def}(U(T_1))$ *and*

$$U(T_1)\,x \in \operatorname{def}(U(T_2)),$$

then,

$$x \in \operatorname{def}(U(T_2 * T_1))$$

and

$$U(T_2 * T_1)\,x = U(T_2)\,U(T_1)\,x\,.$$

Proof. By hypothesis,

$$\int |\mathcal{F}^\star T_1|^2 \,.\, \varpi_{x,x} < +\infty$$

and

$$\int |\mathcal{F}^\star T_2|^2 \,.\, \varpi_{U(T_1)x,\,U(T_1)x} < +\infty\,.$$

The second of these inequalities may be written

$$\int |\mathcal{F}^\star T_2|^2 \,.\, |\mathcal{F}^\star T_1|^2 \,.\, \varpi_{x,x} = \int |\mathcal{F}^\star(T_2 * T_1)|^2 \,.\, \varpi_{x,x} < +\infty\,,$$

so that

$$x \in \text{déf}\,(U(T_2 * T_1)).$$

Consequently, for $y \in H$,

$$\langle\, U(T_2)\,U(T_1)\,x, y \,\rangle = \int \overline{\mathcal{F}^\star T_2} \,.\, \varpi_{U(T_1)x,y} = \int \overline{\mathcal{F}^\star T_2} \,.\, \overline{\mathcal{F}^\star T_1} \,.\, \varpi_{x,y}\,,$$

$$\langle\, U(T_2 * T_1)\,x, y \,\rangle = \int \overline{\mathcal{F}^\star(T_2 * T_1)} \,.\, \varpi_{x,y} = \int \overline{\mathcal{F}^\star T_2} \,.\, \overline{\mathcal{F}^\star T_1} \,.\, \varpi_{x,y}\,.$$

Theorem 13. $\operatorname{def}(U(\tilde{T})) = \operatorname{def} U(T)$. *If* $x, y \in \operatorname{def}(U(T))$, *then*

$$\langle\, U(T)\,x, y \,\rangle = \langle\, x, U(\tilde{T})\,y \,\rangle\,.$$

Proof. The first assertion of the theorem follows immediately from the fact that $\mathcal{F}^\star(\tilde{T}) = \overline{\mathcal{F}^\star(T)}$. Also,

$$\langle\, U(T)\,x, y \,\rangle = \int \overline{\mathcal{F}^\star T} \,.\, \varpi_{x,y}$$

and

$$\langle\, x, U(\tilde{T})\,y \,\rangle = \int \mathcal{F}^\star \tilde{T} \,.\, \varpi_{x,y} = \int \overline{\mathcal{F}^\star T} \,.\, \varpi_{x,y}\,.$$

We shall now apply the preceding results to stationary processes of the second order.

Let X denote a stationary process of the second order. Let $t \to U_t$ denote an associated unitary representation of **R**. For $T \in \mathbf{O}'_c$, we set

$$Z(t) = U(T)\, X(t) = (T * X)\,(t)$$

if $X(t) \in \operatorname{def}(U(T))$, which will be the case if and only if

$$\int |\mathcal{F}^\star T|^2 \,.\, \varpi_{X,X} < +\infty\,.$$

We then have

$$U_\tau Z(t) = U(\delta_\tau)\, U(T)\, X(t) = U(T)\, X(t-\tau) = Z(t-\tau)\,.$$

Consequently, the process $\begin{bmatrix} X \\ Z \end{bmatrix}$ is stationary.

Convolution has the following properties:

$$\begin{aligned}
(S * T) * X &= S * (T * X) && \text{if the right-hand member is defined;}\\
\varphi_{S*X,Y} &= \tilde{S} * \varphi_{X,Y} && \text{if } S * X \text{ is defined;}\\
\varphi_{X,T*Y} &= T * \varphi_{X,Y} && \text{if } T * Y \text{ is defined;}\\
\varphi_{T*X,T*X} &= \tilde{T} * T * \varphi_{X,X} && \text{if } T * X \text{ is defined;}\\
\varpi_{S*X,Y} &= \overline{\mathcal{F}^\star S} \,.\, \varpi_{X,Y} && \text{if } S * X \text{ is defined;}\\
\varpi_{X,T*Y} &= \mathcal{F}^\star T \,.\, \varpi_{X,Y} && \text{if } T * Y \text{ is defined;}\\
\varpi_{T*X,T*X} &= |\mathcal{F}^\star T|^2\, \varpi_{X,X} && \text{if } T * X \text{ is defined.}
\end{aligned}$$

$$m_{T*X} = m_X \,.\, (\mathcal{F}^\star T)\,(0)\,.$$

In particular, we shall call the process $\delta' * X$ the derivative of the process X if this operation is defined. In this case, we shall say that X is differentiable (in mean square).

A necessary and sufficient condition for X to be differentiable is that

$$\int \omega^2 \,.\, \varpi_{XX} < +\infty\,.$$

We then have

$$\begin{aligned}
\varphi_{\delta'*X\,\delta'*X} &= \tilde{\delta}' * \delta' * \varphi_{X,X}\\
&= -\,\varphi''_{X,X}
\end{aligned}$$

$$\varpi_{\delta'*X,\delta'*X} = \omega^2 \,.\, \varpi_{X,X}\,.$$

One should note that the existence of $\delta' * X$ implies that $\varphi_{X,X}$ is twice continuously differentiable. The converse is also true but we shall not prove it here. One can also show that a necessary and sufficient condition for X to be differentiable is that $[X(t) - X(0)]/t$ have a limit in H as $t \to 0$. We then have

$$(\delta' * X)(t) = \lim_{\Delta t \to 0} \frac{X(t + \Delta t) - X(t)}{\Delta t},$$

where the limit is taken in the sense of the topology of H. It is this property that justifies the expression "mean-square differentiable." We note finally that X is infinitely differentiable if $\omega_{X,X}$ has compact support.

6. GENERALIZATIONS. POISSON PROCESSES

Let X denote a stationary process (centered or not) of the second order. For every bounded measure μ, we can define the random variable X_μ by

$$X_\mu = \int X(t) \,.\, \mu^t = (\check{\mu} * X)(0) = U(\check{\mu})\, X(0)\,.$$

Let us calculate

$$\mathbf{E}(X_\mu) = \langle 1, X_\mu \rangle = \int \langle 1, X(t) \rangle \,.\, \mu^t = \int \mathbf{E}(X(t)) \,.\, \mu^t = m_X \int \mu\,.$$

We have

$$\begin{aligned}\mathbf{v}(X_\nu, X_\mu) &= \int \varpi_{\check{\nu}*X,\check{\mu}*X} \\ &= \int \overline{\mathcal{F}^* \check{\nu}} \,.\, \mathcal{F}^* \check{\mu} \,.\, \varpi_{X,X} \\ &= \int \overline{\mathcal{F}\nu} \,.\, \mathcal{F}\mu \,.\, \varpi_{X,X}\,.\end{aligned}$$

We can also calculate the same expression with the aid of the correlation function $\varphi_{X,X}$. We have

$$\begin{aligned}\langle X_\nu, X_\mu \rangle &= \varphi_{\check{\nu}*X,\check{\mu}*X}(0) \\ &= (\bar{\nu} * \check{\mu} * \varphi_{X,X})(0) \\ &= [(\tilde{\nu} * \mu)^{\vee} * \varphi_{X,X}](0) \\ &= \int \varphi_{X,X} \,.\, (\tilde{\nu} * \mu)\,.\end{aligned}$$

Let us make note of these formulas:

$$\boxed{\mathbf{v}(X_\nu, X_\mu) = \int \overline{\mathcal{F}\nu} \,.\, \mathcal{F}\mu \,.\, \varpi_{X,X} = \int \varphi_{X,X} \,.\, (\tilde{\nu} * \mu)}\,.$$

Naturally,

$$\mathbf{E}(X_{\delta_a*\mu}) = \mathbf{E}(X_\mu),$$
$$\mathbf{v}(X_{\delta_a*\nu}, X_{\delta_a*\mu}) = \mathbf{v}(X_\nu, X_\mu) \qquad \text{for every } a \in \mathbf{R}.$$

Let us rewrite these formulas in the case in which μ and ν are replaced by functions $f, g \in \mathbf{L}^1$ by using the following notation, which is more suited for what we shall be doing in this section:

$$\int f \,.\, X = \int f(t)\, X(t)\, \mathrm{d}t\,.$$

We then have

$$\begin{cases} \mathbf{v}\left(\int g \,.\, X, \int f \,.\, X\right) = \int \overline{\mathcal{F}g}\, \mathcal{F}f \,.\, \varpi_{X,X} = \int \tilde{g} * f \,.\, \varphi_{X,X}, \\ \mathbf{E}\left(\int f \,.\, X\right) = m_X \int f. \end{cases}$$

Also,

$$\mathbf{E}\left(\int (\delta_a * f) \,.\, X\right) = \mathbf{E}\left(\int f \,.\, X\right)$$

and

$$\mathbf{v}\left(\int (\delta_a * g) \,.\, X, \int (\delta_a * f) \,.\, X\right) = \mathbf{v}\left(\int g \,.\, X, \int f \,.\, X\right), \quad (\forall a \in \mathbf{R}).$$

An application. Let $z_0(t)$ denote the elementary solution of a differential equation $P(D)\, z = x(t)$ in $\mathcal{D}'_+$. If the support of $x(t)$ is positive, the solution with positive support is

$$z(t) = (z_0 * x)(t) = \int_0^t z_0(t - \theta)\, x(\theta)\, \mathrm{d}\theta, \qquad \text{for} \quad t \geqslant 0\,.$$

Now, let X denote a stationary random process of the second order and let $\mathcal{Y}$ denote Heaviside's function. Similarly, we define the response to a right-hand member equal to $\mathcal{Y}(t)\, X(t)$ by

$$Z(t) = \int_0^t z_0(t - \theta)\, X(\theta)\, \mathrm{d}\theta\,, \qquad \text{for} \quad t \geqslant 0\,.$$

For $t > 0$, we set

$$h_t(\theta) = \begin{cases} z_0(t - \theta) & \text{for} \quad \theta \in [0, t] \\ 0 & \text{for} \quad \theta \notin [0, t] \end{cases}.$$

We have

$$Z(t) = \int X(\theta)\, h_t(\theta)\, d\theta = \int h_t \,.\, X$$

and, consequently, for $t_1, t_2 \geqslant 0$,

$$\mathbf{v}(Z(t_1), Z(t_2)) = \int \overline{\mathcal{F}h_{t_1}} \,.\, \mathcal{F}h_{t_2} \,.\, \varpi_{X,X} \,.$$

Of course, the process Z is not stationary.

Now we shall give a generalization of the concept of a random process. We say above that every stationary process of the second order assigns to every function $f \in \mathbf{L}^1$ a random variable $\int f \,.\, X \in H$. We now obtain a natural generalization by restricting the domain of definition of this mapping, for example, by assuming simply that it is defined on $\mathfrak{D}$. By virtue of the hypothesis that the continuity condition (CC) is satisfied, this amounts to defining the mapping $f \to \int f \,.\, X$ as a distribution with range in H. The fact that $\int f \,.\, X$ belongs to H is in fact a simplifying hypothesis and we shall confine ourselves to the case in which the mapping $f \to \int f \,.\, X$ is defined on $\mathbf{S}$. Thus, we are led to the following definition:

Definition. A *tempered process-distribution* is defined as any tempered distribution X on $\mathbf{R}$ into the space H of square-integrable random variables, that is, any continuous mapping $X : f \to \int f \,.\, X$ of $\mathbf{S}$ into H.

In what follows, we shall use the term "process-distribution" for "tempered process-distribution."

A process-distribution will be called a "stationary process-distribution of the second order" if

$$\mathbf{E}\left(\int (\delta_a * f) \,.\, X\right) = \mathbf{E}\left(\int f \,.\, X\right) \qquad (\forall a \in \mathbf{R})$$

$$\mathbf{v}\left(\int (\delta_a * g) \,.\, X, \int (\delta_a * f) \,.\, X\right) = \mathbf{v}\left(\int g \,.\, X, \int f \,.\, X\right).$$

In what follows, all the process-distributions that we shall consider will be stationary of the second order, and this qualification will be understood.

The stationary processes of the second order studied in the preceding sections are treated as particular cases of process-distributions. To emphasize their special features, we shall call them "process-functions."

Let X denote a process-distribution. The mapping $f \in \mathbf{E}(\int f.X)$ is a continuous linear form on $\mathbf{S}$, that is, a tempered distribution. Furthermore, if X is stationary, this distribution is invariant under translation, which implies that it is proportional to Lebesgue measure. Therefore, we can set

$$\mathbf{E}\left(\int f.X\right) = m_X \int f.$$

The number m_X is called the *mean* of the process-distribution X. If $m_X = 0$, then X is said to be *centered*. In all cases, the process-distribution $X - m_X$ is centered. For a given process-distribution, we show that there exists a tempered positive measure $\varpi_{X,X}$ and a tempered distribution $\varphi_{X,X}$ such that

$$\begin{cases} \varphi_{X,X} = \mathcal{F}\varpi_{X,X} \\ \mathbf{v}\left(\int g\,.\,X, \int f\,.\,X\right) = \int \overline{\mathcal{F}g}\,.\,\mathcal{F}f\,.\,\varpi_{X,X} \\ \qquad = \int (\tilde{g} * f)\,.\,\varphi_{X,X}\,. \end{cases}$$

These formulas generalize those given at the beginning of this section for process-functions. The measure $\varpi_{X,X}$ is called the *spectral measure* of X, and the distribution $\varphi_{X,X}$ is called the *distribution of autocorrelation* of X. If $\varpi_{X,X}$ has a density $\Phi_{X,X}$ with respect to Lebesgue measure, then $\Phi_{X,X}$ is called the *spectral density* of X.

One can define the convolution product of a process-distribution X and a distribution $T \in \mathbf{O}'_c$ as follows: The process $T * X$ is a process-distribution that assigns to every $f \in \mathbf{S}$ the random variable $\int f\,.\,T * X$ defined by

$$\int f\,.\,(T * X) = \int (\check{T} * f)\,.\,X\,.$$

This definition is actually the general definition of the convolution product of a tempered vector distribution and a distribution belonging to $\mathbf{O}'_c$, and it generalizes (with modification in notation) the definition given in Chapter 3 of the convolution product of a scalar tempered distribution and a distribution in $\mathbf{O}'_c$. Let us verify that $T * X$ is stationary. We have

$$\mathbf{E}\left(\int (\delta_a * f)\,.\,(T * X)\right) = \mathbf{E}\left(\int (\check{T} * \delta_a * f)\,.\,X\right)$$

$$= \mathbf{E}\left(\int (\check{T} * f)\,.\,X\right) = \mathbf{E}\left(\int f\,.\,(T * X)\right).$$

Theorem 14.

$$m_{T*X} = (\mathcal{F}^{\star} T)(0)\,.\,m_X\,.$$

Proof. We have

$$\mathbf{E}\left(\int f \,.\, (T * X)\right) = \mathbf{E}\left(\int (\check{T} * f) \,.\, X\right) = m_X \,.\, \int \check{T} * f$$
$$= m_X \,.\, [\mathcal{F}(\check{T} * f)](0) = m_X \,.\, (\mathcal{F}^\star T)(0) \,.\int f \,.$$

Theorem 15.

$$\varpi_{T*X,T*X} = |\,\mathcal{F}^\star T\,|^2 \,.\, \varpi_{X,X}$$

and

$$\varphi_{T*X,T*X} = \tilde{T} * T * \varphi_{X,X} \,.$$

Proof. For $f, g \in \mathbf{S}$, we have

$$\int \overline{\mathcal{F}g} \,.\, \mathcal{F}f \,.\, \varpi_{T*X,T*X} = \mathbf{v}\left(\int g \,.\, (T * X), \int f \,.\, (T * X)\right)$$
$$= \mathbf{v}\left(\int (\check{T} * g) \,.\, X, \int (\check{T} * f) \,.\, X\right)$$
$$= \int \overline{\mathcal{F}(\check{T} * g)} \,.\, \mathcal{F}(\check{T} * f) \,.\, \varpi_{X,X}$$
$$= \int \overline{\mathcal{F}\check{T}} \,.\, \overline{\mathcal{F}g} \,.\, \mathcal{F}\check{T} \,.\, \mathcal{F}f \,.\, \varpi_{X,X}$$
$$= \int |\,\mathcal{F}^\star T\,|^2 \,.\, \overline{\mathcal{F}g} \,.\, \mathcal{F}f \,.\, \varpi_{X,X} \,.$$

From this, we conclude that

$$\mathcal{F}f \,.\, \varpi_{T*X,T*X} = \mathcal{F}f \,.\, |\,\mathcal{F}^\star T\,|^2 \,.\, \varpi_{X,X} \qquad (\forall f \in \mathbf{S}) \,.$$

Therefore, if we take, for example, $(\mathcal{F}f)(\omega) = e^{-\omega^2}$ and multiply by e^{ω^2}, we get the first formula of the theorem. The second formula is obtained by means of the Fourier transformation.

Remark. We can easily obtain formulas analogous to formulas (3)–(6) of Theorem 9 by introducing "two-dimensional" process-distributions.

Suppose that $\xi \in H$. The mapping $f \to \langle \xi, \int f \,.\, X \rangle$ is a scalar tempered distribution. Let us denote it by X_ξ. By definition, we have

$$\langle \xi, \int f \,.\, X \rangle = \int f \,.\, X_\xi \,.$$

Lemma 2. *For* $T \in \mathbf{O}'_c$, *we have*

$$(T * X)_\xi = T * X_\xi .$$

Proof. Suppose that $f \in \mathbf{S}$. We have

$$\int f \,.\, (T * X)_\xi = \langle \xi, \int f \,.\, (T * X) \rangle = \langle \xi, \int (\check{T} * f) \,.\, X \rangle$$

$$= \int (\check{T} * f) \,.\, X_\xi = \int f \,.\, (T * X_\xi),$$

which proves the lemma.

For an application of this lemma, let us study the convolution product of a process-distribution and a function $h \in \mathbf{S}$.

Theorem 16. *If* $h \in \mathbf{S}$, *then* $h * X$ *is a process-function, and we have*

$$(h * X)(t) = \int h(t - \theta) \,.\, X^\theta .$$

Proof. For every $\xi \in H$ and $f \in \mathbf{S}$, we have

$$\langle \xi, \int f(t) \left[\int h(t - \theta) \,.\, X^\theta\right] dt \rangle = \int f(t) \langle \xi, \int h(t - \theta) \,.\, X^\theta \rangle \, dt =$$

$$= \int f(t) \left[\int h(t - \theta) \,.\, X^\theta_\xi\right] dt = \int f(t) \, (h * X_\xi)(t) \, dt$$

$$= \int f(t) \, (h * X)_\xi (t) \, dt = \langle \xi, \int f \,.\, h * X \rangle .$$

Therefore,

$$\int f(t) \left[\int h(t - \theta) \, X_\theta\right] dt = \int f \,.\, (h * X),$$

which completes the proof of the theorem.

Theorem 16 shows that the convolution of a fixed element of $\mathbf{S}$ and any process-distribution maps the latter into a process-function. For a given process-distribution, there will be other elements with this property. One can show that a necessary and sufficient condition for $T * X$ to be a process-function is that

$$\int | \mathcal{F}^* T |^2 \,.\, \varpi_{X,X} < \infty .$$

That the condition is necessary is obvious since the left-hand member is equal to $\int \varpi_{T*X, T*X}$. One can show that it is sufficient by using the following theorem, which we shall state without proof:

A necessary and sufficient condition for a process-distribution Y *to be a process function is that*

$$\int \varpi_{Y,Y} < +\infty.$$

The reason for the difficulty that we encountered in section 5 in defining the convolution of a process-function and an element $T \in \mathbf{O}_c'$ is that we were seeking to define $T * X$ as a process-function. This fact also explains the possibility condition that we have found.

We shall apply these concepts to the study of a certain type of random processes known as *Poisson processes.* We shall consider phenomena of the same nature (for example, telephone calls on a line, emission of particles, etc.) that take place in a period of time such that the instants at which these phenomena take place depend on chance. We shall denote by $n(E)$ the number of occurrences of the phenomenon in question in an interval E. This number is a random variable. The precise definition of a Poisson process consists of a certain number of hypotheses on the random variables $n(E)$.

In what follows, the intervals E in question are of the form $[a, b[$, where a and b are both finite.

Definition. A *Poisson process* is a mapping $E \to n(E)$ that assigns to every interval E a random variable $n(E)$ (on an unspecified probabilized space) such that

(1) The probability distribution of $n(E)$ is a Poisson distribution with parameter lc, where l is the length of the interval E and c is a positive constant. We then have

$$\text{prob}\{\, n(E) = k \,\} = e^{-cl}\,\frac{(cl)^k}{k\,!}.$$

(2) If E_1 and E_2 are two disjoint intervals, then $n(E_1)$ and $n(E_2)$ are independent random variables.

The constant c is called the density of the Poisson process. It follows from the definitions that

$$\mathbf{E}(n(E)) = cl, \qquad \sigma^2(n(E)) = cl.$$

We shall now use a Poisson process to construct a process-distribution.

Let us denote by F the vector space of functions that assume only finitely many values, each nonzero value being taken over the union of finitely many intervals. Suppose that $f, g \in F$. There exist finitely many intervals E_i such that (see Fig. 1)

$$f(x) = \begin{cases} \xi_i & \text{for} \quad x \in E_i\,, \\ 0 & \text{for} \quad x \notin \bigcup E_i\,. \end{cases} \qquad g(x) = \begin{cases} \eta_i & \text{for} \quad x \in E_i\,, \\ 0 & \text{for} \quad x \notin \bigcup E_i\,. \end{cases}$$

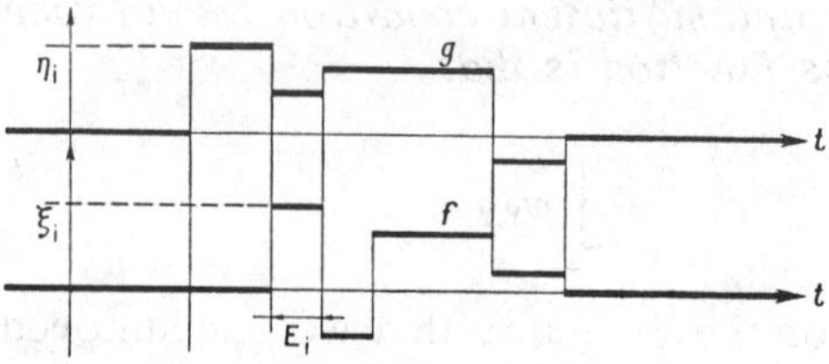

FIG. 1.

Let l_i denote the length of the interval E_i. To f and g, let us assign the random variables

$$X_f = \sum_i \xi_i \, n(E_i) , \qquad X_g = \sum_i \eta_i \, n(E_i) .$$

Then,

$$\mathbf{E}(X_f) = \sum_i \xi_i \, \mathbf{E}(n(E_i)) = \sum_i \xi_i \, cl_i = c \sum_i \xi_i \, l_i = c \int f$$

and, similarly,

$$\mathbf{E}(X_g) = c \int g .$$

Furthermore,

$$\begin{aligned} \mathbf{v}(X_g, X_f) &= \sum_{i,j} \bar{\eta}_j \, \xi_i \, \mathbf{v}(n(E_i), n(E_j)) \\ &= \sum_i \bar{\eta}_i \, \xi_i \, \boldsymbol{\sigma}^2(n(E_i)) \\ &= \sum_i \bar{\eta}_i \, \xi_i \cdot cl_i = c \int \bar{g} f . \end{aligned}$$

In particular, $\boldsymbol{\sigma}^2(X_f) = c \int |f|^2$. Thus, the norm of X_f in H is given by the formula

$$\| X_f \|^2 = c^2 \left| \int f \right|^2 + c \int |f|^2 .$$

Now, let us consider the Banach space $\mathbf{L}^1 \cap \mathbf{L}^2$ equipped, for example, with the norm

$$f \to \| f \|_{1,2} = \| f \|_{\mathbf{L}^1} + \| f \|_{\mathbf{L}^2} .$$

We have $F \subset \mathbf{L}^1 \cap \mathbf{L}^2$, and the mapping $f \to X_f$ is a continuous linear mapping of F into H, where F is equipped with the topology induced by $\mathbf{L}^1 \cap \mathbf{L}^2$. Consequently, it can be extended as a continuous mapping X of $\mathbf{L}^1 \cap \mathbf{L}^2$ into H. This mapping X is, in particular, defined and continuous on **S**. Thus, it is a process-distribution. Therefore, let us set $X_f = \int f \,.\, X$ for $f \in F$. The formulas

$$\mathbf{E}\left(\int f \,.\, X\right) = c \int f$$

$$\mathbf{V}\left(\int g \,.\, X, \int f \,.\, X\right) = c \int \bar{g} f \,,$$

which are valid for $f, g \in F$, remain valid when these are extended as continuous mappings for $f, g \in \mathbf{L}^1 \cap \mathbf{L}^2$. Let us verify that the process-distribution thus defined is stationary. We have

$$\mathbf{E}\left(\int (\delta_a * f) \,.\, X\right) = c \int \delta_a * f = c \int f = \mathbf{E}\left(\int f \,.\, X\right),$$

$$\mathbf{V}\left(\int (\delta_a * g) \,.\, X, \int (\delta_a * f) \,.\, X\right) = c \int \overline{(\delta_a * g)} \,.\, (\delta_a * f) = c \int \bar{g} f$$

$$= \mathbf{V}\left(\int g \,.\, X, \int f \,.\, X\right).$$

Let us calculate the spectral measure $\varpi_{X,X}$ and the autocorrelation distribution. We have

$$\mathbf{V}\left(\int g \,.\, X, \int f \,.\, X\right) = c \int \bar{g} f = \frac{c}{2\pi} \int \overline{\mathcal{F} g} \,.\, \mathcal{F} f.$$

When we compare this with the general definition, we see that the spectral measure is equal to the product of $c/2\pi$ and the Lebesgue measure; in other words, the spectral measure has constant density $\Phi_{X,X}(\omega) = c/2\pi$. The autocorrelation distribution is therefore equal to $c\delta$. Finally, the mean m_X is equal to c. Thus we have

Theorem 17. *Every Poisson process of density c can be considered as a process-distribution X of constant spectral density $c/2\pi$, of autocorrelation distribution $c\delta$, and of mean c.*

The fact that a Poisson distribution has a constant spectral density is expressed by saying that *the process is a white noise.*

We shall study now the convolution of a Poisson process and a distribution or a function. Of course, one can apply the general formulas for convolution with an element $T \in \mathbf{O}'_c$. We shall simply prove a classical result concerning convolution with a function.

For $h \in \mathbf{S}$, we have

$$(h * X)(t) = \int h(t - \theta) \,.\, X^{\theta}.$$

The right-hand member remains meaningful for $h \in \mathbf{L}^1 \cap \mathbf{L}^2$, so that the preceding formula enables us to define $h * X$ for $h \in \mathbf{L}^1 \cap \mathbf{L}^2$. If $h = \lim_{n\to\infty} h_n$, where each $h_n \in \mathbf{S}$, we have

$$(h * X)(t) = \lim_{n\to\infty} (h_n * X)(t).$$

Theorem 18 (Campbell). *The convolution product $h * X$ of a Poisson process X and a function $h \in \mathbf{L}^1 \cap \mathbf{L}^2$ is a process-function with spectral density $c\,|\mathcal{F}^* h|^2/2\pi$, with autocorrelation function $c\tilde{h} * h$, and with mean $c\int h$.*

Proof. If $h \in \mathbf{S}$, the conclusions follow from Theorems 14 and 15. In the general case, let us set $h = \lim_{n\to\infty} h_n$ in $\mathbf{L}^1 \cap \mathbf{L}^2$, where each $h_n \in \mathbf{S}$. Suppose that $Y = h * X$ and $Y_n = h_n * X$. We have $\lim_{n\to\infty} Y_n(t) = Y(t)$ (in H) and consequently,

$$\begin{aligned}
\varphi_{Y,Y}(t) = \mathbf{v}(Y(0), Y(t)) &= \lim_{n\to\infty} \mathbf{v}(Y_n(0), Y_n(t)) \\
&= \lim_{n\to\infty} \varphi_{Y_n,Y_n}(t) \\
&= \lim_{n\to\infty} c \int \tilde{h}_n(t-\theta)\, h_n(\theta)\, d\theta \\
&= c \int \tilde{h}(t-\theta)\, h(\theta)\, d\theta = (\tilde{h} * h)(t).
\end{aligned}$$

Thus, $\varphi_{Y,Y} = c\tilde{h} * h$ and, hence,

$$\varpi_{Y,Y} = \frac{c}{2\pi} |\mathcal{F}^* h|^2.$$

Also,

$$m_Y = \langle 1, Y(t) \rangle = \lim_{n\to\infty} \langle 1, Y_n(t) \rangle = \lim_{n\to\infty} c \int h_n = c \int h.$$

Poisson processes have many interesting properties. For information concerning these we refer the reader to specialized treatises. We simply state the following result, expressed in the language of our initial definition.

Theorem 19. *Let $T_1, T_2, \ldots, T_n, \ldots$ denote instants, all $\geqslant 0$, arranged in order of occurrence of phenomena to which they correspond. The random variables $T_1, T_2 - T_1, \ldots, T_n - T_{n-1}, \ldots$ are independent and they have the same probability density, namely,*

$$\theta(u) = \begin{cases} c\,e^{-cu} & \text{for} \quad u \geqslant 0, \\ 0 & \text{for} \quad u < 0. \end{cases}$$

8

Linear Servo Systems

1. DESCRIPTION OF STATIONARY LINEAR SYSTEMS

By "system" we mean any device S, physical or otherwise, that, in the course of time, transforms a magnitude $e(t)$, known as the *input*, into a magnitude $s(t)$, known as the *output*.

The input and the output may be either scalar or vector quantities. We shall say that a system is *linear* if the output s depends linearly on the input. We shall say that it is *stationary* if every translation in time on the input implies the same translation on the output, in other words, if the transformation $e \to s$ commutes with translations in time.

For the moment, let us assume that the input and the output are scalars.

A stationary linear system S will be represented by a diagram like that shown in Fig. 1.

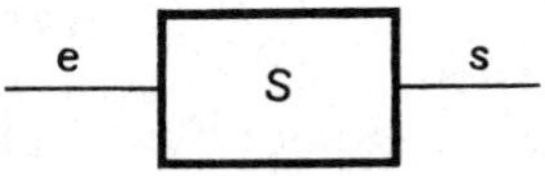

FIG. 1.

With the aid of certain hypotheses regarding admissible inputs (for which the transformation $e \to s$ is defined) and regarding the continuity of that transformation, one can show that every stationary linear system is defined by a convolution $e \to s = f * e$, where f may be a distribution. Thus, we may consider the transformation

$e \to s$ defined for certain "nonfunction" inputs such as measures or distributions. Conversely, every convolution $e \to s = f * e$ can be thought of as defining a stationary linear system.

One can characterize a stationary linear system with the aid of the following concepts:

Impulse response,
Unit step response,
Transfer function,
Harmonic response,
Nyquist locus.

I. Impulse response. The *impulse response* is the element f (a function, measure, or distribution) such that

$$s = f * e .$$

Thus, it is the value of s corresponding to $e = \delta$ (unit impulse). For physical reasons, we always assume that f has positive support and that the system cannot react before the phenomenon represented by the input is produced.

The set of inputs e for which the transformation $e \to f * e = s$ is defined depends, of course, on f.*

II. Unit step response. We designate by *unit step function* the function $\mathcal{Y}$, whose definition, we recall, is

$$\mathcal{Y}(t) = \begin{cases} 1 & \text{for} \quad t \geqslant 0 , \\ 0 & \text{for} \quad t < 0 . \end{cases}$$

The unit step response is $f * \mathcal{Y}$. Its derivative in the distribution sense is f.

In certain cases, $\mathcal{Y}$ is also called the position step. The functions $\mathcal{Y} * \mathcal{Y}$ and $\mathcal{Y} * \mathcal{Y} * \mathcal{Y}$ are often called, respectively, the *velocity step* and the *acceleration step*. We have

$$(\mathcal{Y} * \mathcal{Y})(t) = \begin{cases} t & \text{for } t \geqslant 0 , \\ 0 & \text{for } t < 0 ; \end{cases}$$

$$(\mathcal{Y} * \mathcal{Y} * \mathcal{Y})(t) = \begin{cases} t^2/2 & \text{for } t \geqslant 0 , \\ 0 & \text{for } t < 0 . \end{cases}$$

III. The transfer function. If f has a Laplace transform $F(p)$, this transform is called the *transfer function*.

If e has positive support and has a Laplace transform $E(p)$, then s has positive support and has a Laplace transform $S(p)$ given by the following formula:

$$S(p) = F(p)\, E(p) .$$

* We recall that if f "is a function," this function is defined modulo a function that vanishes almost everywhere.

Example. Suppose that the relationship between e and s is given by a differential equation with constant coefficients:

$$(\alpha) \qquad a_0 D^n s + a_1 D^{n-1} s + \cdots + a_{n-1} Ds + a_n s = e$$

or, in more concise form,

$$A(D)\, s = e\,,$$

where

$$A(p) = a_0 p^n + a_1 p^{n-1} + \cdots + a_{n-1} p + a_n\,.$$

If we set $e = \delta$, there exists only one solution, which we denote by f, with positive support. The Laplace transform $F(p)$ of f is given by the formula

$$A(p)\, F(p) = 1\,,$$

that is,

$$F(p) = \frac{1}{A(p)}\,.$$

This solution f is simply the elementary solution of equation (α) in $\mathfrak{D}'_+$.

If e has positive support and has a Laplace transform $E(p)$, equation (α) has a unique solution s with positive support. The Laplace transform of s is

$$S(p) = \frac{E(p)}{A(p)}\,.$$

If e has arbitrary support, let us agree to choose from among the solutions of equation (α) (which are defined up to an arbitrary solution of the corresponding homogeneous equation) the solution defined by

$$s = f * e\,,$$

at least if the right-hand member is defined. Under these conditions, equation (α) defines a stationary linear system whose impulse response is f and whose transfer function is

$$F(p) = \frac{1}{A(p)}\,.$$

More generally, if the relationship between e and s is given by the equation

$$A(D)\, s = B(D)\, e\,,$$

where A and B are polynomials, the impulse response f will be the solution in $\mathfrak{D}'_+$ of the equation

$$A(D)f = B(D)\,\delta$$

and the transfer function will be

$$F(p) = \frac{B(p)}{A(p)}\,.$$

In general (for physical reasons), we have deg $B \leqslant$ deg A.

IV. Harmonic response. Let us suppose that the impulse response is a bounded measure (or a function $f \in \mathbf{L}^1$). Then, we can take as input

$$e(t) = \mathrm{e}^{\mathrm{i}\omega t}.$$

For output, we obtain

$$\begin{aligned} s(t) &= (\mu * e)(t) \\ &= \int \mathrm{e}^{\mathrm{i}\omega(t-\theta)}\,.\,\mu^\theta \\ &= \mathrm{e}^{\mathrm{i}\omega t}\int \mathrm{e}^{-\mathrm{i}\omega\theta}\,.\,\mu^\theta \\ &= \mathrm{e}^{\mathrm{i}\omega t}(\mathcal{F}^*\mu)(\omega)\,. \end{aligned}$$

Let $M(p)$ denote the Laplace transform of μ, that is, the transfer function. We have $(\mathcal{F}^*\mu)(\omega) = M(\mathrm{i}\,\omega)$, from which we get the result

$$e(t) = \mathrm{e}^{\mathrm{i}\omega t} \quad \Rightarrow \quad s(t) = M(\mathrm{i}\omega)\,\mathrm{e}^{\mathrm{i}\omega t}\,.$$

The function $M(\mathrm{i}\,\omega)$ is called the *harmonic response* of the system. If μ is real, we have $M(\mathrm{i}\,\omega) = \overline{M(-\mathrm{i}\,\omega)}$. If

$$\begin{aligned} e(t) &= e_0 \cos(\omega t + \varphi) \\ &= \frac{e_0}{2}\left[\mathrm{e}^{\mathrm{i}\varphi}\,\mathrm{e}^{\mathrm{i}\omega t} + \mathrm{e}^{-\mathrm{i}\varphi}\,\mathrm{e}^{-\mathrm{i}\omega t}\right], \end{aligned}$$

we obtain

$$\begin{aligned} s(t) &= \frac{e_0}{2}\left[\mathrm{e}^{\mathrm{i}\varphi}\,M(\mathrm{i}\,\omega)\,\mathrm{e}^{\mathrm{i}\omega t} + \mathrm{e}^{-\mathrm{i}\varphi}M(-\mathrm{i}\,\omega)\,\mathrm{e}^{-\mathrm{i}\omega t}\right] \\ &= s_0 \cos(\omega t + \psi)\,, \end{aligned}$$

where

$$s_0 = e_0\,|\,M(\mathrm{i}\,\omega)\,|\,, \qquad \psi = \varphi + \operatorname{Arg} M(\mathrm{i}\,\omega)\,.$$

Thus, for a sinusoidal input, we have a sinusoidal output with amplitude equal to $|\,M(\mathrm{i}\,\omega)\,|$ and phase equal to Arg $M(\mathrm{i}\,\omega)$.

The definition of a harmonic response is *a priori* impossible for an impulse response.

However, let us consider a system defined by a differential equation

$$A(D)\, s = e\, .$$

For $t > 0$, the impulse response is a particular solution of the equation

$$A(D)\, s = 0\, .$$

If all the roots of $A(p) = 0$ have negative real parts, the impulse response is of rapid decrease as $t \to +\infty$ and hence is integrable. The preceding results are applicable, and we have

$$e(t) = \mathrm{e}^{\mathrm{i}\omega t} \quad \Rightarrow \quad s(t) = \frac{1}{A(\mathrm{i}\,\omega)}\,\mathrm{e}^{\,\mathrm{i}\omega t}\, .$$

In the case in which certain roots of $A(p) = 0$ have nonnegative real parts, the convolution product $e * f$ is no longer defined if $e(t) = \mathrm{e}^{\mathrm{i}\omega t}$. But there always exists among the solutions of $A(D)\ s = e$ a particular solution equal to

$$\frac{1}{A(\mathrm{i}\,\omega)}\,\mathrm{e}^{\,\mathrm{i}\omega t}$$

(except when $\mathrm{i}\,\omega$ is one of the roots of the polynomial A). For this reason, the quantity $1/A(\mathrm{i}\,\omega)$ will be called the harmonic response of the system in all cases. We note that, if $F(p)$ is the Laplace transform of the impulse response f, then

$$\frac{1}{A(\mathrm{i}\,\omega)} = F(\mathrm{i}\,\omega)\, .$$

More generally, if the relationship between the input and output is given by the formula

$$A(D)\, s = B(D)\, e\, ,$$

where A and B are polynomials, the harmonic response is, by definition, the function

$$\frac{B(\mathrm{i}\,\omega)}{A(\mathrm{i}\,\omega)}\, .$$

Let f denote the impulse response and let $F = \mathcal{L}f$ denote the transfer function. If the zeros of A all have negative real parts, then

$$e(t) = \mathrm{e}^{\mathrm{i}\omega t} \quad \Rightarrow \quad s(t) = (e * f)\,(t) = \frac{B(\mathrm{i}\,\omega)}{A(\mathrm{i}\,\omega)}\,\mathrm{e}^{\mathrm{i}\omega t} = F(\mathrm{i}\,\omega)\,\mathrm{e}^{\mathrm{i}\omega t}\, .$$

In the general case, we can say only that the solutions of the equation $A(D)\, s = B(D)\, e$ where $e(t) = \mathrm{e}^{i\omega t}$, include the particular solution

$$s(t) = \frac{B(\mathrm{i}\,\omega)}{A(\mathrm{i}\,\omega)} \mathrm{e}^{i\omega t} = F(\mathrm{i}\,\omega)\, \mathrm{e}^{i\omega t}.$$

Finally, if the transfer function F is a meromorphic function in a half-plane of the form $\mathcal{R}(p) > \alpha$, where $\alpha < 0$, we can again define the harmonic response in terms of the function $F(\mathrm{i}\,\omega)$.

Here, we should point out that taking an input $e(t) = \mathrm{e}^{i\omega t}$ is not a physically realizable operation since all the inputs need to have positive support if 0 is the earliest instant at which the system is at the disposal of the experimenter. Therefore, it is useful to study the output corresponding to an input $e(t) = \mathcal{Y}(t)\, \mathrm{e}^{i\omega t}$. We then have the following result:

Theorem 1. *If the impulse response is a bounded measure, then the output $s(t)$ corresponding to the input $e(t) = \mathcal{Y}(t)\, \mathrm{e}^{i\omega t}$ is of the form*

$$s(t) = \mathcal{Y}(t)\,.\, M(\mathrm{i}\,\omega)\, \mathrm{e}^{i\omega t} + \varepsilon(t),$$

where $M = \mathcal{L}\mu$ and $\lim_{t \to +\infty} \varepsilon(t) = 0$.

Proof. We set

$$s_1(t) = \mathrm{e}^{i\omega t}, \qquad s_2(t) = \mathcal{Y}(t)\, \mathrm{e}^{i\omega t}.$$

Then, $s_1 - s_2$ has negative support. The theorem is immediately derived from the following lemma.

Lemma 1. *If μ has bounded measure and if h is a bounded function with negative support, then*

$$\lim_{t \to +\infty} (\mu * h)(t) = 0.$$

Proof. We have

$$(\mu * h)(t) = \int h(t - \theta)\,.\, \mu^{\theta} = \int_{[t, +\infty} h(t - \theta)\,.\, \mu^{\theta},$$

so that

$$|(\mu * h)(t)| \leqslant \sup_t |h(t)|\,.\, \mu([t, +\infty\,|).$$

Now,

$$\lim_{t \to +\infty} \mu([t, +\infty\,|) = 0,$$

from which the conclusion of the lemma follows.

Thanks to the preceding theorem, the quantity $M(\mathrm{i}\,\omega)$ can easily be measured by application of an input $e(t) = \mathcal{Y}(t)\,e^{\mathrm{i}\omega t}$ if μ is a bounded measure. In particular, if $A(D)\,s = B(D)\,e$, then $\varepsilon(t)$ (in the notation of Theorem 1) is, for $t > 0$, a particular solution of the equation $A(D)\,Y = 0$. If the real parts of all the zeros of A are negative, the theorem is applicable. This is not the case if A has zeros with nonnegative real parts.

V. The Nyquist locus. Consider a system whose transfer function $F(p)$ is a meromorphic function for $\mathcal{R}(p) > \alpha$, where $\alpha < 0$. The locus of the point $F(\mathrm{i}\,\omega)$ in the complex plane as ω varies from $-\infty$ to $+\infty$ is called the *Nyquist locus.* Thus, the Nyquist locus is a graphical representation of the harmonic response. For real systems, we have $F(-\mathrm{i}\,\omega) = \overline{F(\mathrm{i}\,\omega)}$. The portions of the Nyquist locus corresponding to $\omega \geqslant 0$ and $\omega \leqslant 0$ are symmetric about the real axis. In the examples that follow, the part corresponding to $\omega \geqslant 0$ is shown by a solid curve and that corresponding to $\omega \leqslant 0$ by a dashed curved. For a Nyquist locus to be useful, one should indicate for a sufficient number of points the values of ω to which they correspond.

Examples. (1) $F(p) = \dfrac{1}{1 + Tp}$.

This is the transfer function of a system for which

$$(Tp + 1)\,s(p) = E(p)\,,$$

that is,

$$TDs + s = e\,.$$

We have

$$F(\mathrm{i}\,\omega) = \frac{1}{1 + T\mathrm{i}\,\omega}.$$

The Nyquist locus is the circle on which 0 and +1 are diametrically opposite points (see Fig. 2).

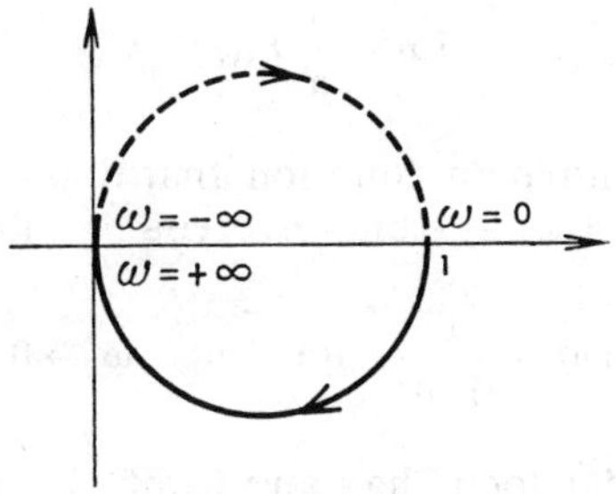

FIG. 2.

(2) $F(p) = \frac{1}{p}\frac{1}{1+Tp}$.

This is the transfer function of a system for which

$$TD^2 s + Ds = e .$$

We have

$$F(\mathrm{i}\,\omega) = \frac{1}{\mathrm{i}\,\omega(1+T\mathrm{i}\,\omega)} .$$

As $\omega \to 0$, we have an infinite branch asymptotic to a vertical line. If we perform a finite expansion of this quotient, we obtain

$$F(\mathrm{i}\,\omega) = \frac{1}{\mathrm{i}\,\omega} - T + \varepsilon(\omega), \qquad \text{where } \varepsilon(\omega) \to 0 \text{ as } \omega \to 0 .$$

The asymptote of the Nyquist locus is the vertical line of abscissa $-T$. Since $F(\mathrm{i}\,\omega) = -1/\omega^2$ as $\omega \to \infty$, the origin is a cusp (see Fig. 3).

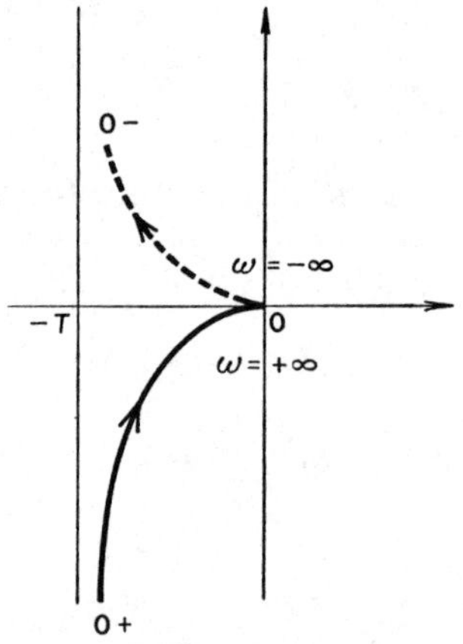

FIG. 3.

(3) Let us examine the influence of a pole of the transfer function at $p = 0$. Suppose that

$$F(p) = \frac{1}{p^k} F_1(p) ,$$

where F_1 is a holomorphic function that does not vanish at $p = 0$. Then, we say that the system involves k integrations. We have

$$F(\mathrm{i}\omega) \sim \frac{1}{(\mathrm{i}\,\omega)^k} F_1(0) \quad \text{as} \quad \omega \to 0 .$$

Therefore, the Nyquist locus has an infinite branch. If $F_1(0) > 0$, the aymptotic direction is $-k\pi/2$ as $\omega \to +0$ and $k\pi/2$ as $\omega \to -0$.

(4) $F(p) = \dfrac{1}{(1 + T_1 p)(1 + T_2 p)}$.

Here, we have a second-order system involving no integration. We have

$$F(i\,\omega) = \frac{1}{(1 + T_1\, i\,\omega)(1 + T_2\, i\,\omega)}.$$

The Nyquist locus has no infinte branch. As $\omega \to \infty$, we have

$$F(i\,\omega) \sim \frac{-1}{T_1 T_2 \omega^2},$$

so that there is a cusp at the origin (see Fig. 4).

(5) $F(p) = e^{-\tau p}\ (\tau > 0)$.

In this case, $s = \delta_\tau * e$. Such a system is called a *pure lag system*. We have

$$F(i\,\omega) = e^{-i\tau\omega}.$$

The Nyquist locus is a circle of radius 1 with center at the origin (see Fig. 5).

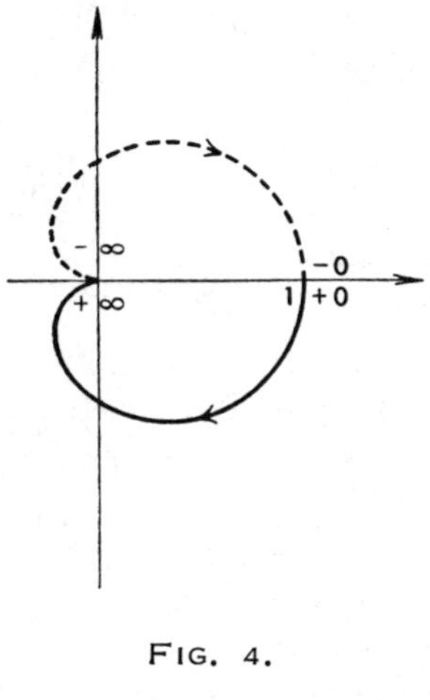

FIG. 4.

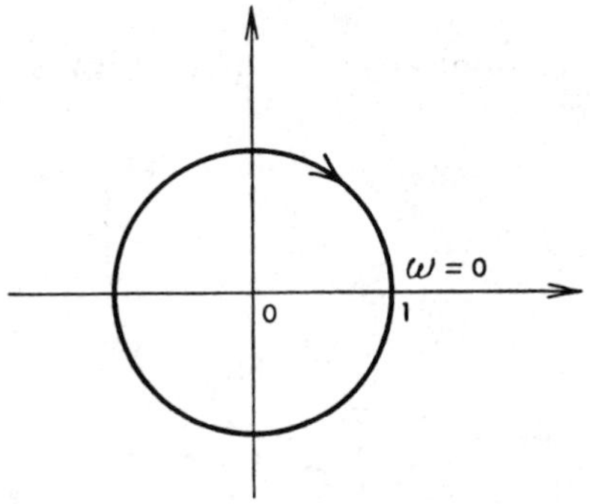

FIG. 5.

In conclusion, let us say a few words about stationary linear systems with several inputs and outputs (that is, with vector-valued input and output).

Let $e_i(t)$ (for $i = 1, \ldots, p$) denote the inputs and let $s_j(t)$ (for $j = 1, \ldots, q$) denote the outputs. Suppose that there exist elements $f_{i,j}$ (functions, measures, or distributions) with positive support such that

$$s_j = \sum_i f_{i,j} * e_i .$$

The elements $f_{i,j}$ are called *impulse responses.* If they have Laplace transforms $F_{i,j}$, these transforms are called the *transfer functions.* If the inputs e_i have positive support and if they have Laplace transforms E_i, then the outputs s_j have positive support and they have Laplace transforms S_j. Then,

$$S_j(p) = \sum_i F_{i,j}(p)\, E_i(p) .$$

One can extend the definition of harmonic responses to such vector-valued inputs and outputs.

2. CONTROL SYSTEMS

Let $c(t)$ denote the input and $s(t)$ the output of the system S. Suppose that we are given a quantity $e(t)$.

Upstream from the system S, let us place a system $\sum$ whose output is $c(t)$. We wish the system $\sum$ to calculate or to determine $c(t)$ so that $s(t)$ is as close as possible to $e(t)$ (in a sense that we shall make precise) (see Fig. 6). Two cases arise according to the nature of the inputs of $\sum$:

(1) $\sum$ has only one entry e. The entire system is then called an *open-loop system* (see Fig. 7).

(2) $\sum$ has inputs e and s. The entire system is then called a *closed-loop system* (see Fig. 8). We also say that it constitutes a *servo system.*

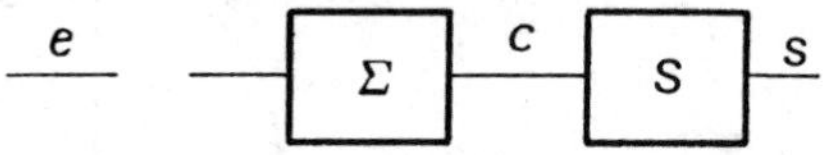

FIG. 6.

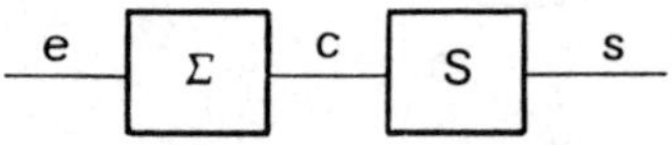

FIG. 7.

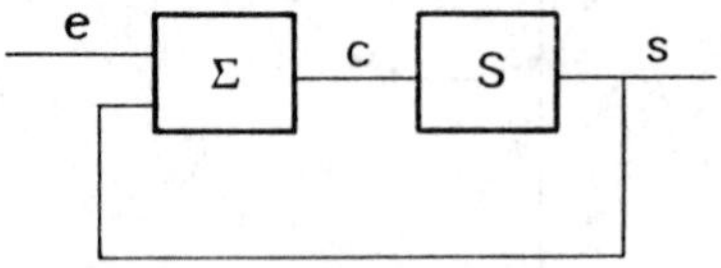

FIG. 8.

In both cases, the system $\sum$ is called a *control system.* The quantity $c(t)$ is called a control.

An important case of a closed-loop system is that of an error correction system. This is set up as follows: Two systems A and B have inputs e and s, respectively. Their outputs are compared with each other by a so-called *differential system*, and the difference is fed into a third system C the output of which is the control $c(t)$ (see Fig. 9 and 9A). Figure 9A is a functional diagram of the entire system as it is usually represented. The system A is called an *anticipation system*, the system C is called a *compensation system*, and the system B is called a *feedback system*. In the particular case in which B performs the identity transformation, that is, when the impulse response is δ and the transfer function is 1, the system is called a *unit-feedback system*.

We note that the anticipation system is in series with the rest of the system. Hence we can study the systems without anticipation components since systems with anticipation components can be derived easily from them.

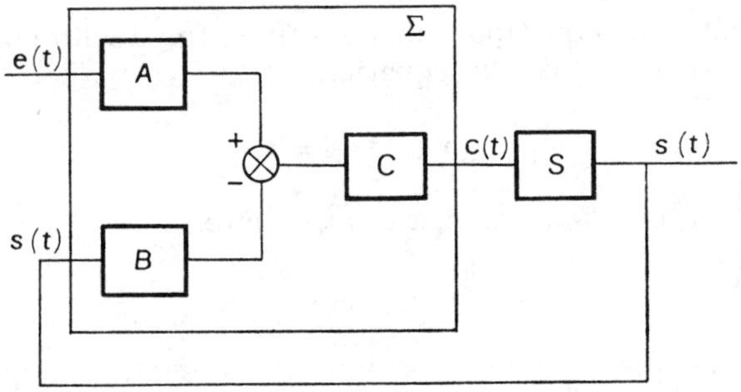

FIG. 9.

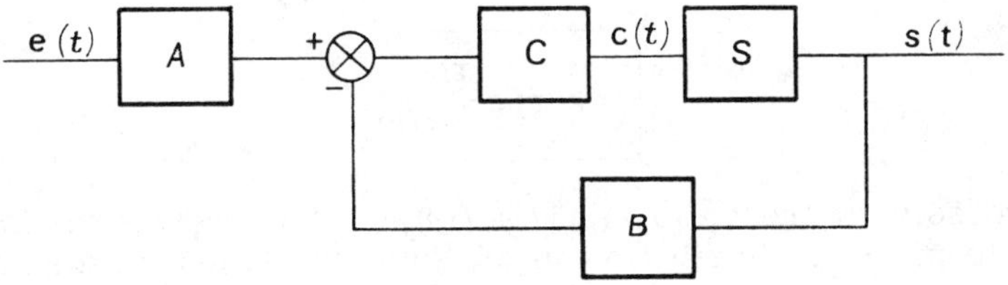

FIG. 9A.

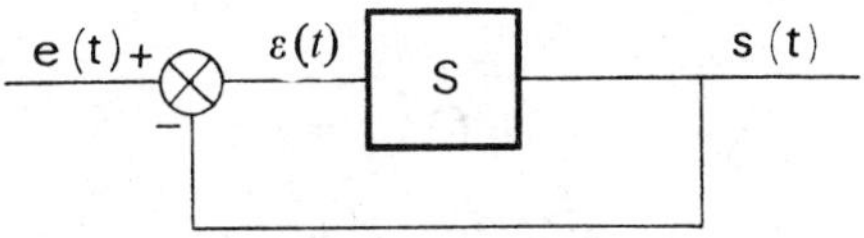

FIG. 10.

The simplest system that we shall need to study consists of a closed-loop system without anticipation system and with unit feedback. If we consider the compensation system and the control system as a single system, we obtain the functional diagram shown in Fig. 10, in which $\varepsilon(t)$ represents the difference $e(t) - s(t)$ between the input and the output.

It should be noted that although the above considerations seem natural from a physical point of view, it is not obvious that they lead to the construction of a stationary linear system. We still need to verify that they lead to a well-defined relation between the input $e(t)$ and the output $s(t)$ and this relation is of the type $s = g * e$.

Let us consider first of all the case of the unit feedback system shown in Fig. 10. Let f denote the impulse response of the system S and let $F = \mathfrak{L}f$ denote its transfer function. The relation between the input and the output is

$$f * (e - s) = s \;;$$

that is,

$$(\delta + f) * s = f * e .$$

Suppose that this equation can be solved for s with solution in $\mathfrak{D}'_+$ when $e = \delta$, that is, that the equation

$$(\delta + f) * g = f$$

has a solution g in $\mathfrak{D}'_+$. Then, we may write

$$s = g * e ,$$

and g will be the impulse response of the system in question. If f and g have Laplace transforms F and G, they must satisfy the relationship

$$\boxed{G(p) = \frac{F(p)}{1 + F(p)}} \; .$$

Conversely, if the ratio $F(p)/[1 + F(p)]$ on the right is the Laplace transform of a distribution in $\mathfrak{D}'_+$, this transform is the impulse response of the system.

In general, it can be shown* that, if f is a locally integrable function, the equation

$$(\delta + f) * g = f$$

has a locally integrable solution g, which is the impulse response of the system.

Now, let us consider a nonunit feedback system. Let h denote the impulse response of the feedback system. The relation between input and output is then written

$$f * (e - h * s) = s ,$$

that is,

$$(\delta + f * h) * s = f * e .$$

---

*[64], Proposition 16, p. 143.

If the equation

$$(\delta + f * h) * g = f$$

has a solution $g \in \mathfrak{D}'_+$, we have

$$s = g * e$$

and g is the impulse response of the system.

If f, h and g have Laplace transforms F, H and G, respectively, then they must satisfy the relation

$$\boxed{G(p) = \frac{F(p)}{1 + F(p)\,H(p)}}\,.$$

Conversely, if the ratio on the right is the Laplace transform of a distribution in $\mathfrak{D}'_+$, this distribution is the impulse response of the system.

More generally, if $f * h$ is a locally summable function, the equation

$$(\delta + f * h) * g = f$$

has a locally summable solution g, which represents the impulse response of the system.

3. STABILITY

Definitions. Suppose that a system has an impulse response f that is a function with positive support.

(1) If $f(t)$ tends to 0 as $t \to +\infty$, we say that the system is *stable*.
(2) If $f(t)$ is bounded, we say that the system is *quasistable*.
(3) If $f(t)$ is unbounded, we say that the system is *unstable*.*

---

*Since f is defined only modulo functions that vanish almost everywhere, these definitions should be made more precise, as follows:

(1) If the set of possible determinations of the impulse response includes a function f such that $f(t)$ tends to 0 as $t \to +\infty$, the system is *stable*.

(2) If the set of possible determinations of the impulse response includes a function f such that $f(t)$ is bounded, the system is *quasistable*.

(3) If condition (2) is not satisfied, the system is *unstable*.

Another wording would be as follows: For every t, let $\alpha(t)$ denote the greatest lower bound of the set of numbers M such that $f(\theta) \leqslant M$ for $\theta \geqslant t$ except on a set of Lebesgue measure 0. The number $\alpha(t)$ is independent of the determination chosen for the impulse response. Then,

(1) If $\lim_{t \to +\infty} \alpha(t) = 0$, the system is *stable*.

(2) If $\alpha(t)$ is bounded, the system is *quasistable*.

(3) If $\alpha(t)$ is unbounded, the system is *unstable*.

An example. Suppose that a system is defined by the differential equation

$$A(D)\,s = B(D)\,e\,,$$

where A and B are polynomials such that $\deg B < \deg A$. For $t > 0$, the impulse response f is a solution of the equation $A(D)f = 0$. We have the following results:

If the real parts of all the roots of the polynomial $A(p)$ are negative, the system is *stable*.

If at least one of the roots of the polynomial $A(p)$ has positive real parts, the system is *unstable*.

If the real parts of all the roots of the polynomial $A(p)$ are nonpositive, the system will be

quasistable if the purely imaginary roots are simple
unstable otherwise.

Let us now consider a closed-loop system with transfer function

$$G(p) = \frac{F(p)}{1 + F(p)}\,.$$

Let us set $p = \xi + i\,\omega$.

The contour $\gamma = \gamma_1 + \gamma_2 + \gamma_3 + \gamma_4$ (see Fig. 11), where the γ's are the curves defined as follows, is called a *Nyquist contour*:

γ_1 is the segment $[i\,\varepsilon,\ i\,R]$ of the imaginary axis,

γ_2 is the semicircle of radius R with center at the origin and lying in the half-plane $\xi \geqslant 0$,

γ_3 is the segment $[-\,i\,R,\ -\,i\,\varepsilon]$ of the imaginary axis,

γ_4 is the semicircle of radius ε with center at the origin and lying in the half-plane $\xi \geqslant 0$.

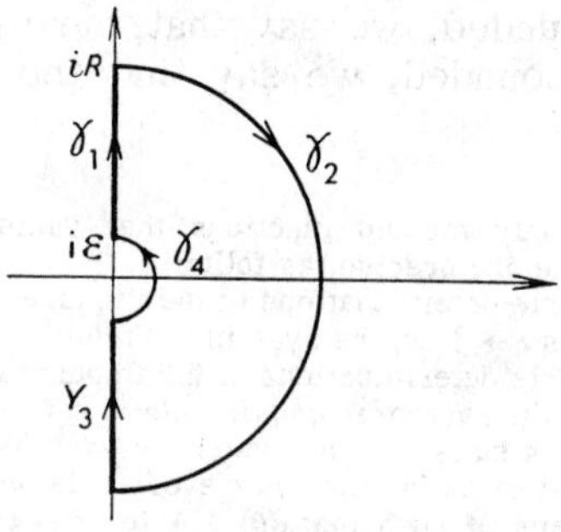

FIG. 11.

The Nyquist contour is assumed to be described in the negative sense (Fig. 11).

Let P denote the number of poles of $F(p)$ in the interior of the Nyquist contour, where a pole of order n is counted n times. (P is also the number of poles $1 + F(p)$.)

Let Z denote the number of zeros of $1 + F(p)$ in the interior of the Nyquist contour, where a zero of order n is counted n times. Then,

$$P - Z = \Delta_\gamma \operatorname{Arg}[1 + F(p)] ,$$

where the right-hand member represents the variation, expressed in the number of times the origin is encircled, in the argument of $1 + F(p)$ as p describes the Nyquist contour.

Let us also assume that $F(p)$ approaches 0 as $p \to \infty$, where $\Re(p) \geqslant 0$. We can then make R approach $+\infty$ in the preceding formula. This gives us the

Rule. *Let P denote the number of poles of $F(p)$ (= the number of poles of $1 + F(p)$) other than 0 in the half-plane $\Re(p) \geqslant 0$ and let Z denote the number of zeros of $1 + F(p)$ in this same half-plane. We have*

$$P - Z = \Delta_\Gamma \operatorname{Arg}[1 + F(p)] ,$$

where the right-hand member is the variation (expressed in the number of times the origin is encircled) in the argument of $1 + F(p)$ as $p = \xi + i\omega$ describes the path Γ (see Fig. 12) in the sense of increasing ω, where Γ consists of the ray $\omega \leqslant \varepsilon$ of the imaginary axis, the semicircle of radius ε with center at the origin and lying in the half-plane $\xi \geqslant 0$, and the ray $\omega \geqslant \varepsilon$ of the imaginary axis.

If 0 is not a pole of $F(p)$, we can replace the above path with the imaginary axis described in the sense of increasing ω.

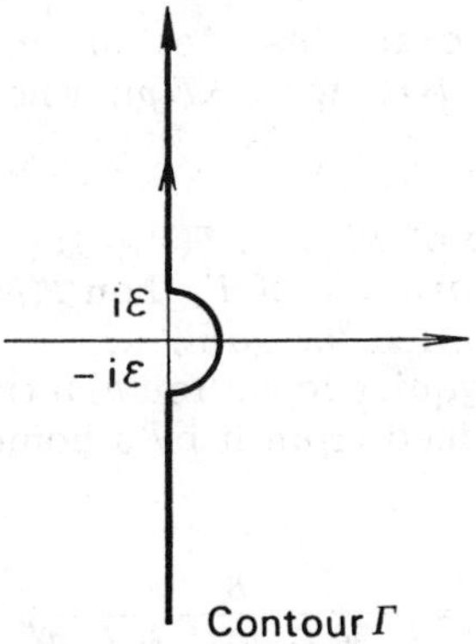

FIG. 12.

Corollary 1. *Let $F(p)$ denote a rational fraction that approaches 0 as $p \to \infty$ and that has no nonzero purely imaginary poles. Then, a necessary and sufficient condition for the system with transfer function*

$$G(p) = \frac{F(p)}{1 + F(p)}$$

to be stable is that

$$\Delta_\Gamma[\mathrm{Arg}\,(1 + F(p))] = P\ ,$$

where P denotes the number of poles $F(p)$ (= the number of poles of $1 + F(p)$) in the half-plane $\Re(p) > 0$ and $\Delta_\Gamma[\mathrm{Arg}\,(1 + F(p))]$ is the variation of the argument of $1 + F(p)$ as p describes the path Γ (or the imaginary axis if 0 is not a pole of F).

We shall refer to the locus of $F(p)$ as p describes a semicircle of radius ε with center at the origin as the connecting arc. We obtain it by taking an approximation of $F(p)$ as $p \to 0$. If 0 is a pole of order k of F, we have

$$F(p) = \frac{1}{p^k} F_1(p)\,,$$

where F_1 is a holomorphic function that does not vanish at 0. Therefore,

$$F(p) \sim \frac{F_1(0)}{p^k}\,.$$

Consequently, $F(p)$ varies approximately on a circle of radius $1/\varepsilon^k$ and encircles the origin approximately $\pm k/2$ times.

The remainder of the locus of $F(p)$ as p describes Γ is a portion of the Nyquist locus.

Before turning to examples, let us make a slight change of notation. We replace $F(p)$ with $KF(p)$, where K is defined by the requirements

(1) If 0 is not a pole of F, then $F(0) = 1$;
(2) if 0 is a pole of order k of F, then $F(p) \sim 1/p^k$ as $p \to 0$.

The constant K is called the *gain.*

Let us trace the Nyquist locus for arbitrary $K = 1$ (the Nyquist locus for K can be obtained from it by a homothetic transformation of ratio K).

Examples. (1) $F(p) = \dfrac{K}{(1 + T_1\ p)(1 + T_2\ p)}$, where $T_1, T_2 > 0$. The Nyquist locus is shown in Fig. 13. We have $P = 0$ and

$$\Delta_\Gamma[\mathrm{Arg}\,(1 + F(p))] = 0$$

(this is the variation in Arg $(1 + F(i\,\omega))$ as ω varies from $-\infty$ to $+\infty$. Therefore, the system is stable for every K.

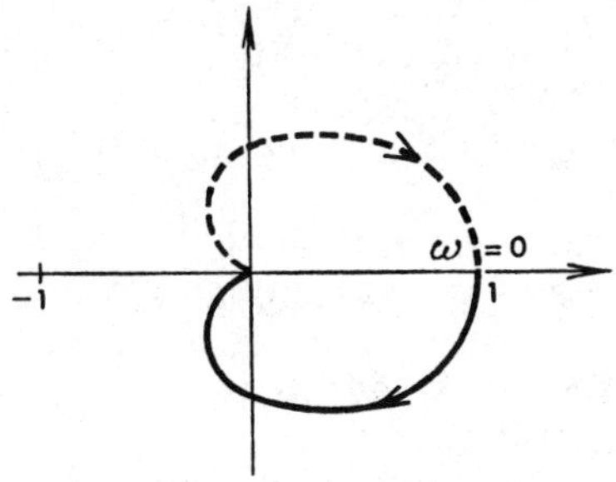

FIG. 13.

(2) $F(p) = \dfrac{K}{p(p^2 + 2zp + 1)}$, for $\begin{cases} z > 0 \\ K > 0. \end{cases}$

We set

$$F_1(p) = \frac{1}{p(p^2 + 2zp + 1)}.$$

The Nyquist locus has been drawn for $K = 1$. The two figures 14 and 14A differ in the position of the point -1 with respect to the intersection of the Nyquist locus and the real axis. The locus has been completed by the locus of $F_1(p)$ as p describes the semicircle of radius r. As $p \to 0$, we have

$$F_1(p) \sim \frac{1}{p}$$

and $F_1(p)$ describes approximately a semicircle of radius $1/r$ as its argument varies from $+\pi/2$ to $-\pi/2$.

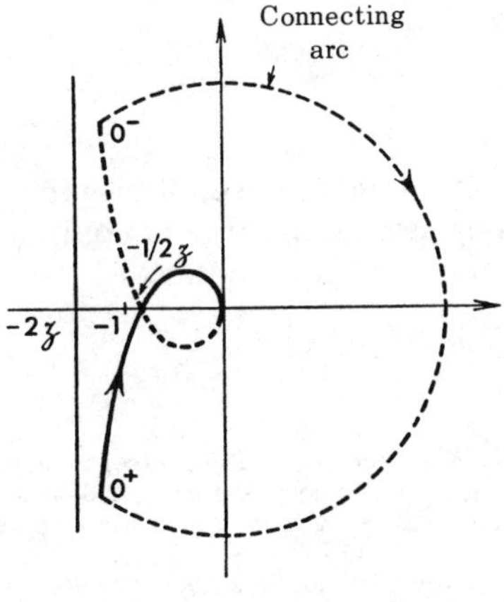

FIG. 14.

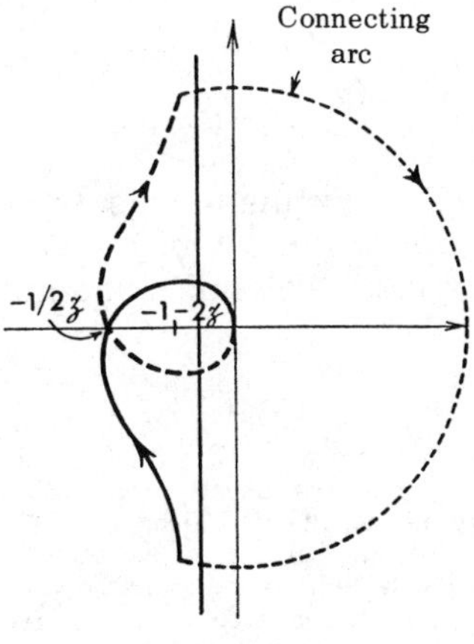

FIG. 14A.

Since, in addition, $P = 0$, we see that the system is stable for $K = 1$ in the case represented by Fig. 14 but unstable in the case represented by Fig. 14A.

In both cases, there exists a value K_0 of the gain such that

$$K < K_0 \longrightarrow \text{the system is stable}$$
$$K > K_0 \longrightarrow \text{the system is unstable.}$$

For a chosen value of K less than K_0, the ratio K_0/K is called the *gain stability margin.**

The reader can verify that the system is unstable for $z < 0$ no matter what the value of K.

$$(3)\ F(p) = \frac{K}{p(1 + Tp)}, \quad (T > 0, K > 0).$$

The Nyquist locus is shown for $K = 1$ in Fig. 15.

This system is stable** for every value of K.

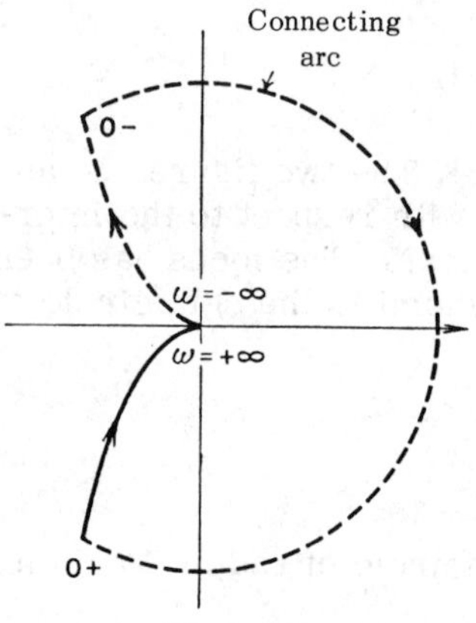

FIG. 15.

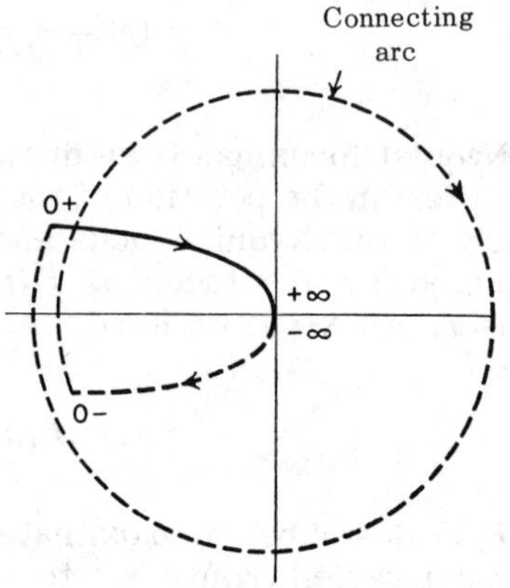

FIG. 16.

$$(4)\ F(p) = \frac{K}{p^2(1 + Tp)}, \quad (T > 0, K > 0).$$

The Nyquist locus is shown for $K = 1$ in Fig. 16. The variation in the argument of $1 + F(p)$ on the connecting arc is approximately $\pm 2\pi$. The system is unstable for every value of K since

$$P = 0, \ \Delta_\Gamma \mathrm{Arg}\,(1 + F(p)) = -2.$$

* Engineers usually measure the ratio K_0/K in decibels. The decibel measure of K_0/K is the quantity $20 \log_{10} (K_0/K)$. If $K_0/K = 1.3$, its decibel measure is 2.3. The gain stability margin enables us to be sure that there will be no loss in stability as a result of an accidental variation in the gain.

** For this reason, engineers feel that real systems cannot be conveniently described by such an equation. It is, in fact, observed that physical systems become unstable when the gain increases.

Now, we shall give a second corollary enabling us to apply the rule to a system whose transfer function is not a rational function.

Corollary 2. *Suppose that $F(p)$ is a meromorphic function in a half-plane of the form $\Re(p) > \alpha$, where $\alpha < 0$. Suppose that $F(p)$ has no nonzero purely imaginary poles. Suppose, finally, that $F(p) = O(1/p^2)$ as $p \to \infty$ for $\Re(p) \geqslant 0$. Then, a necessary and sufficient condition for the system with transfer function*

$$G(p) = \frac{F(p)}{1 + F(p)}$$

to be stable is that

$$\Delta_\Gamma[\operatorname{Arg}(1 + F(p))] = P,$$

where P denotes the number of poles of $F(p)$ (= number of poles of $1 + F(p)$) in the half-plane $\Re(p) > 0$ and $\Delta_\Gamma[\operatorname{Arg}(1 + F(p))]$ denotes the variation of the argument of $1 + F(p)$ expressed in the encirclements of the origin as p describes the path Γ (or the imaginary axis if 0 is not a pole of F).

Proof. If

$$\Delta_\Gamma[\operatorname{Arg}(1 + F(p))] = P,$$

$G(p)$ has no pole in the half-plane $\Re(p) > 0$. Therefore, $G(p)$ is holomorphic in that half-plane. Furthermore, $G(p) = O(1/p^2)$ as $p \to \infty$ with $\Re(p) \geqslant 0$. Consequently (see Chapter 4, Theorem 11), $G(p)$ is the Laplace transform of a function g that approaches 0 as its argument becomes infinite.

On the other hand, if $\Delta_\Gamma[\operatorname{Arg}(1 + F(p))] \neq P$, the function $G(p)$ is not holomorphic in the half-plane $\Re(p) > 0$.

Example.

$$F(p) = \frac{K\,e^{-Tp}}{p(p^2 + zp + 1)}, \qquad (\text{where } z > 0 \text{ and } K > 0).$$

Let us draw the Nyquist locus for $K = 1$. We set

$$F_1(p) = \frac{e^{-Tp}}{p(p^2 + zp + 1)} = H(p)\,e^{-Tp},$$

where

$$H(p) = \frac{1}{p(p^2 + zp + 1)}.$$

We have

$$F_1(i\,\omega) = H(i\,\omega)\,e^{-iT\omega}.$$

The Nyquist locus for $F(p)$ is obtained from the Nyquist locus for a transfer function $H(p)$ by transforming the point corresponding to the parameter ω by a rotation through the angle $-T\omega$. (See Fig. 17. We assume that the system with transfer function $H(p)/[1 + H(p)]$ is stable.)

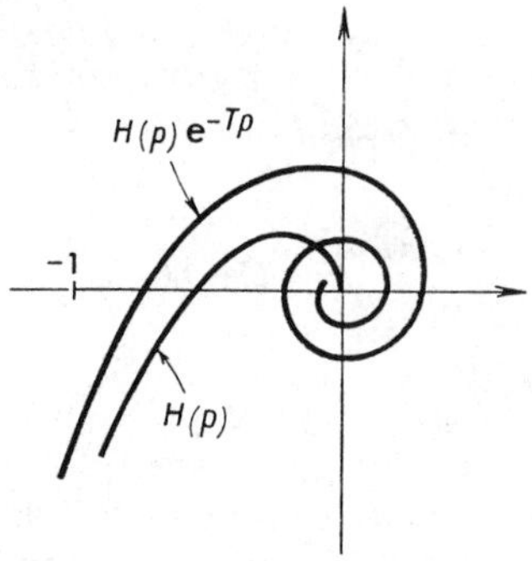

An example of Nyquist loci for $H(p)$ and $H(p)e^{-Tp}$.

FIG. 17.

Since $P = 0$, the stability condition is

$$\Delta_\Gamma[\operatorname{Arg}\ (1 + F(p))] = 0.$$

We see that there exists a value K_T for the gain such that the inequality $K < K_T$ implies that the system is stable and the inequality $K > K_T$ implies that it is unstable. We note that $K_T < K_0$ (for $T > 0$). In particular, there exist values of K (between K_T and K_0) such that the system with transfer function $KH(p)/[1 + H(p)]$ is stable, whereas the system with transfer function

$$\frac{KH(p)\,e^{-Tp}}{1 + H(p)\,e^{-Tp}}$$

is unstable. This fact is expressed by saying that the introduction of a pure lag into the transfer function in closed-loop reduces or destroys the stability of a system.*

4. GAIN CONTROL

Suppose that a servo system with unit feedback has no anticipation system. If $KF(p)$ is the transfer function in an open-loop system, the transfer function in a closed-loop system is

$$G(p) = \frac{KF(p)}{1 + KF(p)}.$$

* The mathematical result finds experimental verification in the oscillations observed in the behavior of a motorist who is intoxicated or half-asleep.

The system would be perfect if $s(t)$ were equal to $e(t)$ no matter what $e(t)$ might be. For this it would be necessary to have $G(p)$ equal to 1. This condition cannot be achieved perfectly, but we can get a good approximation by taking K as great as possible. In practice, we are limited by stability conditions, which, in general, compel us to take K less than a critical value K_0 and even to take K_0/K at least equal to some imposed stability margin. This fact is expressed by stating that every servo system must represent a compromise between stability and precision, this last quality consisting in the fact that the output $s(t)$ is close to the input $e(t)$.

There are various criteria for precision. Of these, we cite the quantity

$$\int_0^\infty \varepsilon^2(t)\,dt\,, \text{ where } \varepsilon(t) = s(t) - e(t)\,,$$

which can be evaluated for a particular input, for example, $e(t) = \mathcal{Y}(t)$.*

We shall study responses with unit position step, velocity step and acceleration step. We set $\varepsilon(t) = e(t) - s(t)$ and $\mathcal{E} = \mathfrak{L}\varepsilon$.

We have

$$\mathcal{E}(p) = E(p) - S(p) = [1 - G(p)]\,E(p) = \frac{1}{1 + KF(p)}\,E(p)\,.$$

Suppose that

$$e_0(t) = \mathcal{Y}(t), \quad e_1(t) = t\mathcal{Y}(t), \quad e_2(t) = \frac{t^2}{2}\,\mathcal{Y}(t),$$

where $s_0(t)$, $s_1(t)$ and $s_2(t)$ are the corresponding outputs.

We set

$$\varepsilon_i = e_i - s_i\,, \qquad \mathfrak{L}e_i = E_i\,, \qquad \mathfrak{L}s_i = S_i\,, \qquad \mathfrak{L}\varepsilon_i = \mathcal{E}_i\,(i = 0, 1, 2)\,.$$

Then,

$$E_0(p) = \frac{1}{p}, \quad E_1(p) = \frac{1}{p^2}, \quad E_2(p) = \frac{1}{p^3},$$

so that

$$\mathcal{E}_0(p) = \frac{1}{1 + KF(p)}\,\frac{1}{p},$$

$$\mathcal{E}_1(p) = \frac{1}{1 + KF(p)}\,\frac{1}{p^2},$$

* The criterion for precision thus obtained is very conventional. It has the advantage that it can be calculated mathematically. However, in practice, it is preferable to use an input $e(t)$ that represents as faithfully as possible the inputs that will need to be fed into the system.

$$\mathcal{E}_2(p) = \frac{1}{1 + KF(p)} \frac{1}{p^3} .$$

If $\varepsilon_i(t)$ has a limit as $t \to +\infty$, it follows from the theorem on initial and final values that

$$\lim_{t \to +\infty} \varepsilon_i(t) = \lim_{p \to 0} p\mathcal{E}_i(p) .$$

Consequently, for ε_0 to approach 0 as $t \to +\infty$, it is necessary that

$$\lim_{p \to 0} \frac{1}{1 + KF(p)} = 0 ,$$

and for this, it is necessary that 0 be a pole of $F(p)$. Similarly, for $\varepsilon_1(t)$ to approach 0 as $t \to +\infty$, it is necessary that

$$\lim_{p \to 0} \frac{1}{1 + KF(p)} \frac{1}{p} = 0 ,$$

for which it is necessary that 0 be a pole of order at least 2 of $F(p)$. For $\varepsilon_2(t)$ to approach 0 as $t \to +\infty$, it is necessary that 0 be a pole of order at least 3 of $F(p)$.

Conversely, let us suppose successively that 0 is a nonpole of F, then that it is a pole of order 1, then a pole of order 2, . . . , and let us examine the limits of $\varepsilon_0(t)$, $\varepsilon_1(t)$ and $\varepsilon_2(t)$ as $t \to +\infty$.

We shall confine ourselves to the case in which F is a rational fraction and the system is stable. In this case, $\mathcal{E}_i(p)$ is a rational fraction, all the nonzero poles of which have negative real parts, and

$$\lim_{t \to +\infty} \varepsilon_i(t) = \lim_{p \to 0} p\mathcal{E}_i(p) .$$

Case (A). The function F does not have a pole at 0 (there is no integration in the transfer function of the open loop). Therefore,

$$\lim_{t \to +\infty} \varepsilon_0(t) = \frac{1}{1 + K} ,$$

$$\lim_{t \to +\infty} \varepsilon_1(t) = \lim_{t \to +\infty} \varepsilon_2(t) = \infty .$$

The first of these equations is expressed by saying that there exists an error in position equal to $1/(1 + K)$ (see Fig. 18).

Case (B). The function F has a simple pole at 0 (the transfer function of the open loop involves an integration). Therefore,

$$\lim_{t \to +\infty} \varepsilon_0(t) = 0 ,$$

$$\lim_{t \to +\infty} \varepsilon_1(t) = \frac{1}{K} ,$$

$$\lim_{t \to +\infty} \varepsilon_2(t) = \infty .$$

There is no error in position. We express the second of the equations above by saying that there is an error in lag equal to $1/K$ (see Fig. 19).

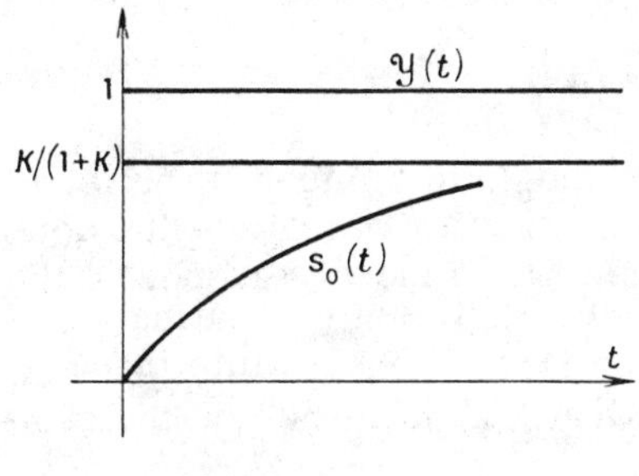

FIG. 18.

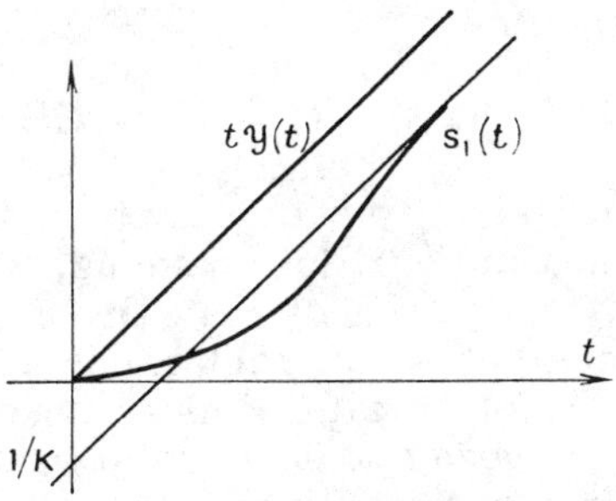

FIG. 19.

Case (C). The function F has a double pole at 0 (the transfer function of the open loop involves two integrations). Then,

$$\lim_{t \to +\infty} \varepsilon_0(t) = \lim_{t \to +\infty} \varepsilon_1(t) = 0\,,$$

$$\lim_{t \to +\infty} \varepsilon_2(t) = \frac{1}{K}\,.$$

Case (D). The function F has a pole or order 3 or more at 0. Then,

$$\lim_{t \to +\infty} \varepsilon_0(t) = \lim_{t \to +\infty} \varepsilon_1(t) = \lim_{t \to +\infty} \varepsilon_2(t) = 0\,.$$

This case is quite rare in practice.

In the case in which there exists at least one integration in the transfer function of the open loop, we can use the quantity

$$\int_0^{+\infty} \varepsilon_0^2(t)\, \mathrm{d}t$$

as stability criterion. We have

$$(\mathcal{F}^\star\ \varepsilon_0)(\omega) = \mathcal{E}(\mathrm{i}\,\omega) = \frac{1}{\mathrm{i}\,\omega + KF(\mathrm{i}\,\omega)}\,,$$

so that, in accordance with Plancherel's theorem,

$$\int_0^{+\infty} \varepsilon_0^2(t)\, \mathrm{d}t = \frac{1}{2\pi} \int_{-\infty}^{+\infty} \frac{\mathrm{d}\omega}{|\mathrm{i}\,\omega + KF(\mathrm{i}\,\omega)|^2}\,.$$

Remark. In the case of a feedback system with transfer function $H(p)$, the results regarding responses to position, velocity, and acceleration steps remain unchanged if $H(0) = 1$, which is the usual case in practice. The formula giving $\int_0^{+\infty} \varepsilon_0^2(t)\,dt$ takes a more complicated form, as the reader can show without difficulty, by use of Plancherel's theorem.

5. COMPENSATION AND ANTICIPATION

For a given control system, the choice of compensating feedback and anticipation systems, which together constitute the control system, is a delicate problem, in which the experience of the engineer plays a considerable role. We shall confine ourselves to pointing out certain transfer functions that are used at the present time for these systems and to indicating their influence on the performance of the system.

We denote by $KF(p)$ the transfer function of the system to be controlled.

The simplest compensating system is obviously a system involving multiplication by a constant K_1. The gain in an open-loop system is then $K_1 K$. For stability reasons, one will in general need to take smaller values of K_1 in order to obtain sufficient precision. In particular, if the system to be controlled includes no integration, the servo system will produce a position error.

Compensation by derived control. We introduce a transfer function compensator

$$F_1(p) = 1 + \tau p\,.$$

Figure 20 show the Nyquist locus of this system. We note that, for $\omega > 0$, we have

$$\operatorname{Arg} F_1(i\,\omega) = \operatorname{Arctan} \tau\omega\,.$$

Consequently, the Nyquist locus of F_1 is obtained from that of F by transforming every point corresponding to positive values of the parameter ω by a rotation through the positive angle Arctan $\tau\omega$. This transformation is such as would stabilize the system. An example is given by Fig. 21.

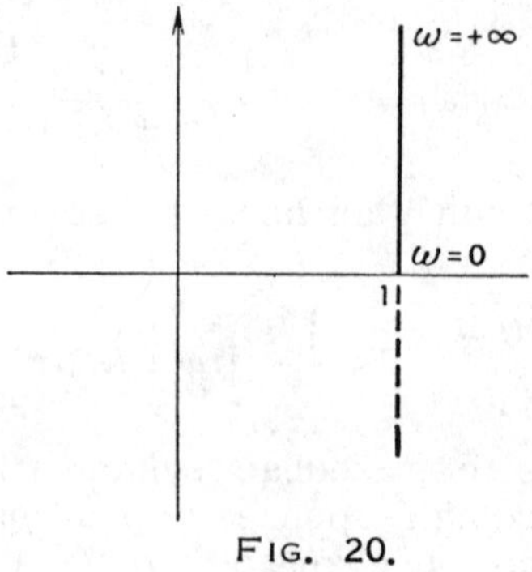

FIG. 20.

In practice, we often replace the above transfer function with a transfer function of the form $F_1(p) = \dfrac{1 + \tau_1 p}{1 + \tau_2 p}$, with $\tau_1 > \tau_2$. The Nyquist locus is shown in Fig. 22.

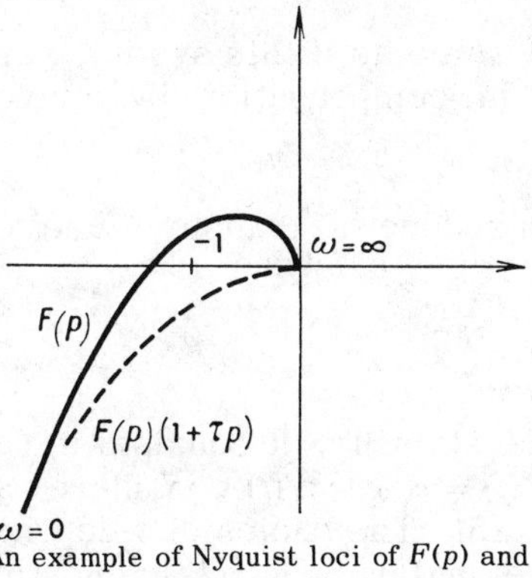

An example of Nyquist loci of $F(p)$ and of $(1 + \tau p)F(p)$.

FIG. 21.

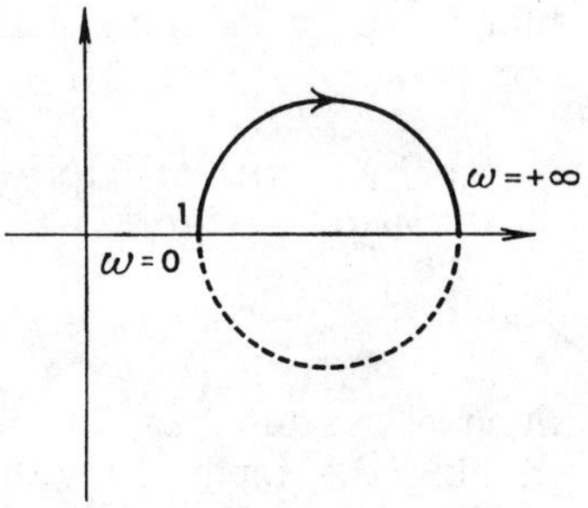

FIG. 22.

This type of compensator, which does not involve any integration, will be used primarily in the case in which the system to be controlled does involve an integration.

Compensation by pure integral control. We introduce a transfer function compensator

$$F_1(p) = 1 + \frac{1}{\tau p}.$$

Figure 23 shows the Nyquist locus of this system.

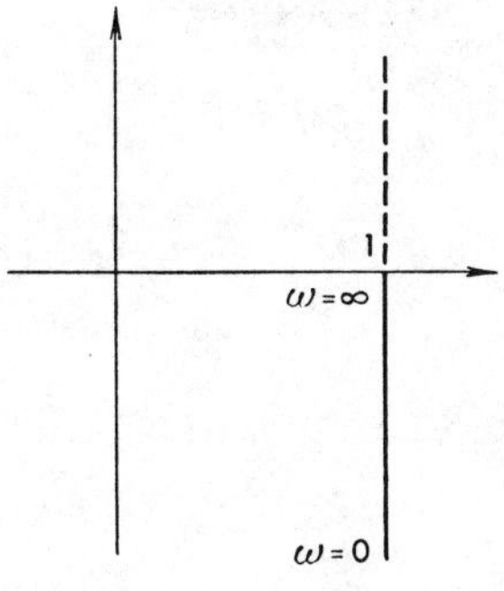

FIG. 23.

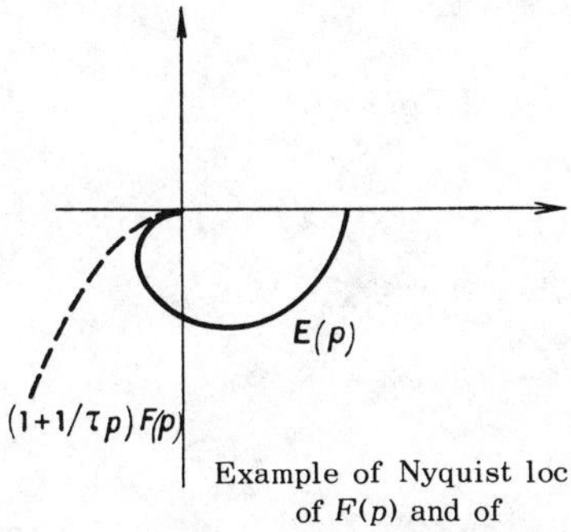

Example of Nyquist loci of $F(p)$ and of $(1 + 1/\tau p)F(p)$.

FIG. 24.

Figure 24 shows an example for comparing the Nyquist locus of F and that of $F_1 F$.

Note that compensation by pure integral control amounts to the introduction of an integration (which increases the precision). This should be compared with compensation by derived control (which avoids loss of stability).

Tachymetric feedback. We introduce a transfer function feedback system

$$F_2(p) = 1 + \tau p .$$

The influence on the stability is the same as if this system were placed in a compensating role as in compensation by derived control.

Tachymetric anticipation. We introduce a transfer function anticipation system

$$F_3(p) = 1 + \tau p .$$

Such a system has no influence on the stability. It consists in replacing the input $e(t)$ with the value $e^*(t) = e(t) + \tau e'(t)$ considered as anticipated value of $e(t + \tau)$ (see Fig. 25). The choice of τ depends on the characteristics of the input. We shall return to this question in Chapter 9.

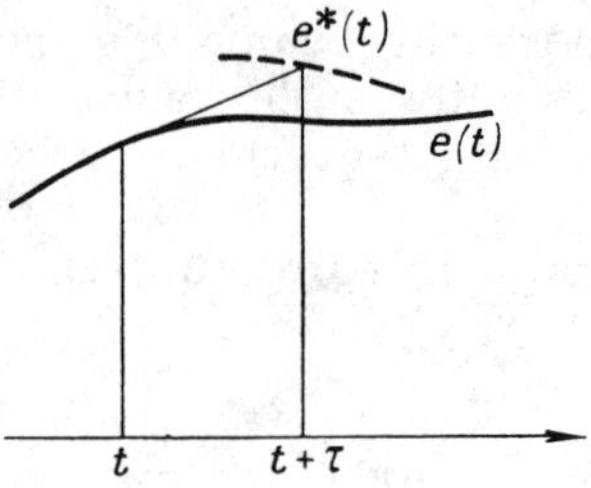

FIG. 25.

9

Filtering, Prediction, Anticipation

1. SYSTEMS WITH RANDOM INPUTS

We shall make the hypotheses that we can take as inputs of a stationary linear system stationary random processes of the second-order and that the relations between the inputs and the outputs

$$s_j = \sum_i f_{i,j} * e_i$$

remain valid in this case. However, it will be convenient to confine ourselves on occasion to inputs such that $f_{i,j} * e_i$ is defined for every i and j. This will be the case with no condition on the entries e_i, if, in particular, the impulse responses $f_{i,j}$ are bounded measures and, in particular, integrable functions.

Suppose that we have a system with impulse response $f \in \mathbf{M}$, with transfer function $F = \mathfrak{L}f$, and with harmonic response

$$F(\mathrm{i}\,\omega) = (\mathcal{F}^\star f)(\omega) .$$

If we consider a second-order stationary random entry $e(t)$, the output $s(t)$ will also be a stationary random process and we shall have (in the notations of Chapter 7)

$$\varphi_{e,s} = \varphi_{e,f*e} = f * \varphi_{e,e} ,$$

$$\varphi_{s,s} = \tilde{f} * f * \varphi_{e,e}$$

$$\varpi_{e,s} = \mathcal{F}^{\star} f \, . \, \varpi_{e,e}$$

$$\varpi_{s,s} = | \mathcal{F}^{\star} f |^2 \, \varpi_{e,e} \, .$$

We can also consider systems of which some of the entries are random and others certain. We can decompose each output s_j into two terms, one representing the response to the certain entries, the other the response to the random entries. This last term (where the e_i are assumed to be stationary and second-order) will be a second-order stationary random process.

Example. The system in Fig. 1 has two inputs $e(t)$ and $b(t)$ and a single output $s(t)$. The input $e(t)$, called the ***principal input***, represents the ideal output. The input $b(t)$ is a random perturbation, called ***noise***.

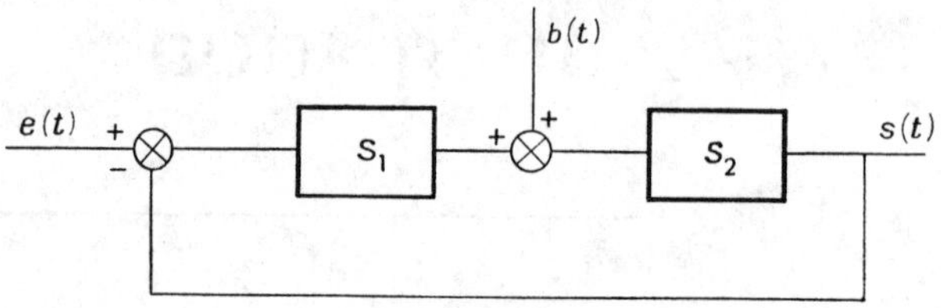

FIG. 1.

Returning for the moment to an input $b(t)$ with positive support, we write the relation between the Laplace transforms E, B, and S of e, b, and s. We denote by KF_1 and F_2 the transfer functions of S_1 and S_2 (where K is the gain of S_1). Let us suppose that these are rational fractions. We then have

$$S(p) = F_2(p) \left[B(p) + KF_1(p) \left[E(p) - S(p) \right] \right] ,$$

from which we conclude

$$S(p) = \frac{KF_1(p) \, F_2(p)}{1 + KF_1(p) \, F_2(p)} E(p) + \frac{F_2(p)}{1 + KF_1(p) \, F_2(p)} B(p) \, .$$

By setting $b = 0$ so that $B = 0$, we again obtain, as one can easily verify, the usual formula for servo systems. We have two transfer functions: first, the principal-input–output transfer function

$$G_{e,s}(p) = \frac{KF_1(p) \, F_2(p)}{1 + KF_1(p) \, F_2(p)}$$

and second, the noise-output transfer function

$$G_{b,s}(p) = \frac{F_2(p)}{1 + KF_1(p) \, F_2(p)} \, .$$

Let us suppose that the system is stable for $b = 0$. Then, the functions

$$\frac{KF_1 F_2}{1 + KF_1 F_2}$$

and

$$\frac{1}{1 + KF_1 F_2}$$

are holomorphic at every point of the half-plane $\Re(p) \geqslant 0$. Let us suppose also that the system S_2 is stable, in which case F_2 is also holomorphic at every point in the half-plane $\Re(p) \geqslant 0$. Then, the function

$$\frac{F_2}{1 + KF_1 F_2}$$

is holomorphic for $\Re(p) \geqslant 0$. Since, in addition,

$$\lim_{p \to \infty} \frac{F_2(p)}{1 + KF_1(p) F_2(p)} = 0\,,$$

we see that the system is stable for $e = 0$. Consequently, it is stable for the two entries e and b.

We note that the preceding conclusion would no longer be valid if S_2 were merely quasistable, in particular, if S_2 involved an integration. Let us now take for b a centered stationary random process of the second order. We denote its spectral power distribution by $\varpi_{b,b}$. We have

$$s(t) = s_1(t) + s_2(t)\,,$$

where $s_1(t)$ is the response to the input e for $b = 0$ and $s_2(t)$ is the response to the input b for $e = 0$. We have

$$\mathfrak{L}s_1 = S_1\,, \quad \text{where} \quad S_1(p) = \frac{KF_1(p) F_2(p)}{1 + KF_1(p) F_2(p)} E(p)\,.$$

Furthermore, s_2 is a centered stationary random process of the second order with spectral power distribution

$$\varpi_{s_2,s_2} = \left| \frac{F_2(i\omega)}{1 + KF_1(i\omega) F_2(i\omega)} \right|^2 \varpi_{b,b}\,.$$

For the system to have as weak as possible a "sensitivity to noise," we should make

$$\int \varpi_{s_2,s_2} = \int \left| \frac{F_2(i\omega)}{1 + KF_1(i\omega) F_2(i\omega)} \right|^2 \cdot \varpi_{b,b}$$

as small as possible. This means that we should take K as large as possible. In general, we shall be limited by the necessity of maintaining the stability of the system and even keeping a certain stability margin.

2. FILTERING AND PREDICTION. THE WIENER-HOPF EQUATION

The problem of filtering consists in finding a stationary linear system with the following properties.

(1) The input is the sum $e = i + b$ of two centered random processes, where the pair $\begin{bmatrix} i \\ b \end{bmatrix}$ is stationary of the second-order;

(2) the output s is as close as possible to i.

More precisely, we shall seek to minimize the power $\int \varpi_{\varepsilon,\varepsilon}$ of the error $\varepsilon = s - i$. We shall call the term b the *noise*.

The *prediction* problem consists in finding a stationary linear system with the following two properties:

(1) The input e is a centered stationary random process of the second order;

(2) the output s should be such that $s(t)$ is as close as possible to $e(t + T)$.

More precisely, we shall seek to minimize the power of the error $\varepsilon(t) = s(t) - e(t + T)$. We have $\varepsilon = s - (e * \delta_{-T})$.

We can combine these two problems by considering, for example, the filtering problem and taking $\varepsilon(t) = s(t) - i(t + T)$. The problem is then the filtering problem with prediction.

Also, the two problems can be considered as particular cases of the following problem:

For a given centered stationary process of the second-order $t \to \begin{bmatrix} x(t) \\ z(t) \end{bmatrix}$, *find a linear system admitting* x *as input such that* $\int \varpi_{y-z,y-z}$ *is minimized, where* y *denotes the output.*

For the straight filtering problem, we have $x = i + b$, $z = i$.
For the straight prediction problem, we have $x = e$, $z = e * \delta_{-T}$.
For the filtering problem with prediction, we have $x = i + b$, $z = i * \delta_{-T}$.

We propose to solve this general problem. Let us denote by H the Hilbert space of random variables ξ such that $\mathbf{E}(|\xi|^2) < +\infty$.

Let us consider a stationary linear system with impulse response $\mu \in \mathbf{M}$ (with positive support). We have

$$y(t) = \int_0^{+\infty} x(t - \theta) \,.\, \mu^\theta \,.$$

We denote by $\mathfrak{M}_t$ the closed subspace of H generated by the elements $x(\tau)$ for $\tau \leqslant t$. We have $y(t) \in \mathfrak{M}_t$ $(\forall t)$. This is true because, for every $a \in \mathfrak{M}_t^\perp$, we have $\langle a, x(\tau) \rangle = 0$ $(\forall \tau \leqslant t)$, so that

$$\langle a, y(t) \rangle = \int_0^{+\infty} \langle a, x(t-\theta) \rangle \,.\, \mu^\theta = 0\,.$$

Now,

$$\int \varpi_{y-z,y-z} = \mathbf{E}(|\, y - z\,|^2) = \|\, y(t) - z(t)\,\|_H^2 \quad (\forall t)\,.$$

A necessary and sufficient condition for $\int \varpi_{y-z,y-z}$ to be minimal is that $y(t)$ be the projection of $z(t)$ onto $\mathfrak{M}_t$ (see Fig. 2).

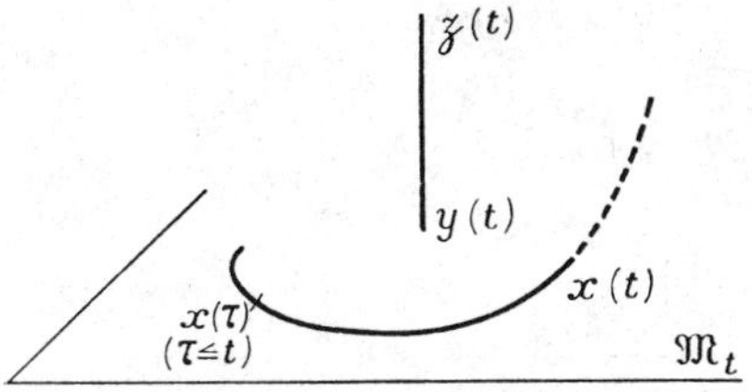

FIG. 2.

We remark that since $\|\, y(t) - z(t)\,\|_H^2$ is independent of t, if this condition is satisfied for one value of t, it is satisfied for every value of t. One can also see this by introducing the unitary representation $\theta \to U_\theta$ associated with the process $t \to \begin{bmatrix} x(t) \\ z(t) \end{bmatrix}$ and remarking that

$$\begin{aligned} x(t-\theta) &= U_\theta\, x(t) \\ \mathfrak{M}_{t-\theta} &= U_\theta(\mathfrak{M}_t) \\ y(t-\theta) &= U_\theta\, y(t)\,, \\ z(t-\theta) &= U_\theta\, z(t)\,. \end{aligned}$$

Let us write the condition

$$z(t) - y(t) \perp \mathfrak{M}_t$$

for $t = 0$. We have

$$z(0) - y(0) \perp x(\theta) \qquad (\forall \theta \leqslant 0)\,,$$

that is,

$$\langle x(\theta), z(0) \rangle = \langle x(\theta), y(0) \rangle, \qquad (\forall \theta \leqslant 0)$$

or

$$\langle U_{-\theta}\, x(0), z(0) \rangle = \langle U_{-\theta}\, x(0), y(0) \rangle \quad (\forall \theta \leqslant 0)$$

or

$$\langle U_\theta x(0), z(0) \rangle = \langle U_\theta x(0), y(0) \rangle \qquad (\forall \theta \geqslant 0)$$

or, finally,

$$\varphi_{x,z}(\theta) = \varphi_{x,y}(\theta) \qquad (\forall \theta \geqslant 0) ,$$

where $\varphi_{x,z}$ represents the intercorrelation function of x and z and $\varphi_{x,y}$ represents the intercorrelation function of x and y. Now, we have

$$\varphi_{x,y} = \varphi_{x,x} * \mu ,$$

where $\varphi_{x,x}$ is the autocorrelation function of x.

Thus, we obtain the relation

$$\varphi_{x,z}(\theta) = \int \varphi_{x,x}(\theta - t) \,.\, \mu^t \qquad (\forall \theta \geqslant 0)$$

or, by change of notation,

$$\boxed{\varphi_{x,z}(t) = \int \varphi_{x,x}(t - \theta) \,.\, \mu^\theta , \qquad \forall t \geqslant 0} \,.$$

This equation is known as the *Wiener-Hopf equation*. It constitutes a necessary and sufficient condition for μ to be the impulse response of an optimal stationary linear system for the criterion in question.

3. SOLUTION OF THE WIENER-HOPF EQUATION

The case in which the spectral densities are rational fractions. In the complex plane, we shall designate by upper half-plane (resp. lower half-plane) the set of complex numbers with nonnegative (resp. nonpositive) imaginary parts and we shall designate by open upper half-plane (resp. open lower half-plane) the set of complex numbers with positive (resp. negative) imaginary parts.

Let μ denote a member of $\mathbf{M}$ with positive support. We have

$$(\mathcal{F}^\star \mu)(\omega) = (\mathcal{L}\mu)(i\,\omega) .$$

Since $(\mathcal{L}\mu)(p)$ is holomorphic in the half-plane $\mathcal{R}(p) > 0$, the function $\mathcal{F}^\star \mu$ is the restriction to the real axis of a function that is continuous in the lower half-plane and holomorphic in the open lower half-plane. We shall express this fact by saying that $\mathcal{F}^\star \mu$ can be extended analytically into the lower half-plane. We note that, since $\mathcal{L}\mu(p)$ is bounded for $\mathcal{R}(p) > 0$, this extension of $\mathcal{F}^\star \mu$ to the lower half-plane is a bounded function. We obtain analogous results by replacing $\mathcal{F}$ with $\mathcal{F}^\star$ and taking functions with negative support. The following table summarizes these results:

$\mu \in \mathbf{M}$	$\mathfrak{F}\mu$ can be extended analytically	$\mathfrak{F}^*\mu$ can be extended analytically
With *positive* support	to the *upper* half-plane	to the *lower* half-plane
With *negative* support	to the *lower* half-plane	to the *upper* half-plane

Conversely, we have the following proposition:

Proposition 1. *Let F denote a continuous function belonging to $\mathbf{L}^2$ that can be continued analytically to the lower half-plane as the function with slow increase, that is, a function satisfying a condition of the form*

$$F(p) = O(p^k) \quad \text{as} \quad p \to \infty .$$

for some k. Then, $\mathfrak{F}F$ has positive support.

Proof. Let us set $F(\omega) = F_1(\mathrm{i}\,\omega)$ for $\omega \in \mathbf{R}$. Then, F_1 can be continued analytically to the half-plane $\mathcal{R}(p) \geqslant 0$ and we have $F_1(p) = O(p^k)$ in that half-plane. Thus, $F_1 = \mathfrak{L}T$, where T is a distribution with positive support. Furthermore, $T = D^{k+2}\,g$, where $\mathfrak{L}g$ has a non-positive abscissa of absolute convergence.

For every $\xi > 0$, we have

$$[\mathfrak{F}^*(\mathrm{e}^{-\xi t}\,T)](\omega) = F_1(\xi + \mathrm{i}\,\omega) .$$

The function $\omega \to F_1(\xi + \mathrm{i}\,\omega)$ tends in $\mathbf{S}'$ to the function $\omega \to F_1(\mathrm{i}\,\omega)$, that is, the function F, as ξ approaches 0. Consequently, the distribution $\mathrm{e}^{-\xi\cdot}\,T$ tends to $(1/2\pi)\,\mathfrak{F}F$, which implies that $T = (1/2\pi)\,\mathfrak{F}F$. Thus, $\mathfrak{F}F$ has positive support. We note, finally, that $T \in \mathbf{L}^2$.

In the following table, we present results analogous to those obtained by replacing $\mathfrak{F}$ with $\mathfrak{F}^*$ or the lower half-plane with the upper half-plane:

$F \in \mathbf{L}^2$ can be continued analytically as a function with slow increase	$\mathfrak{F}F$	$\mathfrak{F}^*F$
In the *upper* half-plane	with *negative* support	with *positive* support
In the *lower* half-plane	with *positive* support	with *negative* support

We shall now give a method of solving the Wiener-Hopf equation. Consider the equation

$$(\alpha) \qquad \varphi_{x,z}(t) - (\varphi_{x,x} * \mu)(t) = 0 \quad \text{for} \quad t \geqslant 0 .$$

We set

$$(\beta) \qquad \varphi_{x,z}(t) - (\varphi_{x,x} * \mu)(t) = h(t) \quad (\forall t \in \mathbf{R}) .$$

The function h as negative support. We have

$$\varphi_{x,z} = \mathcal{F}\varpi_{x,z}$$

$$\varphi_{x,x} = \mathcal{F}\varpi_{x,x}\,.$$

We shall suppose that $\varpi_{x,x}$ and $\varpi_{x,z}$ have densities $\Phi_{x,x}$ and $\Phi_{x,z}$ with respect to Lebesgue measure, that $\Phi_{x,x}$ is a rational fraction, and that $\Phi_{x,z}$ is a rational fraction possibly multiplied by a factor of the form $e^{i\omega T}$. These hypotheses are satisfied in the case of the general filtering problem with prediction if $\varpi_{s,s}$, $\varpi_{s,i}$ and $\varpi_{i,i}$ have densities that are rational fractions.

Then, we shall have

$$\varphi_{x,z} = \mathcal{F}\Phi_{x,z}\,,$$

$$\varphi_{x,x} = \mathcal{F}\Phi_{x,x}\,.$$

Since $\Phi_{x,z}$ and $\Phi_{x,x}$ are infinitely differentiable and since their derivatives are integrable, the functions $\varphi_{x,z}$ and $\varphi_{x,x}$ are of rapid decrease. In particular, $\varphi_{x,z}$ and $\varphi_{x,x}$ are integrable. Therefore, the same is true of $\varphi_{x,x} * \mu$ and hence of h.

Let us apply $\mathcal{F}^\star$ to both sides of (β). If we set $\mathcal{F}^\star \mu = M$ and $\mathcal{F}^\star h = H$, we obtain

$$(\gamma) \qquad \Phi_{x,z}(\omega) - \Phi_{x,x}(\omega)\, M(\omega) = H(\omega)\,,$$

where $M(\omega)$ can be continued analytically to the lower half-plane and $H(\omega)$ can be continued analytically to the upper half-plane. Let us set

$$\Phi_{x,x} = \Phi_+\,\Phi_-\,,$$

where Φ_+ and Φ_- are rational fractions such that

$$\begin{cases} \Phi_+ \text{ has neither pole nor zero in the open upper half-plane,} \\ \Phi_- \text{ has neither pole nor zero in the open lower half-plane,} \\ \Phi_-(\omega) = \overline{\Phi_+(\overline{\omega})}\,. \end{cases}$$

Such a decomposition is always possible because $\Phi_{x,x}(\omega) \geqslant 0$ for $\omega \in \mathbf{R}$ and hence the real zeros of $\Phi_{x,x}$ are of even order.

Therefore, equation (γ) can be written

$$\Phi_{x,z}(\omega) - \Phi_+(\omega)\,\Phi_-(\omega)\, M(\omega) = H(\omega)$$

or

$$(\delta) \qquad \frac{\Phi_{x,z}(\omega)}{\Phi_+(\omega)} - \Phi_-(\omega)\, M(\omega) = \frac{H(\omega)}{\Phi_+(\omega)}\,.$$

We shall recall from the final remark of Chapter 7, section 2, that

$$\Phi_{x,z} = \Psi_{z|x}\,\Phi_{x,x}\,,$$

where

$$\int_{-\infty}^{+\infty} |\Psi_{z|x}(\omega)|^2 \cdot \Phi_{x,x}(\omega)\,d\omega < +\infty;$$

that is,

$$\int_{-\infty}^{+\infty} \frac{|\Phi_{x,z}(\omega)|^2}{\Phi_{x,x}(\omega)}\,d\omega < +\infty$$

or

$$\int_{-\infty}^{+\infty} \frac{|\Phi_{x,z}(\omega)|^2}{\Phi_+(\omega)\,\Phi_-(\omega)}\,d\omega < +\infty.$$

Since

$$|\Phi_+(\omega)| = |\Phi_-(\omega)| \quad \text{for} \quad \omega \in \mathbf{R},$$

we have finally

$$\frac{\Phi_{x,z}}{\Phi_+} \in \mathbf{L}^2.$$

Furthermore,

$$\int_{-\infty}^{+\infty} \Phi_+(\omega)\,\Phi_-(\omega)\,d\omega = \int_{-\infty}^{+\infty} |\Phi_-(\omega)|^2\,d\omega,$$

so that $\Phi_- \in \mathbf{L}^2$. Since M is bounded, we have $M\Phi_- \in \mathbf{L}^2$ and, consequently,

$$\frac{H}{\Phi_+} \in \mathbf{L}^2.$$

Let us now apply the transformation $\overset{-1}{\mathcal{F}}\mathcal{Y}\mathcal{F}$ to both sides of equation (δ). The function H/Φ_+ can be continued analytically to the upper half-plane. In that half-plane, H is bounded and hence there exists a k such that

$$\frac{H(p)}{\Phi_+(p)} = O(p^k).$$

Consequently, $\mathcal{F}(H/\Phi_+)$ has negative support. Therefore,

$$\overset{-1}{\mathcal{F}}\mathcal{Y}\mathcal{F}\left(\frac{H}{\Phi_+}\right) = 0.$$

Similarly, $\mathcal{F}(\Phi_- M)$ has positive support and hence,

$$\overset{-1}{\mathcal{F}}\mathcal{Y}\mathcal{F}(\Phi_- M) = \Phi_- M.$$

If we set

$$\overset{-1}{\mathcal{F}}\, \mathcal{Y} \mathcal{F} \left(\frac{\Phi_{x,z}}{\Phi_+} \right) = \Psi ,$$

we obtain the relation

$$\Psi(\omega) - \Phi_-(\omega)\, M(\omega) = 0 ,$$

which determines $M(\omega)$.

Conversely, let us suppose that μ is a bounded measure such that

$$\Psi(\omega) - \Phi_-(\omega)\, M(\omega) = 0 .$$

We set

$$\frac{\Phi_{x,z}}{\Phi_+} - \Psi = \frac{H}{\Phi_+} .$$

The left-hand member is the image under $\overset{-1}{\mathcal{F}} = \dfrac{\mathcal{F}^*}{2\pi}$ of the function

$$(1 - \mathcal{Y})\, \mathcal{F} \left(\frac{\Phi_{x,z}}{\Phi_+} \right)$$

which has negative support. Since $\mathcal{F} \left(\dfrac{\Phi_{x,z}}{\Phi_+} \right)$ is of rapid decrease and hence is integrable, the function $\dfrac{\Phi_{x,z}}{\Phi_+} - \Psi$ can be extended as a bounded function to the closed upper half-plane and as a holomorphic function to the open upper half-plane. Consequently, H is holomorphic and bounded in the upper half-plane. Thus,

$$\Phi_{x,z} - \Phi_{x,x}\, M = H$$

and consequently,

$$\varphi_{x,z} - \varphi_{x,x} * \mu = h ,$$

where $h = \mathcal{F}H$ has negative support. Therefore, the measure μ is indeed a solution of the Wiener-Hopf equation.

Example. Let $t \to \begin{bmatrix} i(t) \\ b(t) \end{bmatrix}$ denote a centered process stationary of the second order. Find a stationary linear system with input $i(t) + b(t)$ and output $s(t)$ that will minimize $\int \varpi_{\varepsilon,\varepsilon}$, where $\varepsilon(t) = s(t) - i(t + T)$. Suppose that the spectral densities

$$\Phi_{i,i}(\omega) = \frac{1}{\omega^2 + 1}$$

$$\Phi_{b,b}(\omega) = \frac{3\,\omega^2}{\omega^4 + \omega^2 + 1}$$

$$\Phi_{i,b} = \Phi_{b,i} = 0$$

are given. Define

$$x = i + b\,, \qquad y = s\,, \qquad z = i * \delta_{-T}\,.$$

We have

$$\Phi_{x,x} = \Phi_{i,i} + \Phi_{b,b} \quad (\text{since } \Phi_{i,b} = \Phi_{b,i} = 0)\,,$$

so that

$$\Phi_{x,x}(\omega) = \frac{(2\,\omega^2 + 1)^2}{(\omega^2 + 1)\,(\omega^4 + \omega^2 + 1)}\,.$$

Also,

$$\begin{aligned} \Phi_{x,z}(\omega) &= \Phi_{i,i*\delta_{-T}}(\omega) \\ &= \Phi_{i,i}(\omega)\,(\mathcal{F}^{\star}\,\delta_{-T})\,(\omega) \\ &= \frac{e^{iT\omega}}{\omega^2 + 1}\,. \end{aligned}$$

We have

$$\Phi_{x,x}(\omega) = \frac{(\omega\sqrt{2} + i)^2\,(\omega\sqrt{2} - i)^2}{(\omega + i)\,(\omega - i)\,(\omega + \alpha)\,(\omega + \bar{\alpha})\,(\omega + \beta)\,(\omega + \bar{\beta})}$$

$$\text{with} \quad \begin{cases} \alpha = \dfrac{1 + i\sqrt{3}}{2} \\ \beta = \dfrac{-1 + i\sqrt{3}}{2}\,; \end{cases}$$

that is,

$$\Phi_{x,x}(\omega) = \Phi_{+}(\omega)\,\Phi_{-}(\omega)\,,$$

where

$$\Phi_{+}(\omega) = \frac{(\omega\sqrt{2} + i)^2}{(\omega + i)\,(\omega + \alpha)\,(\omega + \beta)}$$

and

$$\Phi_{-}(\omega) = \frac{(\omega\sqrt{2} - i)^2}{(\omega - i)\,(\omega + \bar{\alpha})\,(\omega + \bar{\beta})}\,.$$

Calculation yields

$$\frac{\Phi_{x,z}(\omega)}{\Phi_+(\omega)} = \frac{e^{iT\omega}(\omega + i)(\omega + \alpha)(\omega + \beta)}{(\omega^2 + 1)(\omega\sqrt{2} + i)^2}$$

$$= \frac{e^{iT\omega}(\omega^2 + i\,\omega\sqrt{3} - 1)}{(\omega - i)(\omega\sqrt{2} + i)^2}$$

$$= e^{iT\omega}\left[\frac{A}{\omega - i} + \frac{B}{(\omega\sqrt{2} + i)^2} + \frac{C}{\omega\sqrt{2} + i}\right],$$

where

$$A = \frac{2 + \sqrt{3}}{2\sqrt{2} + 3} \cdot$$

(The precise values of B and C are unimportant.)

Now, we need to calculate

$$\Psi = \overset{-1}{\mathcal{F}}\,\mathcal{Y}\mathcal{F}\left(\frac{\Phi_{x,z}}{\Phi_+}\right).$$

The function $e^{iT\omega}$ is bounded in the upper half-plane. Consequently, the functions

$$\frac{e^{iT\omega}}{(\omega\sqrt{2} + i)^2} \quad \text{and} \quad \frac{e^{iT\omega}}{\omega\sqrt{2} + i}$$

are holomorphic and bounded in the upper half-plane. Consequently, their Fourier transforms have negative support. The operator $\overset{-1}{\mathcal{F}}\mathcal{Y}\mathcal{F}$ maps them into 0.

Now, let us calculate the Fourier transform of

$$\frac{e^{iT\omega}}{\omega - i} = \theta(\omega).$$

We have

$$\frac{1}{\omega - i} = \frac{i}{i\,\omega + 1} = i\,G(i\,\omega), \qquad \text{with } G(p) = \frac{1}{p + 1};$$

that is,

$$G(p) = \mathfrak{L}(\mathcal{Y}(t)\,e^{-t}).$$

Therefore, we conclude that

$$\left[\mathcal{F}\left(\frac{1}{\omega - i}\right)\right](t) = 2\,i\,\pi\mathcal{Y}(t)\,e^{-t}$$

and, consequently,

$$(\mathcal{F}\theta)(t) = 2\,i\,\pi\mathcal{Y}(t + T)\,e^{-(t+T)},$$

so that

$$\mathcal{Y}(t)\left[(\mathcal{F}\theta)(t)\right] = 2\,\mathrm{i}\,\pi\,\mathrm{e}^{-T}\,\mathcal{Y}(t)\,\mathrm{e}^{-t}$$

and

$$\frac{1}{2\pi}(\mathcal{F}^{\star}\,\mathcal{Y}\mathcal{F}\theta)(\omega) = \mathrm{i}\,\mathrm{e}^{-T}\left[\mathcal{L}(\mathcal{Y}(t)\,\mathrm{e}^{-t})\right](\mathrm{i}\,\omega) = \mathrm{i}\,\mathrm{e}^{-T}\frac{1}{\mathrm{i}\,\omega+1} = \frac{\mathrm{e}^{-T}}{\omega-\mathrm{i}}\,.$$

Thus,

$$\Psi(\omega) = \left[(\overset{-1}{\mathcal{F}}\mathcal{Y}\mathcal{F})\left(\frac{\Phi_{x,z}}{\Phi_{+}}\right)\right](\omega) = \frac{A\,\mathrm{e}^{-T}}{\omega-\mathrm{i}}$$

and, consequently,

$$M(\omega) = \frac{\Psi(\omega)}{\Phi_{-}(\omega)} = A\,\mathrm{e}^{-T}\frac{\omega^2 - \mathrm{i}\sqrt{3}\,\omega - 1}{(\omega\sqrt{2}-\mathrm{i})^2}\,.$$

The transfer function of the system sought is therefore

$$F(p) = M\left(\frac{p}{\mathrm{i}}\right) = A\,\mathrm{e}^{-T}\frac{p^2 + p\sqrt{3}+1}{(p\sqrt{2}+1)^2}\,.$$

The impulse response will be of the form

$$\mu = \alpha\delta + \mathcal{Y}(t)\left[\beta t\,\mathrm{e}^{-t/\sqrt{2}} + \gamma\,\mathrm{e}^{-t/\sqrt{2}}\right].$$

We note in conclusion that, with the aid of supplementary hypotheses, the method given for solving the Wiener-Hopf equation can be applied to the case in which the spectral densities are not rational fractions.

4. ANTICIPATION

We now return to the problem of the choice of τ in an anticipation system with transfer function $1 + \tau p$ (cf. Chapter 8, section 5). Let G denote the transfer function of the system to which we intend to add the anticipation system. After the anticipation system is joined onto it, the complete system will have as its transfer function $(1+\tau p)\,G(p)$. If we take for input a centered random process $e(t)$ (stationary of the second order), the error $\varepsilon = s - e$ will have spectral power distribution

$$\varpi_{\varepsilon,\varepsilon} = \left|\,1 - (1+\mathrm{i}\,\tau\omega)\,G(\mathrm{i}\,\omega)\,\right|^2 . \,\varpi_{e,e}\,.$$

We can choose τ in such a way as to minimize the integral

$$\int \varpi_{\varepsilon,\varepsilon} = \int \left|\,1 - (1+\mathrm{i}\,\tau\omega)\,G(\mathrm{i}\,\omega)\,\right|^2 \varpi_{e,e}\,.$$

The right-hand member is a second-degree polynomial in τ. Therefore, it will be easy to calculate the optimum value of τ.

10

Discrete Systems, Part I

1. DEFINITIONS AND NOTATIONS

We shall call any mapping $\mathbf{a}(n \to \mathbf{a}(n))$ defined on $\mathbf{Z}$ a *bilateral sequence*.* In this chapter, we shall call any complex bilateral sequence simply a sequence. We define

$$\langle \mathbf{a}, \mathbf{b} \rangle = \sum_{n=-\infty}^{+\infty} \bar{\mathbf{a}}(n)\, \mathbf{b}(n),$$

if the series on the right is absolutely convergent.

We denote by l^1 the space of summable sequences. Equipped with the norm

$$\mathbf{a} \to \|\mathbf{a}\| = \sum_{n=-\infty}^{+\infty} |\mathbf{a}(n)|,$$

l^1 is a Banach space.

We denote by l^2 the space of square-summable sequences. Equipped with the norm

*In this chapter and the following one, we denote by $\mathbf{a}(n)$, instead of $\mathbf{a}_n$, the value of the sequence a at the point $n \in \mathbf{Z}$. The purpose of this convention is to render uniform the notations dealing with the Fourier transformation in $\mathbf{R}$, $\mathbf{T}$, and $\mathbf{Z}$. If the reader wishes, he may go back to the more common notation. We shall do this ourselves in the remaining chapters of the book. Also, to make the reading of Chapters 10 and 11 easier, we denote the sequences in these chapters by boldface letters.

$$\mathbf{a} \to \| \mathbf{a} \| = \left(\sum_{n=-\infty}^{+\infty} | \mathbf{a}(n) |^2 \right)^{1/2}$$

and the scalar product

$$\mathbf{a}, \mathbf{b} \to \langle \mathbf{a}, \mathbf{b} \rangle = \sum_{n=-\infty}^{+\infty} \overline{\mathbf{a}(n)}\, \mathbf{b}(n) \,,$$

l^2 is a Hilbert space, an orthonormal basis for which is formed by the sequences δ_i such that

$$\delta_i(n) = 0 \quad \text{for } n \neq i \quad \text{and} \quad \delta_i(i) = 1 \,.$$

We recall that $l^1 \subset l^2$. In fact, if $\mathbf{a} \in l^1$, we have

$$\mathbf{a}(n) \to 0 \text{ as } n \to \infty$$

and, consequently, the sequence $\mathbf{a}$ is bounded. Let M denote a number such that $| \mathbf{a}(n) | \leqslant M$ $(\forall n)$. We have

$$\sum_{n=-\infty}^{+\infty} | \mathbf{a}(n) |^2 \leqslant M \sum_{n=-\infty}^{+\infty} | \mathbf{a}(n) | \,.$$

We denote by s the space of sequences "of rapid decrease." By definition, we have $\mathbf{a} \in s$ if $n^k \mathbf{a}(n) \to 0$ as $n \to \pm \infty$ for every $k > 0$. Let us equip s with the topology defined by the increasing sequence of norms

$$\mathbf{a} \to \| \mathbf{a} \|_k = \left(\sum_{n=-\infty}^{+\infty} (1 + n^2)^k \, | \mathbf{a}(n) |^2 \right)^{1/2} \qquad (k = 0, 1, 2, \dots) \,.$$

Equipped with this topology, s is a locally convex metrizable space. It can be shown that s is complete and hence is a Fréchet space. This sequence of norms can be replaced by the equivalent sequence of seminorms

$$\mathbf{a} \to \| \mathbf{a} \|_k' = \left(\sum_{n=-\infty}^{+\infty} | n^k \mathbf{a}(n) |^2 \right)^{1/2} .$$

Let us denote by s' the space of sequences "of slow increase." By definition, $\mathbf{a} \in s'$ if there exists a $k > 0$ such that $\mathbf{a}(n) = O(n^k)$ as $n \to \infty$.

Theorem 1. *s' is the dual of s.*

Proof. (*a*) Let a denote *a* member of s and let *b* denote a member of s'. We define

$$\langle \mathbf{b}, \mathbf{a} \rangle = \sum_{n=-\infty}^{+\infty} \overline{\mathbf{b}(n)}\, \mathbf{a}(n) \,.$$

The series on the right is absolutely convergent since

$$\lim_{n\to\infty} n^2\, \overline{\mathbf{b}(n)}\, \mathbf{a}(n) = 0\,.$$

The mapping $\mathbf{a} \to \langle \mathbf{b}, \mathbf{a} \rangle$ is a linear form on s. Let us show that it is continuous. Suppose that $|\,\mathbf{b}(n)\,| \leqslant M(1 + n^2)^k$. Then,

$$|\langle \mathbf{b}, \mathbf{a} \rangle| \leqslant \sum_{n=-\infty}^{+\infty} M(1 + n^2)^k\, \mathbf{a}(n) = M \sum_{n=-\infty}^{+\infty} \frac{1}{1 + n^2} \cdot (1 + n^2)^{k+1}\, |\,\mathbf{a}(n)\,|.$$

Let us apply Schwarz' inequality (in l^2) to the second member. We get

$$|\langle \mathbf{b}, \mathbf{a} \rangle| \leqslant M \left(\sum_{n=-\infty}^{+\infty} \frac{1}{(1 + n^2)^2}\right)^{1/2} \left(\sum_{k=-\infty}^{+\infty} (1 + n^2)^{2(k+1)}\, |\,\mathbf{a}(n)\,|^2\right)^{1/2}$$
$$= k\, \|\,\mathbf{a}\,\|_{2(k+1)}\,.$$

Consequently, the mapping $\mathbf{a} \to \langle \mathbf{b}, \mathbf{a} \rangle$ is continuous for one of the norms defining the topology of s. Therefore, this mapping is continuous.

(*b*) Let u denote a continuous linear form on s. We note that, for every k, the norm $\mathbf{a} \to \|\,\mathbf{a}\,\|_k$ is associated with the Hilbert structure on s defined by the scalar product

$$\mathbf{b}, \mathbf{a} \to \langle \mathbf{b}, \mathbf{a} \rangle_k = \sum_{k=-\infty}^{+\infty} (1 + n^2)^k\, \overline{\mathbf{b}(n)}\, \mathbf{a}(n)\,.$$

There exists a k such that the linear form u is continuous with respect to the norm $\|\,.\,\|_k$. Consequently, there exists a sequence $\mathbf{b}'$ such that

$$\begin{cases} \displaystyle\sum_{n=-\infty}^{+\infty} (1 + n^2)^k\, |\,\mathbf{b}'(n)\,|^2 < +\infty \\ \displaystyle u(\mathbf{a}) = \langle \mathbf{b}', \mathbf{a} \rangle_k = \sum_{k=-\infty}^{+\infty} (1 + n^2)^k\, \overline{\mathbf{b}'(n)}\, \mathbf{a}(n)\,. \end{cases}$$

Let us set $\mathbf{b}(n) = (1 + n^2)^k\, \mathbf{b}'(n)$. We then have

$$\begin{cases} \displaystyle\sum_{n=-\infty}^{+\infty} \frac{|\,\mathbf{b}(n)\,|^2}{(1 + n^2)^k} < +\infty \\ \displaystyle u(\mathbf{a}) = \sum_{n=-\infty}^{+\infty} \overline{\mathbf{b}(n)}\, \mathbf{a}(n)\,. \end{cases}$$

The first relation implies that $\mathbf{b}(n) = O((1 + n^2)^{k/2})$. Therefore, $\mathbf{b} \in s'$.

Thus, we have shown that there is a one-to-one correspondence between s' and the dual of s such that every $\mathbf{b} \in s'$ defines a continuous linear form on s by

$$\langle \mathbf{b}, \mathbf{a} \rangle = \sum_{n=-\infty}^{+\infty} \overline{\mathbf{b}(n)}\, \mathbf{a}(n)\,,$$

and, conversely, every continuous linear form on s can be defined on the basis of any element $\mathbf{b} \in s'$ by the above formula.

Topology of s'. Let us equip s' with the topology of simple convergence on s.

We denote by $\mathfrak{D}^0(\mathbf{T})$ the space of continuous functions on $\mathbf{T}$ equipped with the topology of uniform convergence defined by the norm

$$\varphi \to \| \varphi \|_u = \sup_{\theta \in \mathbf{T}} \| \varphi(\theta) \| .$$

Equipped with this norm, $\mathfrak{D}^0(\mathbf{T})$ is a Banach space.

We denote by $\mathbf{M}(\mathbf{T})$ the space of measures on $\mathbf{T}$, in other words, the dual of $\mathfrak{D}^0(\mathbf{T})$, the space of continuous linear forms μ: $\varphi \to \int \varphi . \mu$ on $\mathfrak{D}^0(\mathbf{T})$. The space $\mathbf{M}(\mathbf{T})$ is a Banach space for the norm

$$\mu \to \| \mu \| = \sup_{\|f\|_u \leqslant 1} \left| \int f . \mu \right| .$$

If μ is a positive measure, we have

$$\| \mu \| = \int \mu .$$

We note that it is possible to associate with every function φ on $\mathbf{T}$ a function φ on $\mathbf{R}$ that is periodic with period 2π by

$$\dot{\varphi}(t) = \varphi(\dot{t}) ,$$

where $\dot{t}$ is the residue-class of t modulo 2π. We shall call the function $\dot{\varphi}$ the evolute of $\mathbf{R}$ of the function φ.

Conversely, to every function $\dot{\varphi}$ that is periodic with period 2π on $\mathbf{R}$ is assigned a function φ on $\mathbf{T}$ by the preceding formula. In particular, if φ is continuous, so is $\dot{\varphi}$ and vice versa. The mapping

$$\varphi \to \int_{\alpha}^{\alpha + 2\pi} \dot{\varphi}(t) \, \mathrm{d}t \qquad (\varphi \in \mathfrak{D}^0(\mathbf{T}))$$

defines a measure on $\mathbf{T}$ that is independent of α. We call this measure the *Lebesgue measure* on $\mathbf{T}$. We define

$$\int_{\mathbf{T}} \varphi(\theta) \, \mathrm{d}\theta = \int_{-\pi}^{+\pi} \dot{\varphi}(t) \, \mathrm{d}t .$$

Lebesgue measure on $\mathbf{T}$ is invariant under translation. A function φ on $\mathbf{T}$ is Lebesgue-measurable on $\mathbf{T}$ if and only if $\dot{\varphi}$ is Lebesgue-measurable on $\mathbf{R}$. In what follows, we shall say "measurable" for "Lebesgue-measurable."

Let us denote by $\mathbf{L}^1(\mathbf{T})$ the space of measurable functions φ on $\mathbf{T}$ such that

$$\int_{\mathbf{T}} |\varphi(\theta)|\,d\theta = \int_{-\pi}^{+\pi} |\dot{\varphi}(t)|\,dt < +\infty .$$

The space $\mathbf{L}^1(\mathbf{T})$ is a Banach space for the norm

$$\varphi \to \|\varphi\| = \int_{\mathbf{T}} |\varphi(\theta)|\,d\theta$$

if we identify elements φ_1 and φ_2 such that $\varphi_1(\theta) = \varphi_2(\theta)$ almost everywhere.

Similarly, we shall denote by $\mathbf{L}^2(\mathbf{T})$ the space of measurable functions on $\mathbf{T}$ such that

$$\int_{\mathbf{T}} |\varphi(\theta)|^2\,d\theta = \int_{-\pi}^{+\pi} |\dot{\varphi}(t)|^2\,dt < +\infty .$$

Equipped with the scalar product

$$\varphi, \psi \to \langle \varphi, \psi \rangle = \int_{\mathbf{T}} \overline{\varphi}(\theta)\,\psi(\theta)\,d\theta = \int_{-\pi}^{+\pi} \overline{\dot{\varphi}}(t)\,\dot{\psi}(t)\,dt ,$$

the space $\mathbf{L}^2(\mathbf{T})$ is a Hilbert space if we identify two elements φ_1 and φ_2 such that $\varphi_1(\theta) = \varphi_2(\theta)$ almost everywhere. The associated norm is

$$\varphi \to \|\varphi\| = \left(\int_{\mathbf{T}} |\varphi(\theta)|^2\,d\theta\right)^{1/2} = \left(\int_{-\pi}^{+\pi} |\dot{\varphi}(t)|^2\,dt\right)^{1/2} .$$

We note that $\mathbf{L}^2(\mathbf{T}) \subset \mathbf{L}^1(\mathbf{T})$. To see this, note that $1 \in \mathbf{L}^2(\mathbf{T})$ and, consequently, if $\varphi \in \mathbf{L}^2(\mathbf{T})$, we have

$$\langle 1, |\varphi| \rangle = \int_{\mathbf{T}} |\varphi(\theta)|\,d\theta ,$$

which shows that $\varphi \in \mathbf{L}^1(\mathbf{T})$.

A necessary and sufficient condition for a function φ on $\mathbf{T}$ to be k times differentiable (resp. infinitely differentiable) is that $\dot{\varphi}$ be k times differentiable (resp. infinitely differentiable). We have $(\dot{\varphi})' = \dot{\widehat{\varphi'}}$.

We denote by $\mathfrak{D}(\mathbf{T})$ the space of infinitely differentiable functions on $\mathbf{T}$. One might also denote this space by $\mathbf{S}(\mathbf{T})$. (On $\mathbf{R}$, the difference between $\mathfrak{D}$ and $\mathbf{S}$ has to do with the behavior at infinity, and this difference no longer exists on $\mathbf{T}$.)

It is possible to equip $\mathfrak{D}(\mathbf{T})$ with a locally convex metrizable topology defined by the sequence of seminorms

$$\varphi \to \|\varphi^{(k)}\|_{\mathbf{L}^2(\mathbf{T})}$$

or any equivalent sequence. One can show that $\mathfrak{D}(\mathbf{T})$ equipped with this topology is a complete space and hence a Fréchet space.

We denote by $\mathfrak{D}'(\mathbf{T})$ the dual of $\mathfrak{D}(\mathbf{T})$. Its elements are called *distributions* on **T**. (One might also denote this space by $\mathfrak{S}'(\mathbf{T})$.) It will be equipped with the topology of simple convergence on $\mathfrak{D}(\mathbf{T})$.

2. THE FOURIER TRANSFORMATION

Let θ denote an element of **T** and let t denote a real number such that $\dot{t} = \theta$. We define $e^{i\theta} = e^{it}$. Therefore, the function $\theta \to e^{i\theta}$ is the function on **T** whose evolute on **R** is $t \to e^{it}$.

Definitions. I. For a sequence $\mathbf{a} \in l^1$, the Fourier transform of **a** is defined as the function $\mathcal{F}_{\mathbf{Z}}\,\mathbf{a}$ defined on **T** by

$$(\mathcal{F}_{\mathbf{Z}}\,\mathbf{a})(\theta) = \sum_{n=-\infty}^{+\infty} \mathbf{a}(n)\, e^{in\theta} .$$

The hypothesis $\mathbf{a} \in l^1$ implies that the series on the right is uniformly convergent. Consequently, $\mathcal{F}_{\mathbf{Z}}\,\mathbf{a}$ is a continuous function on **T** whose evolute on **R** is the function

$$t \to \sum_{n=-\infty}^{+\infty} \mathbf{a}(n)\, e^{int} .$$

Similarly, we define

$$(\mathcal{F}^{\star}_{\mathbf{Z}}\,\mathbf{a})(\theta) = \sum_{n=-\infty}^{+\infty} \mathbf{a}(n)\, e^{-in\theta} .$$

When there is no ambiguity, we may write $\mathcal{F}$ (resp. $\mathcal{F}^{\star}$) for $\mathcal{F}_{\mathbf{Z}}$ (resp. $\mathcal{F}^{\star}_{\mathbf{Z}}$).

II. For a measure $\mu \in \mathbf{M}(\mathbf{T})$, the Fourier transform of μ is defined as the sequence $\mathcal{F}_{\mathbf{T}}\,\mu$ defined by

$$(\mathcal{F}_{\mathbf{T}}\,\mu)(n) = \int e^{in\theta} \,.\, \mu^{\theta} \qquad (\theta \in \mathbf{T}) .$$

The sequence $\mathcal{F}_{\mathbf{T}}\,\mu$ is a bounded sequence. We have

$$|(\mathcal{F}_{\mathbf{T}}\,\mu)(n)| \leqslant \|\mu\| .$$

If $\varphi \in \mathbf{L}^1(\mathbf{T})$, we set

$$(\mathcal{F}_{\mathbf{T}}\,\varphi)(n) = \int_{\mathbf{T}} \varphi(\theta)\, e^{in\theta}\, d\theta = \int_{-\pi}^{+\pi} \dot{\varphi}(t)\, e^{int}\, dt .$$

Similarly,

$$(\mathcal{F}^{\star}_{\mathbf{T}}\,\mu)(n) = \int_{\mathbf{T}} e^{-in\theta}\, \mu^{\theta} \text{ (pour } \mu \in \mathbf{M}(\mathbf{T}))$$

and

$$(\mathcal{F}_{\mathbf{T}}^{\star}\varphi)(n) = \int_{\mathbf{T}} \varphi(\theta)\, e^{-in\theta}\, d\theta \quad (\text{ for } \varphi \in \mathbf{L}^1(\mathbf{T}))$$

$$= \int_{-\pi}^{+\pi} \dot{\varphi}(t)\, e^{-int}\, dt .$$

When there is no ambiguity, we may write $\mathcal{F}$ (resp. $\mathcal{F}^{\star}$) for $\mathcal{F}_{\mathbf{T}}$ (resp. $\mathcal{F}_{\mathbf{T}}^{\star}$).

Let us recall the following fundamental result:

Theorem 2 (Parseval-Bessel). *If* $\varphi \in \mathbf{L}^2(\mathbf{T})$, *then* $\mathcal{F}_{\mathbf{T}}\varphi \in l^2$. *If* $\varphi, \psi \in \mathbf{L}^2(\mathbf{T})$, *then*

$$\langle \mathcal{F}_{\mathbf{T}}\varphi, \mathcal{F}_{\mathbf{T}}\psi \rangle = 2\pi \langle \varphi, \psi \rangle ;$$

that is,

$$\sum_{n=-\infty}^{+\infty} \overline{(\mathcal{F}_{\mathbf{T}}\varphi)(n)}\, (\mathcal{F}_{\mathbf{T}}\psi)(n) = 2\pi \int_{\mathbf{T}} \overline{\varphi(\theta)}\, \psi(\theta)\, d\theta .$$

In particular,

$$\| \mathcal{F}_{\mathbf{T}}\varphi \|^2 = 2\pi \| \varphi \|^2 ;$$

that is,

$$\sum_{n=-\infty}^{+\infty} | (\mathcal{F}_{\mathbf{T}}\varphi)(n) |^2 = 2\pi \int_{\mathbf{T}} | \varphi(\theta) |^2\, d\theta .$$

The same results hold if we replace $\mathcal{F}_{\mathbf{T}}$ *by* $\mathcal{F}_{\mathbf{T}}^{\star}$.

Theorem 3. *If* $a \in l^1$, *then* $\mathcal{F}_{\mathbf{T}}^{\star}\mathcal{F}_{\mathbf{Z}}\, \mathbf{a} = 2\pi\, \mathbf{a}$.

Proof.

$$(\mathcal{F}\mathbf{a})(\theta) = \sum_{n=-\infty}^{+\infty} \mathbf{a}(n)\, e^{in\theta} = \varphi(\theta)$$

$$(\mathcal{F}^{\star}\varphi)(k) = \int_{\mathbf{T}} \varphi(\theta)\, e^{-ik\theta}\, d\theta$$

$$= \sum_{n=-\infty}^{+\infty} \mathbf{a}(n) \int_{\mathbf{T}} e^{i(n-k)\theta}\, d\theta = 2\pi\, \mathbf{a}(k) .$$

Corollary. *If* $\varphi \in \mathbf{L}^2(\mathbf{T})$ *and* $\mathcal{F}^{\star}\varphi \in l^1$, *then* $\mathcal{F}\mathcal{F}^{\star}\varphi = 2\pi\varphi$.

Proof. We have $\mathbf{a} = \mathcal{F}^{\star}\varphi$. Also, $\mathbf{a} = \frac{1}{2\pi}\mathcal{F}^{\star}\mathcal{F}\mathbf{a}$, so that $\varphi = \frac{1}{2\pi}\mathcal{F}\mathbf{a}$ (since $\mathcal{F}^{\star}$ is bijective on $\mathbf{L}^2(\mathbf{T})$) and $2\pi\varphi = \mathcal{F}\mathbf{a} = \mathcal{F}\mathcal{F}^{\star}\varphi$.

Proposition 1. *If f is a continuously differentiable function on* $\mathbf{T}$, *then*

$$(\mathcal{F}\varphi')(n) = -\,\mathrm{i}\, n(\mathcal{F}\varphi)(n)$$

and

$$(\mathcal{F}^\star \varphi')(n) = \quad \mathrm{i}\, n(\mathcal{F}^\star \varphi)(n)\,.$$

Proof. Let $\dot\varphi$ and $\dot\varphi'$ denote the evolutes on $\mathbf{R}$ of φ and φ', respectively. We have

$$\begin{aligned}(\mathcal{F}\varphi')(n) &= \int_{-\pi}^{+\pi} \mathrm{e}^{\mathrm{i}nt}\,\dot\varphi'(t)\,\mathrm{d}t \\ &= [\mathrm{e}^{\mathrm{i}nt}\,\dot\varphi(t)]_{-\pi}^{+\pi} - \mathrm{i}\,n \int_{-\pi}^{+\pi} \mathrm{e}^{\mathrm{i}nt}\,\dot\varphi(t)\,\mathrm{d}t \\ &= -\,\mathrm{i}\,n(\mathcal{F}f)(n)\,.\end{aligned}$$

Corollary. *If $\varphi \in \mathfrak{D}(\mathrm{T})$, then $\mathcal{F}\varphi \in s$ and*

$$(\mathcal{F}\varphi^{(k)})(n) = (-\,\mathrm{i}\,n)^k\,(\mathcal{F}\varphi)(n)\,.$$

Proposition 2. *If $\mathbf{a} \in l^1$ and the sequence $n \to n\,\mathbf{a}(n)$ also belongs to l^1, then $\mathcal{F}\mathbf{a}$ is differentiable and*

$$(\mathcal{F}\mathbf{a})'(\theta) = \sum_{n=-\infty}^{+\infty} \mathrm{i}\,n\,\mathbf{a}(n)\,.\,\mathrm{e}^{\mathrm{i}n\theta}\,.$$

In other words, the Fourier transform of the sequence $n \to \mathrm{i}\,n\,\mathbf{a}(n)$ is $(\mathcal{F}\mathbf{a})'$.

Proof. Suppose that

$$\varphi = \mathcal{F}\mathbf{a}\,, \qquad \dot\varphi(t) = \sum_{n=-\infty}^{+\infty} \mathbf{a}(n)\,\mathrm{e}^{\mathrm{i}nt}\,.$$

Then,

$$\dot\varphi'(t) = \sum_{n=-\infty}^{+\infty} \mathrm{i}\,n\,\mathbf{a}(n)\,\mathrm{e}^{\mathrm{i}nt}$$

(the hypotheses assure legitimacy of term-by-term differentiation), and hence

$$\varphi'(\theta) = \sum_{n=-\infty}^{+\infty} \mathrm{i}\,n\,\mathbf{a}(n)\,\mathrm{e}^{\mathrm{i}n\theta}\,.$$

Corollary. *If $\mathbf{a} \in s$, then $\mathcal{F}\mathbf{a} \in \mathfrak{D}(\mathrm{T})$ and $(\mathcal{F}\mathbf{a})^{(k)}$ is the Fourier transform of the sequence $n \to (\mathrm{i}\,n)^k\,\mathbf{a}(n)$.*

Theorem 4. *The Fourier transformations $\mathcal{F}_{\mathrm{Z}}$ of s in $\mathfrak{D}(\mathrm{T})$ and $\mathcal{F}^\star_{\mathrm{T}}$ of $\mathfrak{D}(\mathrm{T})$ in s are continuous isomorphisms such that*

$$\mathcal{F}_{\mathrm{Z}}\mathcal{F}^\star_{\mathrm{T}} = 2\pi \quad \text{and} \quad \mathcal{F}^\star_{\mathrm{T}}\mathcal{F}_{\mathrm{Z}} = 2\pi.$$

Proof. The formulas $\mathcal{F}_{\mathbf{Z}} \overset{\star}{\mathcal{F}}_{\mathbf{T}} = 2\pi$ and $\overset{\star}{\mathcal{F}}_{\mathbf{T}} \mathcal{F}_{\mathbf{Z}} = 2\pi$ follow from Theorem 3 and its corollary.

The topology of s can be defined by the following seminorms:

$$\mathbf{a} \to \| \mathbf{a} \|_k = \left(\sum_{n=-\infty}^{+\infty} | n^k \mathbf{a}(n) |^2 \right).$$

In $\mathfrak{D}(\mathbf{T})$, we shall use the seminorms

$$\varphi \to \| \varphi \|_k = \left[\int | \varphi^{(k)}(\theta) |^2 \, d\theta \right]^{1/2} = \| \varphi^{(k)} \|_{\mathbf{L}^2(\mathbf{T})}.$$

According to the Parseval-Bessel theorem, the equation $\mathbf{a} = \mathcal{F}_{\mathbf{T}} \varphi$ (that is, the equation $\varphi = \frac{1}{2\pi} \overset{\star}{\mathcal{F}}_{\mathbf{Z}} \mathbf{a}$) implies

$$\| \mathbf{a} \|_k = \sqrt{2\pi} \, \| \varphi \|_k,$$

which shows that $\mathcal{F}_{\mathbf{Z}}$ and $\overset{\star}{\mathcal{F}}_{\mathbf{T}}$ are continuous isomorphisms.

Definition. The Fourier transform of a distribution $U \in \mathfrak{D}'(\mathbf{T})$ is defined as the sequence $\mathcal{F}_{\mathbf{T}} U \in s'$ defined by

$$(\alpha) \qquad \langle \mathbf{a}, \mathcal{F}_{\mathbf{T}} U \rangle = \langle \overset{\star}{\mathcal{F}}_{\mathbf{Z}} \mathbf{a}, U \rangle, \qquad \forall \mathbf{a} \in s.$$

Similarly, $\overset{\star}{\mathcal{F}}_{\mathbf{T}} U$ is defined by

$$\langle \mathbf{a}, \overset{\star}{\mathcal{F}}_{\mathbf{T}} U \rangle = \langle \mathcal{F}_{\mathbf{Z}} \mathbf{a}, U \rangle, \qquad \forall \mathbf{a} \in s.$$

The Fourier transform of a sequence $\mathbf{a} \in s'$ is defined as the distribution $\mathcal{F}_{\mathbf{Z}} \mathbf{a} \in \mathfrak{D}'(\mathbf{T})$ defined by

$$(\beta) \qquad \langle \varphi, \mathcal{F}_{\mathbf{Z}} \mathbf{a} \rangle = \langle \overset{\star}{\mathcal{F}}_{\mathbf{T}} \varphi, \mathbf{a} \rangle, \qquad \forall \varphi \in \mathfrak{D}(\mathbf{T}).$$

Similarly, $\mathcal{F}^\star \mathbf{a}$ is defined by

$$\langle \varphi, \overset{\star}{\mathcal{F}}_{\mathbf{Z}} \mathbf{a} \rangle = \langle \mathcal{F}_{\mathbf{T}} \varphi, \mathbf{a} \rangle, \qquad \forall \varphi \in \mathfrak{D}(\mathbf{T}).$$

To legitimize these definitions, it is necessary to verify that (α) is valid for $U \in \mathbf{M}(\mathbf{T})$ and that (β) is valid for $\mathbf{a} \in l_1$. Let $\mathbf{a}$ denote a member of s and let μ denote a member of $\mathbf{M}(\mathbf{T})$. We have

$$\langle \mathbf{a}, \mathcal{F}\mu \rangle = \sum_{n=-\infty}^{+\infty} \overline{\mathbf{a}(n)} \int_{\mathbf{T}} e^{in\theta} \mu^\theta = \int_{\mathbf{T}} \overline{\sum_{n=-\infty}^{+\infty} \mathbf{a}(n) \, e^{-in\theta}} \cdot \mu^\theta = \langle \mathcal{F}^\star \mathbf{a}, \mu \rangle.$$

Suppose that $\varphi \in \mathfrak{D}(\mathbf{T})$ and $\mathbf{a} \in l^1$. Then,

$$\langle \varphi, \mathcal{F}\mathbf{a} \rangle = \int_{\mathbf{T}} \overline{\varphi(\theta)} \left(\sum_{n=-\infty}^{+\infty} \mathbf{a}(n) \, e^{in\theta} \right) d\theta$$

$$= \sum_{n=-\infty}^{+\infty} \mathbf{a}(n) \int_{\mathbf{T}} \overline{\varphi(\theta) \, e^{-in\theta}} \, d\theta = \langle \mathcal{F}^\star \varphi, \mathbf{a} \rangle.$$

Just as on $\mathbf{R}$, we derive the following fundamental result from the definition.

Theorem 5. *The Fourier transformations* $\mathcal{F}_{\mathbf{Z}}$ (*resp.* $\mathcal{F}^{\star}_{\mathbf{Z}}$) *of* s' *in* $\mathcal{D}'(\mathbf{T})$ *and* $\mathcal{F}^{\star}_{\mathbf{T}}$ (*resp.* $\mathcal{F}_{\mathbf{T}}$) *of* $\mathcal{D}'(\mathbf{T})$ *in* s' *are continuous isomorphisms satisfying the equations*

$$\mathcal{F}_{\mathbf{Z}}\mathcal{F}^{\star}_{\mathbf{T}} = \mathcal{F}^{\star}_{\mathbf{Z}}\mathcal{F}_{\mathbf{T}} = 2\pi$$

$$\mathcal{F}^{\star}_{\mathbf{T}}\mathcal{F}_{\mathbf{Z}} = \mathcal{F}_{\mathbf{T}}\mathcal{F}^{\star}_{\mathbf{Z}} = 2\pi .$$

Theorem 6. *If* $U \in \mathcal{D}'(\mathbf{T})$, *we have*

$$(\mathcal{F}U)(n) = \int e^{in\theta} \,.\, U^{\theta}$$

$$(\mathcal{F}^{\star} U)(n) = \int e^{-in\theta} \,.\, U^{\theta} .$$

Proof. We have

$$(\mathcal{F}U)(n) = \langle \delta_n, \mathcal{F}U \rangle = \langle \mathcal{F}^{\star}\delta_n, U \rangle .$$

Now,

$$(\mathcal{F}^{\star}\delta_n)(\theta) = e^{-in\theta} ,$$

so that

$$(\mathcal{F}U)(n) = \int e^{in\theta} \,.\, U^{\theta} .$$

Theorem 7. *If* $\mathrm{a} \in s'$, *then*

$$\mathcal{F}\mathrm{a} \overset{\mathcal{D}'(\mathbf{T})}{=\!=} \sum_{n=-\infty}^{+\infty} \mathrm{a}(n)\, e^{in\theta}, \quad \mathcal{F}^{\star}\mathrm{a} \overset{\mathcal{D}'(\mathbf{T})}{=\!=} \sum_{n=-\infty}^{+\infty} \mathrm{a}(n)\, e^{-in\theta} ,$$

where the symbol $\overset{\mathcal{D}'(\mathbf{T})}{=\!=}$ *means that the series on the right converges in* $\mathcal{D}'(\mathbf{T})$ *and that the sum is equal to the left-hand member.**

Proof. Let P_M^N denote the projection defined in s' by

$$(P_M^N \mathrm{a})(n) = \begin{cases} \mathrm{a}(n) \text{ if } M \leqslant n \leqslant N \\ 0 \quad \text{if } n > N \text{ or } n < M . \end{cases}$$

We have

$$\lim_{\substack{N\to+\infty \\ M\to-\infty}} P_M^N = 1$$

---

*Thus, it is not a matter of convergence of the right-hand member for every value of θ.

(the identity operator in s'). The limit is taken for the topology adopted in s', that is, the topology of simple convergence on s. In fact, for every $\mathbf{b} \in s$, we have

$$\langle \mathbf{b}, \mathbf{a} \rangle = \sum_{n=-\infty}^{+\infty} \overline{\mathbf{b}(n)}\, \mathbf{a}(n) = \lim_{\substack{N \to +\infty \\ M \to -\infty}} \sum_{n=M}^{N} \overline{\mathbf{b}(n)}\, \mathbf{a}(n) = \lim_{\substack{N \to +\infty \\ M \to -\infty}} \langle \mathbf{b}, P_M^N \mathbf{a} \rangle .$$

Since $\mathcal{F}$ is a continuous mapping, we have

$$\mathcal{F}\mathbf{a} = \lim_{\substack{N \to +\infty \\ M \to -\infty}} \mathcal{F} P_M^N \mathbf{a} ,$$

where

$$(\mathcal{F} P_M^N \mathbf{a})(\theta) = \sum_{n=M}^{N} \mathbf{a}(n)\, e^{in\theta} ,$$

from which the conclusion of the theorem follows.

Corollary. *For every* $U \in \mathfrak{D}'(\mathbf{T})$, *we have*

$$U \overset{\mathfrak{D}'(\mathbf{T})}{=\!=} \sum_{n=-\infty}^{+\infty} \mathbf{a}(n)\, e^{in\theta} ,$$

where

$$\mathbf{a}(n) = \frac{1}{2\pi} (\mathcal{F}^* U)(n) = \frac{1}{2\pi} \int_{\mathbf{T}} e^{-in\theta} . U^{\theta} .$$

3. DIFFERENTIATION. CONVOLUTION. INVOLUTIONS

We define the derivative of a distribution $U \in \mathfrak{D}'(\mathbf{T})$, and denote it by DU or U', as the distribution defined by the following formulas:

$$\langle \varphi, DU \rangle = - \langle \varphi', U \rangle , \qquad \forall \varphi \in \mathfrak{D}(\mathbf{T}) .$$

$$\int \varphi . DU = - \int \varphi' . U .$$

One can verify that this definition extends the definition of a differentiable function in the usual sense and that differentiation of distributions is a continuous mapping.

The product ψU of a distribution $U \in \mathfrak{D}'(\mathbf{T})$ and a function $\psi \in \mathfrak{D}(\mathbf{T})$ is defined by one of the following formulas:

$$\langle \varphi, \psi U \rangle = \langle \overline{\psi} \varphi, U \rangle , \qquad \forall \varphi \in \mathfrak{D}(\mathbf{T})$$

$$\int \varphi . \psi U = \int \psi \varphi . U .$$

The set $\mathfrak{D}(\mathbf{T})$ will be considered as the set of multipliers on $\mathfrak{D}'(\mathbf{T})$.

Theorem 8. $(\mathcal{F}_{\mathbf{T}}\, DU)(n) = -\,\mathrm{i}\, n(\mathcal{F}_{\mathbf{T}}\, U)(n) \quad (\forall U \in \mathfrak{D}'(\mathbf{T}))$.

Proof. We have

$$(\mathcal{F} DU)(n) = \int e^{in\theta} \,.\, (DU)^{\theta} = -\int D(e^{in\theta}) \,.\, U^{\theta}$$

$$= -\,\mathrm{i}\, n \int e^{in\theta} \,.\, U^{\theta} = -\,\mathrm{i}\, n(\mathcal{F} U)(n)\,.$$

Corollary. *If* $\mathbf{a}, \mathbf{b} \in s'$, *where* $\mathbf{a}(n) = \mathrm{i}\, n\, \mathbf{b}(n)$, *then*

$$\mathcal{F}_{\mathbf{Z}}\, \mathbf{a} = D\mathcal{F}_{\mathbf{Z}}\, \mathbf{b}\,.$$

Theorem 9. *For every distribution* $U \in \mathfrak{D}'(\mathbf{T})$, *there exists a function* φ *and a positive integer* k *such that* $U = (1 - D^2)^k\, \varphi$.

Proof. Suppose that $U = \mathcal{F}\mathbf{a}$, where $\mathbf{a} \in s'$. There exists a positive integer k and a $\mathbf{b} \in l^1$ such that $\mathbf{a}(n) = (1 + n^2)^k\, \mathbf{b}(n)$. Then, by setting $\varphi = \mathcal{F}\mathbf{b}$, we have

$$U = (1 - D^2)^k\, \varphi\,.$$

Since φ is a continuous function, the theorem is proven.

Let U and V denote two distributions in $\mathfrak{D}'(\mathbf{T})$. We can define the direct product UV by the formula

$$\int \varphi(\theta_1, \theta_2) \,.\, U^{\theta_1}\, V^{\theta_2} = \left[\int \varphi(\theta_1, \theta_2)\, U^{\theta_1}\right] V^{\theta_2}$$

$$= \left[\int \varphi(\theta_1, \theta_2)\, V^{\theta_2}\right] U^{\theta_1}$$

for every φ that is infinitely differentiable on $\mathbf{T}^2$. The equality of the second and third members is proven just as for distributions on $\mathbf{R}$ by beginning with Fubini's theorem in its usual form and using the theorem on the structure of distributions on $\mathbf{T}$ (Theorem 9).

Thus, we can define the convolution product $U * V$ for $U, V \in \mathfrak{D}'(\mathbf{T})$ by

$$\int \varphi(\theta) \,.\, (U * V)^{\theta} = \int \varphi(\theta_1 + \theta_2) \,.\, U^{\theta_1}\, V^{\theta_2}\,.$$

On $\mathbf{T}$, we thus have the advantage that the convolution product of two distributions is always defined. This, of course, stems from the fact that, since $\mathbf{T}$ is compact, the conditions on the supports or on the behavior at infinity that we had to impose when we were discussing $\mathbf{R}$ have no meaning on $\mathbf{T}$.

If $\varphi \in \mathfrak{D}(\mathbf{T})$ and $U \in \mathfrak{D}'(\mathbf{T})$, we see that $\varphi * U$ is a function defined by

$$(\varphi * U)(\theta) = \int \varphi(\theta - \tau) \,.\, U^{\tau}\,.$$

The convolution product of two functions φ and ψ will be defined by

$$(\alpha) \qquad (\varphi * \psi)(\theta) = \int_{\mathbf{T}} \varphi(\theta - \tau)\, \psi(\tau)\, d\tau$$

in the extended cases, for example, $\varphi, \psi \in \mathbf{L}^2(\mathbf{T})$ or $\varphi \in \mathbf{L}^1(\mathbf{T})$, $\psi \in \mathfrak{D}^0(\mathbf{T})$.

Let us define $\chi = \varphi * \psi$. We denote by $\dot{\varphi}, \dot{\psi}$ and $\dot{\chi}$ the evolutes of φ, ψ, and χ on $\mathbf{R}$. We have

$$\dot{\chi}(t) = \int_{-\pi}^{+\pi} \dot{\varphi}(t - u) \,.\, \dot{\psi}(u)\, du \,.$$

For every $\varphi \in \mathfrak{D}(\mathbf{T})$, we denote by $\tilde{\varphi}$ the function defined by

$$\tilde{\varphi}(\theta) = \bar{\varphi}(-\theta)\,.$$

For every $U \in \mathfrak{D}'(\mathbf{T})$, we denote by $\tilde{U}$ the distribution defined by

$$\langle \varphi, \tilde{U} \rangle = \overline{\langle \tilde{\varphi}, U \rangle}\,, \qquad \forall \varphi \in \mathfrak{D}(\mathbf{T})$$

or

$$\int \varphi \,.\, \tilde{U} = \overline{\int \tilde{\varphi} \,.\, U}\,.$$

Theorem 10. *If* $U, V \in \mathfrak{D}'(\mathbf{T})$, *then*

(i) $\mathcal{F}_{\mathbf{T}}(U * V) = \mathcal{F}_{\mathbf{T}} U \,.\, \mathcal{F}_{\mathbf{T}} V$

(ii) $\mathcal{F}_{\mathbf{T}}(\tilde{U}) = \overline{\mathcal{F}_{\mathbf{T}}(U)}$.

These formulas remain valid if we replace $\mathcal{F}$ *by* $\mathcal{F}^*$.

Proof. (I) Define $W = U * V$. Then,

$$\begin{aligned}(\mathcal{F}W)(n) &= \int e^{in\theta}\, W^{\theta} \\ &= \int e^{in(\theta_1+\theta_2)}\, U^{\theta_1}\, V^{\theta_2} \\ &= \int e^{in\theta_1} \,.\, U^{\theta_1} \,.\, \int e^{in\theta_2} \,.\, V^{\theta_2} \\ &= (\mathcal{F}U)(n) \,.\, (\mathcal{F}V)(n)\,.\end{aligned}$$

(II) The result is immediate.

From this theorem, we conclude immediately that $\mathfrak{D}'(\mathbf{T})$ is an algebra with respect to convolution. The unit element of the convolution algebra $\mathfrak{D}'(\mathbf{T})$ is the Dirac measure δ at the point 0 of $\mathbf{T}$.

Let us now look at operations on sequences. Multiplication of sequences is an elementary operation that we have already used.

We note simply that s' and s are algebras with respect to multiplication. If $\mathbf{a} \in s$ and $\mathbf{b} \in s'$, then $\mathbf{ab} \in s$. Conversely, one can show that if $\mathbf{b}$ is a sequence such that $\mathbf{ab} \in s$ for every $\mathbf{a} \in s$, then $\mathbf{b} \in s'$. Thus, s' is the algebra of multipliers on s.

For every sequence $\mathbf{a}$, we define the sequence $\tilde{\mathbf{a}}$ by $\tilde{\mathbf{a}}(n) = \overline{\mathbf{a}(-n)}$.

Let us now turn to convolution in the space of sequences. If we wish to follow the procedure analogous to that followed with $\mathbf{R}$, we denote by d the space of sequences with finite support* and we denote by d' the space of arbitrary sequences. If $\alpha \in d$ and $\mathbf{a} \in d'$, we define

$$\langle \alpha, \mathbf{a} \rangle = \sum_{n=-\infty}^{+\infty} \bar{\alpha}(n)\, \mathbf{a}(n) .$$

For two given sequences $\mathbf{a}, \mathbf{b} \in (d')$, the convolution product $\mathbf{c} = \mathbf{a} * \mathbf{b}$ is defined as

$$\sum_{n=-\infty}^{+\infty} \alpha(n)\, \mathbf{c}(n) = \sum_{p,q=-\infty}^{+\infty} \alpha(p+q)\, \mathbf{a}(p)\, \mathbf{b}(q)$$

if the right-hand member is defined for every $\alpha \in d$. For this, it is necessary and sufficient that it be defined for every $\alpha = \delta_n$ no matter what the value of n. Then,

$$\mathbf{c}(n) = \sum_{p+q=n} \mathbf{a}(p)\, \mathbf{b}(q) = \sum_{p=-\infty}^{+\infty} \mathbf{a}(p)\, \mathbf{b}(n-p) .$$

Let us give some examples where $\mathbf{a} * \mathbf{b}$ is defined:

(1) $\mathbf{a} \in s$, $\mathbf{b} \in s'$,
(2) $\mathbf{a}, \mathbf{b} \in l^2$,
(3) $\mathbf{a} \in l^1$, $\mathbf{b}$ is bounded,
(4) $\mathbf{a}$ and $\mathbf{b}$ have positive support,
(5) $\mathbf{a}$ or $\mathbf{b}$ has finite support.

We consider s as the space of convolvers on s'. We note that s' is also the space of convolvers on s. We have the formula

$$\langle \mathbf{a} * \mathbf{b}, \mathbf{c} \rangle = \langle \mathbf{b}, \tilde{\mathbf{a}} * \mathbf{c} \rangle \quad \text{for} \quad \mathbf{a} \in s, \mathbf{b} \in s', \mathbf{c} \in s .$$

Also,

$$\mathbf{a} * (\mathbf{b} * \mathbf{c}) = (\mathbf{a} * \mathbf{b}) * \mathbf{c}$$

if $\mathbf{a}$, $\mathbf{b}$, and $\mathbf{c}$ have positive support or if $\mathbf{a}, \mathbf{b} \in s$ and $\mathbf{c} \in s'$.

The space s is a convolution algebra.

Theorem 11. *If* $\mathbf{a} \in s$ *and* $\mathbf{b} \in s'$, *then*

*The support of a sequence $\mathbf{a}$ is defined as the set of those n such that $\mathbf{a}(n) \neq 0$.

(i) $\mathcal{F}_Z(\mathbf{a} * \mathbf{b}) = \mathcal{F}_Z \mathbf{a} \,.\, \mathcal{F}_Z \mathbf{b}$,
(ii) $\mathcal{F}_Z \tilde{\mathbf{a}} = \overline{\mathcal{F}_Z \mathbf{a}}$.

Proof. (i) Suppose that

$$\mathbf{c} = \mathbf{a} * \mathbf{b}.$$

We have

$$\begin{aligned}(\mathcal{F}\mathbf{c})(\theta) &= \sum_{n=-\infty}^{+\infty} \mathbf{c}(n)\, e^{in\theta} \\ &= \sum_{n=-\infty}^{+\infty} \left(\sum_{p+q=n} \mathbf{a}(p)\, \mathbf{b}(q) \right) e^{in\theta} \\ &= \sum_{p,q=-\infty}^{+\infty} \mathbf{a}(p)\, \mathbf{b}(q)\, e^{i(p+q)\theta} \\ &= \sum_{p=-\infty}^{+\infty} \mathbf{a}(p)\, e^{ip\theta} \,.\, \sum_{q=-\infty}^{+\infty} \mathbf{b}(q)\, e^{iq\theta} \\ &= (\mathcal{F}\mathbf{a})(\theta) \,.\, (\mathcal{F}\mathbf{b})(\theta)\,.\end{aligned}$$

(ii) The conclusion is immediate.

Remark. One can show that this theorem holds under other hypotheses on $\mathbf{a}$ and $\mathbf{b}$, for example, under the hypothesis that either $\mathbf{a}, \mathbf{b} \in l^2$ or $\mathbf{a} \in l^1$ and $\mathbf{b}$ is bounded.

We conclude this section by showing certain properties of the convolution algebra $\mathbf{d}'_+$ of sequences with positive support (these sequences can be identified with unilateral sequences).

Theorem 12. *d'_+ is an integral domain; that is, for $\mathbf{a}, \mathbf{b} \in d'_+$, the inequalities $\mathbf{a}, \mathbf{b} \neq 0$ imply $\mathbf{a} * \mathbf{b} \neq 0$.*

Proof. Let us suppose that $\mathbf{a}$ and $\mathbf{b}$ are nonzero. Let i be an integer such that $\mathbf{a}(i) \neq 0$ and $\mathbf{a}(p) = 0$ for $p < i$. Also, let j be an integer such that $\mathbf{b}(j) \neq 0$ and $\mathbf{b}(q) = 0$ for $q < j$. Then,

$$(\mathbf{a} * \mathbf{b})(i + j) = \sum_{p+q=i+j} \mathbf{a}(p)\, \mathbf{b}(q) = \mathbf{a}(i)\, \mathbf{b}(j) \neq 0\,.$$

Theorem 13. *If $\mathbf{a}(0) \neq 0$, then a has an inverse in d'_+ with respect to convolution.*

Proof. We seek $\mathbf{b} \in d'_+$ such that $\mathbf{a} * \mathbf{b} = \boldsymbol{\delta}$. For these relationships to be satisfied, we must have

$$\begin{cases} \displaystyle\sum_{p=0}^{n} \mathbf{a}(n-p)\, \mathbf{b}(p) = 0 \quad \text{for } n \geqslant 1 \\ \mathbf{a}(0)\, \mathbf{b}(0) = 1\,. \end{cases}$$

The first relation may be written

$$\mathbf{a}(0)\,\mathbf{b}(n) = -\sum_{p=0}^{n-1} \mathbf{a}(n-p)\,\mathbf{b}(p)\,,$$

and this enables us to find $\mathbf{b}(n)$ by induction.

Remark. With a sequence $\mathbf{a}$ having positive support we often associate a formal power series

$$\sum_{n=0}^{+\infty} \mathbf{a}(n)\,X^n\,,$$

which we denote by $S_{\mathbf{a}}(X)$. If

$$S_{\mathbf{b}}(X) = \sum_{p=0}^{+\infty} \mathbf{b}(p)\,X^p\,,$$

we set

$$\lambda S_{\mathbf{a}}(X) = S_{\lambda\mathbf{a}}(X)$$

$$S_{\mathbf{a}}(X) + S_{\mathbf{b}}(X) = S_{\mathbf{a}+\mathbf{b}}(X)$$

$$S_{\mathbf{a}}(X)\,S_{\mathbf{b}}(X) = \sum_{n,p=0}^{+\infty} \mathbf{a}(n)\,\mathbf{b}(p)\,X^{n+p}\,.$$

This last definition can be written

$$S_{\mathbf{a}}(X)\,S_{\mathbf{b}}(X) = S_{\mathbf{a}*\mathbf{b}}(X)\,.$$

Study of the algebra of formal series (for the laws previously defined*) is thus identified with the study of the algebra $\mathscr{d}'_+$ of the coefficients. Also, the word "formal" indicates that only the coefficients of the power series are of significance and that the nature of the "indeterminate" X is immaterial. If $\mathbf{a}$ has positive and finite support, then $S_{\mathbf{a}}(X)$ is a formal polynomial.**

4. RELATIONSHIPS BETWEEN HARMONIC ANALYSIS ON R AND HARMONIC ANALYSIS ON T OR Z

The evolute on **R** *of a function defined on* **T.** To every function φ on **T**, we have assigned a periodic function defined on **R** by

$$\dot{\varphi}(t) = \varphi(\dot{t}),$$

---

*There is another important law, namely, substitution. Cf. H. Cartan, *Elementary Theory of Analytic Functions of One or Several Complex Variables*, Addison-Wesley, 1963, Chapter I, section 1.

**We mention a tendency that is developing in mathematics instruction where "abstract polynomials" are defined as unilateral sequences that vanish from some point on; the set of these is equipped with the convolution algebra structure, and only then are "specific polynomials" of a specific variable defined.

where $\dot{t}$ is the class of $t \bmod 2\pi$. We shall call the function $\dot{\varphi}$ the *evolute* of **R** of the function φ. We have

$$\int_{\mathbf{T}} \varphi = \int_{-\pi}^{+\pi} \dot{\varphi} \ (\text{si } \varphi \in \mathbf{L}^1(\mathbf{T}))$$

and, consequently,

$$\int_{\mathbf{T}} \varphi\psi = \int_{-\pi}^{+\pi} \dot{\varphi}\dot{\psi} \ (\text{if } \varphi\psi \in \mathbf{L}^1(\mathbf{T}))\,.$$

Reduction to a function on **T** *of a function that is periodic on* **R.** Conversely, with every function f that is periodic on **R** is associated a function f^{T} defined on **T** by

$$f^{\mathsf{T}}(\theta) = f(t)\,, \quad \text{if } \theta = \dot{t}\,.$$

Obviously,

$$\begin{cases} (\dot{\varphi})^{\mathsf{T}} = \varphi \\ \dot{\widehat{f^{\mathsf{T}}}} = f\,. \end{cases}$$

We shall say that a distribution U is *periodic* on **R** with *period* τ if

$$U * \delta_\tau = U\,.$$

In what follows, we shall say "periodic" for "periodic with period 2π." We note that if *either* U or V is periodic, then the convolution product $U * V$ is also periodic if it is defined.

Reduction to a measure on **T** *of a periodic measure defined on* **R.** With every periodic measure μ defined on **R** is associated a measure μ^{T} defined on **T** by

$$\int_{\mathbf{T}} \varphi \cdot \mu^{\mathsf{T}} = \int_{[-\pi,+\pi[} \dot{\varphi} \cdot \mu\,.$$

If $\mu^t = f(t)\,dt$, where f is a periodic function, then $\mu^{\mathsf{T}} = f^{\mathsf{T}}(\theta)\,d\theta$. (The density of the reduction to **T** of a measure of density f is the reduction to **T** of f.)

It should be noted that the definition cannot be applied to a distribution. It should also be noted that the interval of integration chosen has to be the half-open interval $[-\pi, +\pi[$ (or any interval obtained from it by translation) and not the closed interval $[-\pi, +\pi]$. This precaution is superfluous if μ has a density with respect to Lebesgue measure.

Periodification of a bounded measure defined on **R.** *Notation.* We denote by Δ the measure

$$\sum_{k=-\infty}^{+\infty} \delta_{k.2\pi}\,.$$

We point out that $\Delta \in \mathbf{S}'$ and Δ is periodic.

Let f denote a member of $\mathfrak{D}^0(\mathbf{R})$. We have

$$(\Delta * f)(t) = \sum_{k=-\infty}^{+\infty} f(t - k \, . \, 2\pi),$$

and $\Delta * f$ is a continuous periodic function known as the periodification of f. If μ is a bounded measure on $\mathbf{R}$, then $\Delta * \mu$ is defined by the classical formula

$$\int f \, . \, (\Delta * \mu) = \int f(\theta_1 + \theta_2) \, \mu^{\theta_1} \, . \, \Delta^{\theta_2} \, .$$

This is true because the right-hand member is defined and equal to

$$\int \left(\int f(\theta_1 + \theta_2) \, . \, \Delta^{\theta_2} \right) . \, \mu^{\theta_1} = \int \left(\sum_{k=-\infty}^{+\infty} f(\theta_1 + k \, . \, 2\pi) \right) \mu^{\theta_1}$$
$$= \int (\Delta * f) \, . \, \mu \, .$$

In showing this, we have proven the formula

$$\int f \, . \, (\Delta * \mu) = \int (f * \Delta) \, . \, \mu \qquad (\mu \in \mathbf{M}(\mathbf{R}), f \in \mathfrak{D}^0(\mathbf{R}))$$

or

$$\langle f, \Delta * \mu \rangle = \langle \Delta * f, \mu \rangle \, .$$

We shall call the measure $\Delta * \mu$ the periodification of the measure μ.

If μ has a density g (belonging to $\mathbf{L}^1$) with respect to Lebesgue measure, then $\Delta * \mu$ has density $\Delta * g$, where

$$(\Delta * g)(t) \overset{p.p.}{=\!=} \int g(t - u) \, . \, \Delta^u \, ,$$

that is,

$$(\Delta * g)(t) \overset{p.p.}{=\!=} \sum_{k=-\infty}^{+\infty} g(t - k \, . \, 2\pi) \, .$$

Lemma 1. *If μ is a periodic measure on $\mathbf{R}$ and if $f \in \mathfrak{D}^0(\mathbf{R})$, then*

$$\int f \, . \, \mu = \int_{[-\pi, +\pi[} (\Delta * f) \, . \, \mu \, .$$

Proof. We have

$$\int f \cdot \mu = \sum_{k=-\infty}^{+\infty} \int_{I_k} f \cdot \mu, \text{ avec } I_k = [-\pi + k \cdot 2\pi, \pi + k \cdot 2\pi[$$

$$= \sum_{k=-\infty}^{+\infty} \int_{[-\pi,+\pi[} (f * \delta_{k.2\pi}) \cdot \mu$$

$$= \int_{[-\pi,+\pi[} (\Delta * f) \cdot \mu .$$

Piling on **T** *of a bounded measure defined on* **R.** Let μ denote a bounded measure defined on **R.** Its image on **T** under the mapping $t \to \dot{t}$ is a measure $\mathring{\mu}$ defined on **T** by

$$\int \varphi \cdot \mathring{\mu} = \int \varphi(\dot{t}) \cdot \mu^t . \qquad (\forall \varphi \in \mathfrak{D}^0(\mathbf{T})),$$

that is,

$$\int \varphi \cdot \mathring{\mu} = \int \dot{\varphi} \cdot \mu .$$

We shall call the mapping $\mu \to \mathring{\mu}$ a *piling.*

In particular, suppose that $\mu^t = g(t)\,\mathrm{d}t$, where $g \in \mathfrak{D}^0(\mathbf{R})$. We then have

$$\int \varphi \cdot \mathring{\mu} = \int \dot{\varphi}(t)\, g(t)\, \mathrm{d}t = \int_{[-\pi,+\pi[} \dot{\varphi} \cdot (\Delta * g) \qquad \text{(Lemma 1)}$$

$$= \int \varphi \cdot (\Delta * g)^{\mathsf{T}} .$$

Thus, we see that the density of $\mathring{\mu}$ is $(\Delta * g)^{\mathsf{T}}$. We define $\mathring{g} = (\Delta * g)^{\mathsf{T}}$. We then have

Lemma . *If the density of* $\mu(\in \mathbf{M}(\mathbf{R}))$ *is* $g \in \mathfrak{D}^0(\mathbf{R})$, *then the density of* $\mathring{\mu}$ *is*

$$\mathring{g} = (\Delta * g)^{\mathsf{T}} .$$

Remarks. I. We shall show later that $\mathring{\mu} = (\Delta * \mu)^{\mathsf{T}}$ for every bounded measure.

II. $(g * \delta_{2\pi})^\circ = \mathring{g}$.

Lemma 3. *The mapping* $f \to \mathring{f}$ *of* $\mathfrak{D}^0(\mathbf{R})$ (*resp.* $\mathfrak{D}(\mathbf{R})$) *into* $\mathfrak{D}^0(\mathbf{T})$ (*resp.* $\mathfrak{D}(\mathbf{T})$) *is surjective; that is, for every function* $\varphi \in \mathfrak{D}^0(\mathbf{T})$ (*resp.* $\mathfrak{D}(\mathbf{T})$), *there exists a function* $f \in \mathfrak{D}^0(\mathbf{R})$ (*resp.* $\mathfrak{D}(\mathbf{R})$) *such that* $\mathring{f} = \varphi$.

Proof. Let φ denote a member of $\mathfrak{D}(\mathbf{T})$. Suppose that φ vanishes in a neighborhood of 0. Let f denote the function

$$f(t) = \begin{cases} \dot{\varphi}(t) & \text{if } t \in [0, 2\pi[\\ 0 & \text{if } t \notin [0, 2\pi[. \end{cases}$$

This function is infinitely differentiable and

$$\Delta * f = \dot{\varphi}, \quad \text{so that } (\Delta * f)^{\mathrm{T}} = \varphi .$$

Therefore, the set $\mathcal{G}$ of functions $\overset{\circ}{f}$ corresponding to $f \in \mathfrak{D}(\mathbf{R})$ contains all infinitely differentiable functions that vanish in a neighborhood of 0. Also, $\mathcal{G}$ is invariant under translation (since $(f * \delta_a)^{\circ} = \overset{\circ}{f} * \delta_{\dot{a}}$). Consequently, $\mathcal{G}$ contains all infinitely differentiable functions with support different from $\mathbf{T}$. Since every function $\varphi \in \mathfrak{D}(\mathbf{T})$ is the sum of two infinitely differentiable functions with support different from $\mathbf{T}$,* we have $\varphi = \overset{\circ}{f}$ with $f \in \mathfrak{D}(\mathbf{R})$.

An analogous proof shows that every function $\varphi \in \mathfrak{D}^0(\mathbf{T})$ is of the form $\varphi = \overset{\circ}{f}$ with $f \in \mathfrak{D}^0(\mathbf{R})$.

Corollary. *Every continuous periodic function defined on* $\mathbf{R}$ *is of the form* $\Delta * f$, *where* $f \in \mathfrak{D}^0(\mathbf{R})$. *Every infinitely differentiable function periodic defined on* $\mathbf{R}$ *is of the form* $\Delta * f$, *where* $f \in \mathfrak{D}(\mathbf{R})$.

The evolute of $\mathbf{R}$ *of a measure or a distribution on* $\mathbf{T}$. Suppose that $U \in \mathfrak{D}'(\mathbf{T})$. The evolute onto $\mathbf{R}$ of U is defined as the distribution $\dot{U}$ defined by

$$\int f \cdot \dot{U} = \int \overset{\circ}{f} \cdot U \qquad (\forall f \in \mathfrak{D}(\mathbf{R})) .$$

The distribution $\dot{U}$ is periodic. If U is a measure, so is $\dot{U}$.

This definition is compatible with that given for a function. If $\{U_n\}$ converges to U in $\mathfrak{D}'(\mathbf{T})$, then $\{\dot{U}_n\}$ converges to $\dot{U}$ in $\mathfrak{D}'(\mathbf{R})$.

Lemma 4. *The mapping* $U \to \dot{U}$ *is injective; that is, the equation* $\dot{U} = 0$ *implies the equation* $U = 0$.

Proof. If $\dot{U} = 0$, we have

$$\int \overset{\circ}{f} \cdot U = 0 \qquad (\forall f \in \mathfrak{D}(\mathbf{R}))$$

and, consequently,

$$\int \varphi \cdot U = 0, \qquad \forall \varphi \in \mathfrak{D}(\mathbf{T}),$$

and hence, $U = 0$.

---

*To show this, let us take $\theta_0, \theta_1 \in \mathbf{T}$, where $\theta_0 \neq \theta_1$, and $a \in \mathfrak{D}(\mathbf{T})$ where $a(\theta) = \theta$ in a neighborhood of θ_0 and $a(\theta) = 1$ in a neighborhood of θ_1. Then, let us write $\varphi = a\varphi + (1 - a)\,\varphi$.

Lemma 5. *Let μ denote a periodic measure on* **R** *and let ν denote a measure on* **T**. *The two equations*

$$\mu^{\mathsf{T}} = \nu$$
$$\mu = \dot{\nu}$$

are equivalent.

Proof. (*a*) Assume $\mu^{\mathsf{T}} = \nu$. Let f denote a member of $\mathfrak{D}(\mathbf{R})$. We have

$$\int f \cdot \dot{\nu} = \int \mathring{f} \cdot \nu = \int \mathring{f} \cdot \mu^{\mathsf{T}} = \int_{[-\pi,+\pi[} \dot{\mathring{f}} \cdot \mu = \int_{[-\pi,+\pi[} (\Delta * f) \cdot \mu$$
$$= \int f \cdot \mu ,$$

so that $\mu = \dot{\nu}$.

(*b*) Assume $\mu = \nu$. Let φ denote a member of $\mathfrak{D}^0(\mathbf{T})$. Let us set $\varphi = \mathring{f} = (\Delta * f)^{\mathsf{T}}$ with $f \in \mathfrak{D}^0(\mathbf{R})$. We have

$$\int \varphi \cdot \nu = \int \mathring{f} \cdot \nu = \int f \cdot \dot{\nu} = \int f \cdot \mu = \int_{[-\pi,+\pi[} (\Delta * f) \cdot \mu$$
$$= \int_{\mathbf{T}} (\Delta * f)^{\mathsf{T}} \cdot \mu^{\mathsf{T}} = \int_{\mathbf{T}} \varphi \cdot \mu^{\mathsf{T}} ,$$

so that

$$\nu = \mu^{\mathsf{T}} .$$

Lemma 6. *Let μ denote a bounded measure on* **R**. *Then,*

$$\mathring{\mu} = (\Delta * \mu)^{\mathsf{T}} .$$

Proof. Suppose that $\varphi \in \mathfrak{D}^0(\mathbf{T})$. Let us set $\varphi = \mathring{f}$ with $f \in \mathfrak{D}^0(\mathbf{R})$. We have

$$\int \varphi \cdot \mathring{\mu} = \int \mathring{f} \cdot \mathring{\mu} = \int (\Delta * f)^{\mathsf{T}} \cdot \mathring{\mu} = \int (\Delta * f) \cdot \mu = \int f \cdot (\Delta * \mu)$$
$$= \int \mathring{f} \cdot (\Delta * \mu)^{\mathsf{T}} ,$$

from which the desired result follows.

Corollary. $\Delta^{\mathsf{T}} = \delta$.

Embedding. With every sequence $\mathbf{a}$, let us associate the measure

$$\underline{\mathbf{a}} = \sum_{k=-\infty}^{+\infty} \mathbf{a}(k)\, \delta_k .$$

We have the following lemmas, the proofs of which are immediate:

Lemma 7. *The mapping* $\mathbf{a} \to \underline{\mathbf{a}}$ *is injective.*

Lemma 8. *If* $\mathbf{a} \in s'$, *then* $\underline{\mathbf{a}} \in \mathbf{S}'$. *If* $\mathbf{a} \in s$, then $\underline{\mathbf{a}} \in \mathbf{O}'_c$.

Lemma 9. $\underline{\mathbf{a} * \mathbf{b}} = \underline{\mathbf{a}} * \underline{\mathbf{b}}$.

The sampling of a function. The sampling of a function f on $\mathbf{R}$ is defined as the sequence $f^\star$ representing the restriction of f to $\mathbf{Z}$. Thus,

$$f^\star(n) = f(n) \quad \text{for } n \in \mathbf{Z}.$$

Remark. $\underline{f^\star} = f \sum_{n=-\infty}^{+\infty} \delta_n$. We may write $f^{\underline{\star}}$ instead of $\underline{f^\star}$.

Theorem 14.

$$\begin{array}{lll} \text{(i)} & \mathcal{F}\dot{U} = \underline{\mathcal{F}U} & \textit{for } U \in \mathfrak{D}'(\mathbf{T}). \\ \text{(ii)} & \mathcal{F}(\Delta * \mu) = (\mathcal{F}\mu)^{\underline{\star}} & \textit{for } \mu \in \mathbf{M}(\mathbf{R}). \\ \text{(iii)} & \mathcal{F}\overset{\circ}{\mu} = (\mathcal{F}\mu)^\star & \textit{for } \mu \in \mathbf{M}(\mathbf{R}). \\ \text{(iv)} & \underline{\mathcal{F}\mu^{\mathsf{T}}} = \mathcal{F}\mu & \textit{for } \mu \textit{ periodic} \end{array}$$

$$\text{(v)} \qquad \left\{ \begin{array}{l} \mathcal{F}\underline{\mathbf{a}} = \widehat{\dot{\mathcal{F}\mathbf{a}}} \\ \mathcal{F}\underline{\mathbf{a}} \overset{\mathbf{S}'}{=\!=} \sum_{n=-\infty}^{+\infty} \mathbf{a}(n)\, e^{int} \end{array} \right\} \textit{for } \mathbf{a} \in s'.$$

Proof. (v) We need to show that the equation $U = \mathcal{F}\mathbf{a}$ implies $\dot{U} = \mathcal{F}\underline{\mathbf{a}}$. Now (in the notation of the proof of Theorem 7),

$$\underline{\mathbf{a}} = \lim_{\substack{N \to +\infty \\ M \to -\infty}} \underline{P_M^N \mathbf{a}}$$

(in $\mathbf{S}'$, the verification is immediate), so that

$$\mathcal{F}\underline{\mathbf{a}} = \lim_{\substack{N \to +\infty \\ M \to -\infty}} \mathcal{F}\underline{P_M^N \mathbf{a}} = \lim_{\substack{N \to +\infty \\ M \to -\infty}} \sum_{n=M}^{N} \mathbf{a}(n)\, e^{int}.$$

Therefore,

$$\mathcal{F}\underline{\mathbf{a}} \overset{\mathbf{S}'}{=\!=} \sum_{n=-\infty}^{+\infty} \mathbf{a}(n)\, e^{int}.$$

We recall that

$$\mathcal{F}\mathbf{a} \overset{\mathfrak{D}'(\mathbf{T})}{=\!=} \sum_{n=-\infty}^{+\infty} \mathbf{a}(n)\, e^{in\theta}.$$

Now, $\sum_{n=M}^{N} \mathbf{a}(n)\, e^{int}$ is the evolute of $\sum_{n=M}^{N} \mathbf{a}(n)^{in\theta}$. By taking the limit, we conclude that $\mathcal{F}\underline{\mathbf{a}}$ is the evolute of $\mathcal{F}\mathbf{a}$.

(i) We set $U = \mathcal{F}^\star\, \mathbf{a}$, that is, $\mathbf{a} = \dfrac{1}{2\pi} \mathcal{F}U$. Then,

$$\dot{U} = \mathcal{F}^\star\, \underline{\mathbf{a}},$$

so that

$$\mathcal{F}\dot{U} = \mathcal{F}\mathcal{F}^\star\, \underline{\mathbf{a}} = 2\pi\, \underline{\mathbf{a}} = \underline{\mathcal{F}U}\,.$$

(iv) follows from (i) when we set $\mu^{\mathsf{T}} = U$.

(iii) We have

$$(\mathcal{F}\overset{\circ}{\mu})\,(n) = \int_{\mathbf{T}} e^{in\theta}\,.\,\overset{\circ}{\mu}{}^{\theta} = \int e^{int}\,.\,\mu^{t} = (\mathcal{F}\mu)\,(n) = (\mathcal{F}\mu)^\star\,(n)\,.$$

(ii) We have

$$\overset{\circ}{\mu} = (\varDelta * \mu)^{\mathsf{T}},$$

so that by virtue of (iv) and (iii),

$$\mathcal{F}(\varDelta * \mu) = \underline{\mathcal{F}\overset{\circ}{\mu}} = \underline{(\mathcal{F}\mu)^\star}\,.$$

Thus, we have associated with a distribution U on $\mathbf{T}$ a periodic distribution on $\mathbf{R}$. The inverse operation has been defined only for measures. Now, let us study periodic distributions on $\mathbf{R}$.

Theorem 15. *The Fourier transform of every periodic tempered* distribution is a measure of the form*

$$\sum_{k=-\infty}^{+\infty} \mathbf{a}(k)\, \delta_k\,.$$

Proof. Let U denote a tempered periodic distribution. Define

$$h(t) = \frac{1}{\sqrt{2\pi}}\, e^{-t^2/2}\,.$$

Then, $U * h$ is a periodic function. Consequently, there exists a $\mathbf{b} \in \mathbf{s}'$ such that

$$\mathcal{F}(U * h) = \sum_{k=-\infty}^{+\infty} \mathbf{b}(k)\, \delta_k = \mathcal{F}U\,.\,\mathcal{F}h\,.$$

---

*One can show that every periodic distribution is tempered, so that we may drop the word "tempered" in the hypothesis.

Since $(\mathcal{F}h)(\omega) = e^{-\omega^2/2}$, we have

$$\mathcal{F}U = \sum_{k=-\infty}^{+\infty} e^{k^2/2}\, \mathbf{b}(k)\, \delta_k\,.$$

Corollary. *For every periodic tempered distribution U, there exists exactly one distribution V on* $\mathbf{T}$ *such that $\dot{V} = U$. We set $V = U^{\mathrm{T}}$.*

Proof. Suppose that

$$\mathcal{F}U = \sum_{k=-\infty}^{\infty} \mathbf{a}(k)\, \delta_k\,.$$

Let us set

$$V = \frac{1}{2\pi}\mathcal{F}^{\star}\mathbf{a}\,,$$

that is, $\mathbf{a} = \mathcal{F}V$. We have $\mathcal{F}\dot{V} = \underline{\mathbf{a}} = \mathcal{F}U$ and, consequently, $\dot{V} = U$. Conversely, if $\dot{V} = U$, it follows that $\underline{\mathcal{F}V} = \mathcal{F}U$, which shows that V is unique.

Theorem 16. *We have**

$$\mathcal{F}\Delta = \sum_{n=-\infty}^{+\infty} \delta_n$$

Proof. Let $\mu = \delta$ in formula (ii) of Theorem 14.

Corollary. (Poisson's summation formula.) *If $f \in \mathbf{S}$, then*

$$2\pi \sum_{n=-\infty}^{+\infty} f(n\,.\,2\pi) = \sum_{n=-\infty}^{+\infty} (\mathcal{F}f)(n)\,.$$

Proof. The left side of the equation is equal to $2\pi\langle \Delta, f\rangle$ and the right side is equal to

$$\left\langle \sum_{n=-\infty}^{+\infty} \delta_n, \mathcal{F}f\right\rangle = \langle \mathcal{F}\Delta, \mathcal{F}f\rangle = 2\pi\langle \Delta, f\rangle\,.$$

---

*This theorem takes a more desirable form if we adopt the following definition for the Fourier transformation (on $\mathbf{R}$) (cf. formulas at end of Chapter 3):

$$(\mathcal{F}\mu)(x) = \int e^{2\pi i \omega x}\cdot\mu^{\omega}.$$

We then obtain $\mathcal{F}\Delta = \Delta$ if we set

$$\Delta = \sum_{n=-\infty}^{+\infty} \delta_n\,.$$

The reason we have not used this definition is that it would conflict with well-established usage in the field of application that we have in mind.

11

Discrete Systems, Part II

1. THE LAPLACE TRANSFORM OF A UNILATERAL SEQUENCE

A unilateral sequence **a** that maps $\mathbf{Z}_+$ into **R** or **C** can also be considered as a bilateral sequence with positive support by setting $\mathbf{a}(k) = 0$ for $k < 0$.

Definition. The Laplace transform of a unilateral sequence **a** is defined as the function $\mathcal{L}\mathbf{a}$ of a complex variable z, where $\mathcal{L}\mathbf{a}$ is defined by

$$(\mathcal{L}\mathbf{a})(z) = \sum_{n=0}^{\infty} \frac{\mathbf{a}(n)}{z^n}.$$

There exists a nonnegative finite or infinite number R, known as the *radius of convergence*, such that the series on the right converges absolutely for $|z| > R$ (and in certain cases, for $|z| = R$) but diverges for $|z| < R$. The circle of radius R is called the *circle of convergence*. The function $\mathcal{L}\mathbf{a}$ will be assumed defined for $|z| > R$ (at times for $|z| \geqslant R$).

The Laplace transform (at z) of a sequence **a** is often called the transform at z of that sequence.

With the sequence **a** we can associate the measure

$$\underline{\mathbf{a}} = \sum_{n=0}^{+\infty} \mathbf{a}(n)\, \delta_n ,$$

the Laplace transform of which is

$$(\mathcal{L}\underline{a})(p) = \sum_{n=0}^{\infty} \mathbf{a}(n)\, e^{-np} = (\mathcal{L}\mathbf{a})(e^p).$$

It should be noted that $\mathcal{L}\underline{a}$ is a function of period $2i\pi$.

Example. Let $\mathcal{Y}^*$ denote the sequence defined by

$$\begin{cases} \mathcal{Y}^*(n) = 1 & \text{for} \quad n \geqslant 0, \\ \mathcal{Y}^*(n) = 0 & \text{for} \quad n < 0. \end{cases}$$

$\mathcal{Y}^*$ is indeed the sampling of Heaviside's function $\mathcal{Y}$.

We have

$$\mathcal{L}\mathcal{Y}^*(z) = \sum_{n=0}^{\infty} \frac{1}{z^n} = \frac{z}{z-1}.$$

We recall that

$$\underline{\mathcal{Y}}^* = \sum_{n=0}^{+\infty} \delta_n,$$

so that

$$\mathcal{L}\underline{\mathcal{Y}}^*(p) = \sum_{n=0}^{+\infty} e^{-np} = \frac{e^p}{e^p - 1}.$$

Theorem 1.

(i) *The Laplace transform of sequences is a linear mapping*:

$$[\mathcal{L}(\mathbf{a} + \mathbf{b})](z) = (\mathcal{L}\mathbf{a})(z) + (\mathcal{L}\mathbf{b})(z),$$

outside the greater of the circles of convergence of $\mathcal{L}\mathbf{a}$ *and* $\mathcal{L}\mathbf{b}$.

(ii) $\mathcal{L}(\mathbf{a} * \mathbf{b}) = \mathcal{L}\mathbf{a} \,.\, \mathcal{L}\mathbf{b}$ *outside the greater of the circles of convergence of* $\mathcal{L}\mathbf{a}$ *and* $\mathcal{L}\mathbf{b}$.

Proof. The proof of (i) is immediate. To prove (ii), let us refer to the analogous property for the Laplace transform on $\mathbf{R}$. We have

$$\begin{aligned} \mathcal{L}(\mathbf{a} * \mathbf{b})(e^p) &= [\mathcal{L}(\underline{\mathbf{a} * \mathbf{b}})] \quad (p) = [\mathcal{L}(\underline{\mathbf{a}} * \underline{\mathbf{b}})](p) \\ &= [\mathcal{L}\underline{\mathbf{a}}](p) \,.\, [\mathcal{L}\underline{\mathbf{b}}](p) = (\mathcal{L}\mathbf{a})(e^p) \,.\, (\mathcal{L}\mathbf{b})(e^p). \end{aligned}$$

The following theorem shows the relationship between the Laplace and Fourier transforms of sequences:

Theorem 2.

$$(\mathcal{L}\mathbf{a})(\rho\, e^{i\theta}) = \left[\mathcal{F}^* \left\{ \frac{\mathbf{a}(n)}{\rho^n} \right\}\right](\theta), \quad (\rho \geqslant 0, \theta \in \mathbf{T}).$$

Proof. We have

$$(\mathcal{L}\mathbf{a})(\rho\, e^{i\theta}) = \sum_{n=0}^{\infty} \frac{\mathbf{a}(n)}{\rho^n}\, e^{-in\theta}.$$

Corollary. *If* $\mathcal{L}\mathbf{a} = \mathcal{L}\mathbf{b}$, *then* $\mathbf{a} = \mathbf{b}$.

The following table gives the Laplace transforms of certain common sequences:

	$\mathbf{a}$	$\mathcal{L}\mathbf{a}$
	δ	1
	δ_k	$\dfrac{1}{z^k}$
	λ^n	$\dfrac{z}{z-\lambda}$
	1	$\dfrac{z}{z-1}$
	n	$\dfrac{z}{(z-1)^2}$
$n \to \mathcal{Y}^* \times$	n^2	$\dfrac{z(z+1)}{(z-1)^3}$
	$\sin(n\omega)$	$\dfrac{z \sin\omega}{z^2 - 2z\cos\omega + 1}$
	$\cos(n\omega)$	$\dfrac{z^2 - z\cos\omega}{z^2 - 2z\cos\omega + 1}$.

The inverse Laplace transform. Let $A(z)$ denote a function that is holomorphic for $|z| > R$. Let us represent it in its Laurent series:

$$A(z) = \sum_{n=0}^{+\infty} \frac{\mathbf{a}(n)}{z^n}.$$

Then, $A = \mathcal{L}\mathbf{a}$. Thus, every function A that is holomorphic for $|z| > R$ is the Laplace transform of a sequence $\mathbf{a}$. This fact gives us a practical method of evaluating $\mathbf{a}$. We can find the coefficients $\mathbf{a}(n)$ by using the classical formulas

$$(\alpha) \qquad \mathbf{a}(n) = \frac{1}{2\,i\pi} \int_\Gamma A(z)\, z^{n-1}\, dz,$$

where Γ is a circle of radius $\rho > R$ described in the positive direction. On the other hand, the formula for the inverse Fourier transform yields, for $\rho > R$,

$$\frac{\mathbf{a}(n)}{\rho^n} = \frac{1}{2\pi} \mathcal{F} \{ A(\rho\, e^{i\theta}) \}$$

$$= \frac{1}{2\pi} \int_{\mathbf{T}} A(\rho\, e^{i\theta})\, e^{in\theta} d\theta ,$$

so that

$$\mathbf{a}(n) = \frac{\rho^n}{2\pi} \int_{\mathbf{T}} A(\rho\, e^{i\theta})\, e^{in\theta}\, d\theta .$$

This formula is merely the formula obtained if we set $z = \rho\, e^{i\theta}$ in formula (α). Thus, the formula for the inverse Fourier transform does not yield a new method.

There is another method, consisting in decomposing $A(z)$ into a linear combination of functions that one can find in tables. This method is used primarily by engineers who have constant occasion to use the Laplace transform of sequences. Of course, this method requires very complete tables.

Applications of the Laplace transform of sequences. We shall confine ourselves to the application to convolution equations, especially, difference equations.

We note that*

$$(\mathbf{a} * \boldsymbol{\delta}_1)(n) = \mathbf{a}(n-1)$$

and, more generally,

$$(\mathbf{a} * \boldsymbol{\delta}_k)(n) = \mathbf{a}(n-k) .$$

Example. Solve the difference equation

$$\mathbf{a}(n) - 3\,\mathbf{a}(n-1) + 2\,\mathbf{a}(n-2) = 1 \quad \text{for} \quad n \geqslant 2$$

with initial conditions $\mathbf{a}(0) = 1, \mathbf{a}(1) = 0.$

We naturally set $\mathbf{a}(n) = 0$ for $n < 0$. Then,

$$\mathbf{a}(n) - 3\,\mathbf{a}(n-1) + 2\,\mathbf{a}(n-2) = \begin{cases} 1 \text{ for } n \geqslant 2 \\ -3 \text{ for } n = 1 \\ 1 \text{ for } n = 0 \\ 0 \text{ for } n < 0 , \end{cases}$$

which may be written

$$(\boldsymbol{\delta} - 3\,\boldsymbol{\delta}_1 + 2\,\boldsymbol{\delta}_2) * \mathbf{a} = \mathcal{Y}^* - 4\,\boldsymbol{\delta}_1 .$$

If we then set $\mathfrak{L}\mathbf{a} = A$, we have

---

*The operator $\mathbf{a} \to \mathbf{a} * \boldsymbol{\delta}_1$ is often called the unit advance operator and is denoted by $E: \mathbf{a} * \boldsymbol{\delta}_1 = E\mathbf{a}$.

$$\left(1 - \frac{3}{z} + \frac{2}{z^2}\right) A(z) = \frac{z}{z-1} - \frac{4}{z},$$

which yields

$$A(z) = \frac{z(z-2)}{(z-1)^2}.$$

Let us calculate $\mathbf{a}(n)$:

$$\mathbf{a}(n) = \frac{1}{2i\pi} \int_{\Gamma} \frac{z^n(z-2)\,dz}{(z-1)^2},$$

where Γ is a circle of radius exceeding 1 and described in the positive sense. Consequently, $\mathbf{a}(n)$ is the residue of the function

$$\frac{z^n(z-2)}{(z-1)^2}$$

at the point $z = 1$. Let us set $z - 1 = u$. We have

$$\frac{z^n(z-2)}{(z-1)^2} = \frac{(1+u)^n(-1+u)}{u^2} = \frac{[1 + nu + O(u^2)](-1+u)}{u^2}$$

$$= -\frac{1}{u^2} + \frac{1-n}{u} + O(1) \quad \text{as} \quad u \to 0.$$

Consequently,

$$\mathbf{a}(n) = 1 - n \qquad (\text{for } n \geqslant 0).$$

More generally, we can apply the Laplace transform to the solution of convolution equations $\mathbf{a} * \mathbf{x} = \mathbf{b}$ where $\mathbf{a}$, $\mathbf{b}$, and $\mathbf{x}$ have positive supports. If $\mathbf{a}$ and $\mathbf{b}$ have Laplace transforms A and B and if $B(z)/A(z)$ is the Laplace transform of a sequence with positive support, then the equation in question has a unique solution $\mathbf{x}$, the Laplace transform of which is

$$X(z) = \frac{B(z)}{A(z)}.$$

We recall that if $\mathbf{a}(0) \neq 0$, the equation has a solution for every $\mathbf{b}$ and, in particular, an elementary solution $\mathbf{x}_0$ such that

$$\mathbf{a} * \mathbf{x}_0 = \boldsymbol{\delta}$$

and such that the Laplace transform of this solution, if it exists, is

$$X_0(z) = \frac{1}{A(z)}.$$

Thus, the solution of the equation $\mathbf{a} * \mathbf{x} = \mathbf{b}$ may be written

$$\mathbf{x} = \mathbf{x}_0 * \mathbf{b}.$$

If $X(p)$ exists, we may write

$$X(p) = X_0(p)\, B(p)\,.$$

Theorems on initial and final values. One of the two theorems on initial and final values becomes trivial in the case of sequences. It is

Theorem 3. $\lim_{z\to\infty} (\mathfrak{L}\mathbf{a})(z) = \mathbf{a}(0)$.

Suppose now that z is a member of $\mathbf{R}$ greater than 1. We have

$$\sum_{n=0}^{\infty} \frac{1}{z^n} = \frac{z}{z-1},$$

so that the sequence

$$n \to \frac{z-1}{z} \cdot \frac{1}{z^n}$$

may be considered on $\mathbf{Z}_+$ as a positive measure with total mass equal to 1. Consequently, if $\mathbf{a}(n) \in [c, d]$, we also have

$$\frac{z-1}{z} (\mathfrak{L}\mathbf{a})(z) \in [c, d]\,.$$

Theorem 4. *If* $\lim_{n\to+\infty} \mathbf{a}(n)$ *exists, then* $\mathfrak{L}\mathbf{a}$ *has radius of convergence not exceeding 1, and*

$$\lim_{\substack{z\geqslant 1\\ z\to 1}} \frac{z-1}{z} (\mathfrak{L}\mathbf{a})(z) = \lim_{n\to+\infty} \mathbf{a}(n)\,.$$

Proof. The hypothesis implies that the sequence $\mathbf{a}$ is bounded and hence that the series whose general term is $\mathbf{a}(n)/z^n$ is absolutely convergent for $|z| > 1$.

Let us suppose that $\lim_{n\to+\infty} \mathbf{a}(n) = 0$. We then have

$$(\alpha) \qquad \frac{z-1}{z} (\mathfrak{L}\mathbf{a})(z) = \frac{z-1}{z} \sum_{n=0}^{N} \frac{\mathbf{a}(n)}{z^n} + \frac{z-1}{z} \sum_{n=N+1}^{+\infty} \frac{\mathbf{a}(n)}{z^n}\,.$$

Let ε denote any positive number. Let us choose N so that $|\mathbf{a}(n)| \leqslant \varepsilon$ for $n \geqslant N$. We have

$$\left| \frac{z-1}{z} \sum_{n=N+1}^{+\infty} \frac{\mathbf{a}(n)}{z^n} \right| \leqslant \varepsilon\,.$$

Since,

$$\lim_{\substack{z\geqslant 1\\ z\to 1}} \frac{z-1}{z} \sum_{n=0}^{N} \frac{\mathbf{a}(n)}{z^n} = 0\,,$$

we conclude that

$$\lim_{\substack{z \geq 1 \\ z \to 1}} \frac{z-1}{z} (\mathfrak{L}\mathbf{a})(z) = 0 .$$

Let us now suppose that $\lim_{n \to +\infty} \mathbf{a}(n) = \lambda$. Let us apply the preceding result to the sequence $n \to \mathbf{a}(n) - \lambda \mathcal{Y}^\star$, the Laplace transform of which is $(\mathfrak{L}\mathbf{a})(z) - \lambda \dfrac{z}{z-1}$.

We get

$$\lim_{\substack{z \geq 1 \\ z \to 1}} \frac{z-1}{z} \left[(\mathfrak{L}\mathbf{a})(z) - \lambda \frac{z}{z-1} \right] = 0 ,$$

that is,

$$\lim_{\substack{z \geq 1 \\ z \to 1}} \frac{z-1}{z} (\mathfrak{L}\mathbf{a})(z) = \lambda .$$

2. SEQUENCES OF POSITIVE TYPE

Definition. A real or complex sequence $\mathbf{a}$ (on $\mathbf{Z}$) is said to be *of positive type* if, for every $n_1, \ldots, n_k \in \mathbf{Z}$, the matrix of the coefficients $\mathbf{a}(n_i - n_j)$ is positive-semidefinite. If we set $k = 2$, $n_1 = 0$, and $n_2 = n$, we conclude from the definition that, if $\mathbf{a}$ is of positive type, the matrix

$$\begin{bmatrix} \mathbf{a}(0) & \mathbf{a}(n) \\ \mathbf{a}(-n) & \mathbf{a}(0) \end{bmatrix}$$

is positive-semidefinite, so that

$$\begin{cases} \mathbf{a}(n) = \overline{\mathbf{a}(-n)}, \quad \text{that is} \quad \mathbf{a} = \tilde{\mathbf{a}} \\ \mathbf{a}(0) \geq 0 \\ |\mathbf{a}(n)| \leq \mathbf{a}(0) . \end{cases}$$

Theorem 5. *If μ is a positive measure on* $\mathbf{T}$, *then $\mathcal{F}\mu$ is a sequence of positive type.*

Proof. Suppose that $n_1, \ldots, n_k \in \mathbf{Z}$ and $z_1, \ldots, z_k \in \mathbf{C}$. Then,

$$\begin{aligned} \sum_{i,j=1}^{k} \mathbf{a}(n_i - n_j)\, z_i \bar{z}_j &= \sum_{i,j=1}^{k} \int_{\mathbf{T}} \mathrm{e}^{\mathrm{i}(n_i - n_j)\theta}\, z_i \bar{z}_j \,.\, \mu^\theta \\ &= \sum_{i,j=1}^{k} \int_{\mathbf{T}} \mathrm{e}^{\mathrm{i}n_i\theta} z_i\, \mathrm{e}^{-\mathrm{i}n_j\theta}\, \bar{z}_j \,.\, \mu^\theta \\ &= \int_{\mathbf{T}} \left(\sum_{i=1}^{k} \mathrm{e}^{\mathrm{i}n_i\theta} z_i \right) \overline{\left(\sum_{j=1}^{k} \mathrm{e}^{\mathrm{i}n_j\theta} z_j \right)} \,.\, \mu^\theta \\ &= \int_{\mathbf{T}} \left| \sum_{i=1}^{k} \mathrm{e}^{\mathrm{i}n_i\theta} z_i \right|^2 .\, \mu^\theta \geq 0 . \end{aligned}$$

Theorem 6 (Bochner). *Every sequence of positive type is the Fourier transform of a positive measure.*

Proof. Let $\mathbf{a}$ denote a sequence of positive type. Then, $\mathbf{a}$ is bounded and hence a member of s'. Let us set $\mathbf{a} = \mathcal{F}U$, $U = (1/2\pi)\mathcal{F}^{\star}\,\mathbf{a}$, where $U \in \mathfrak{D}'(\mathbf{T})$.

(a) Let us show that, for every $\mathbf{b} \in s$, we have $\langle \tilde{\mathbf{b}} * \mathbf{b}, \mathbf{a} \rangle \geqslant 0$. We have

$$(\tilde{\mathbf{b}} * \mathbf{b})(n) = \sum_{m=-\infty}^{+\infty} \tilde{\mathbf{b}}(n-m)\,\mathbf{b}(m) = \sum_{m=-\infty}^{+\infty} \overline{\mathbf{b}(m-n)\,\mathbf{b}(m)}\,.$$

Consequently,

$$\langle \tilde{\mathbf{b}} * \mathbf{b}, \mathbf{a} \rangle = \sum_{n,m=-\infty}^{+\infty} \mathbf{b}(m-n)\,\overline{\mathbf{b}(m)}\,\mathbf{a}(n)$$

$$= \sum_{m,p=-\infty}^{+\infty} \mathbf{b}(p)\,\overline{\mathbf{b}(m)}\,\mathbf{a}(m-p)$$

(by setting $m - n = p$).

Now,

$$\sum_{m,p=-N}^{+N} \mathbf{b}(p)\,\overline{\mathbf{b}}(m)\,\mathbf{a}(m-p) \geqslant 0$$

from the definition of a sequence of positive type. Therefore,

$$\langle \tilde{\mathbf{b}} * \mathbf{b}, \mathbf{a} \rangle = \sum_{m,p=-\infty}^{+\infty} \mathbf{b}(p)\,\overline{\mathbf{b}}(m)\,\mathbf{a}(m-p) \geqslant 0\,.$$

(b) Let us show that $\int \overline{\psi}\psi\,.\,U \geqslant 0$ for every $\psi \in \mathfrak{D}(\mathbf{T})$. We set $\mathbf{b} = \mathcal{F}\psi$, that is, $\psi = (1/2\,\pi)\mathcal{F}^{\star}\,\mathbf{b}$. Then,

$$4\,\pi^2\,\overline{\psi}\psi = \mathcal{F}^{\star}(\tilde{\mathbf{b}} * \mathbf{b})\,,$$

so that

$$\int \overline{\psi}\psi\,.\,U = \langle \overline{\psi}\psi, U \rangle = \frac{1}{4\,\pi^2} \langle \tilde{\mathbf{b}} * \mathbf{b}, \mathbf{a} \rangle \geqslant 0\,.$$

(c) From this we conclude, just as in the case of Bochner's theorem on $\mathbf{R}$, that $\int \varphi\,.\,U \geqslant 0$ for $\varphi \geqslant 0$ and hence that U is a positive measure.

3. UNITARY REPRESENTATIONS OF Z IN A HILBERT SPACE

A unitary representation of $\mathbf{Z}$ in a Hilbert space H is defined as a mapping $n \to U_n$ of $\mathbf{Z}$ into $L(H, H)$ such that

$$\begin{cases} U_n \text{ is unitary} \\ U_{n+m} = U_n U_m \\ U_{-n} \;= U_n^{-1} \;(= U_n^{\star}) . \end{cases}$$

Necessarily, $U_n = (U_1)^n$. Consequently, every unitary representation of $\mathbf{Z}$ in H is of the form $n \to U^n$, where U is a unitary operator. Conversely, if U is unitary, the mapping $n \to U^n$ is a unitary representation.

Theorem 7. *If $n \to U^n$ is a unitary representation in H, then, for every $x \in H$, the sequence*

$$n \to \varphi_{x,x}(n) = \langle U^n x, x \rangle$$

is a sequence of positive type.

Proof.

$$\varphi_{x,x}(n_i - n_j) = \langle U^{n_i - n_j} x, x \rangle = \langle U^{n_i} x, U^{n_j} x \rangle .$$

Thus, the numbers $\varphi_{x,x}(n_i - n_j)$ are the coefficients of the Gram matrix of the vectors $U^{n_i} x$. Since this matrix is positive-semi-definite, the theorem is proven.

A consequence. We may set

$$\varphi_{x,x} = \mathcal{F} \varpi_{x,x} , \qquad \text{that is,} \qquad \varphi_{x,x}(n) = \int_{\mathbf{T}} e^{in\theta} \varpi^{\theta}_{x,x} ,$$

where $\varpi_{x,x}$ is a positive measure on $\mathbf{T}$.

For $x, y \in H$, if we set $\varphi_{x,y}(n) = \langle U^n x,y \rangle$, we get

$$\varphi_{x,y} = \mathcal{F} \varpi_{x,y} , \qquad \text{that is,} \qquad \varphi_{x,y}(n) = \int_{\mathbf{T}} e^{in\theta} \varpi^{\theta}{}_{x,y} ,$$

where $\varpi_{x,y}$ is a measure on $\mathbf{T}$.

The mapping $x, y \to \varpi_{x,y}$ is sesquilinear. One can easily verify that

$$\begin{cases} \varphi_{y,x} = \tilde{\varphi}_{y,x} \\ \varpi_{y,x} = \overline{\varpi_{x,y}} . \end{cases}$$

Theorem 8. *Let $n \to U^n$ denote a unitary representation in H. With every $\alpha \in l^1$, we can associate a continuous operator $U(\alpha)$ in H such that*

$$\langle x, U(\alpha)\, y \rangle = \sum_{n=-\infty}^{+\infty} \alpha(n) \langle x, U^n y \rangle \qquad (\forall x, y \in H) .$$

The mapping $\alpha \to U(\alpha)$ is a linear mapping of l^1 in $L(H, H)$ and it enjoys the following properties:

(i) $$\| U(\alpha) \| \leqslant \| \alpha \|_{l^1} = \sum_{n=-\infty}^{+\infty} | \alpha(n) | ,$$

(ii) $$U(\tilde{\alpha}) = U^*(\alpha) ,$$

(iii) $$U(\alpha * \beta) = U(\alpha) U(\beta) ,$$

(iv) $$U(\delta_n) = U^n .$$

The operator $U(\alpha)$ is called *Radon's operator.*

Proof. We have

$$\left| \sum_{n=-\infty}^{+\infty} \alpha(n) \langle x, U^n y \rangle \right| \leqslant \| \alpha \|_{l^1} \| x \| \| y \| .$$

Therefore, for every $y \in H$, the mapping

$$x \to \sum_{n=-\infty}^{+\infty} \alpha(n) \langle x, U^n y \rangle$$

is continuous and semilinear on H. Therefore, there exists a $y_1 \in H$ such that

$$\sum_{n=-\infty}^{+\infty} \alpha(n) \langle x, U^n y \rangle = \langle x, y_1 \rangle$$

and we have

$$\| y_1 \| \leqslant \| \alpha \|_{l^1} \| y \| .$$

Therefore, the mapping $y \to y_1$ is continuous and linear. We may set $y_1 = U(\alpha) y$, where $\| U(\alpha) \| \leqslant \| \alpha \|_{l^1}$, which proves the existence of $U(\alpha)$ and the fact that it enjoys property (i).

Let us prove (ii). For every $x, y \in H$,

$$\begin{aligned} \langle x, U(\tilde{\alpha}) y \rangle &= \sum_{n=-\infty}^{+\infty} \tilde{\alpha}(n) \langle x, U^n y \rangle \\ &= \sum_{n=-\infty}^{+\infty} \overline{\alpha(-n)} \langle U^{-n} x, y \rangle \\ &= \sum_{n=-\infty}^{+\infty} \overline{\alpha(n)} \langle U^n x, y \rangle \\ &= \sum_{n=-\infty}^{+\infty} \overline{\alpha(n)} \, \overline{\langle y, U^n x \rangle} \\ &= \overline{\langle y, U(\alpha) x \rangle} = \langle U(\alpha) x, y \rangle = \langle x, U^*(\alpha) y \rangle . \end{aligned}$$

Let us prove (iii). For every $x, y \in H$,

$$\begin{aligned} \langle x, U(\alpha * \beta) y \rangle &= \sum_{n=-\infty}^{+\infty} (\alpha * \beta)(n) \langle x, U^n y \rangle \\ &= \sum_{n,m=-\infty}^{+\infty} \alpha(n) \beta(m) \langle x, U^{n+m} y \rangle \end{aligned}$$

$$\langle x, U(\alpha)\,U(\beta)\,y\rangle = \sum_{n=-\infty}^{+\infty} \alpha(n)\,\langle x, U^n\,U(\beta)\,y\rangle$$

$$= \sum_{n=-\infty}^{+\infty} \alpha(n)\,\langle U^{-n}x, U(\beta)\,y\rangle$$

$$= \sum_{n,m=-\infty}^{+\infty} \alpha(n)\,\beta(m)\,\langle U^{-n}x, U^m\,y\rangle$$

$$= \sum_{n,m=-\infty}^{+\infty} \alpha(n)\,\beta(m)\,\langle x, U^{n+m}\,y\rangle$$

$$= \langle x, U(\alpha * \beta)\,y\rangle .$$

Property (iv) follows from the definition.

Theorem 9. *We have*

$$\begin{cases} \text{(i)} & \varphi_{x,U(\alpha)y} = \varphi_{x,y} * \alpha \\ \text{(ii)} & \varphi_{U(\beta)x,y} = \varphi_{x,y} * \tilde{\beta} \\ \text{(iii)} & \varpi_{x,U(\alpha)y} = \mathcal{F}^\star \alpha \,.\, \varpi_{x,y} \\ \text{(iv)} & \varpi_{U(\beta)x,y} = \overline{\mathcal{F}^\star \beta} \,.\, \varpi_{x,y} \end{cases} .$$

Proof. Let us prove (i). We have

$$\varphi_{x,U(\alpha)y}(n) = \langle U^n x, U(\alpha)\,y\rangle = \sum_{p=-\infty}^{+\infty} \alpha(p)\,\langle U^n x, U^p\,y\rangle$$

$$= \sum_{p=-\infty}^{+\infty} \alpha(p)\,\langle U^{n-p}\,x, y\rangle$$

$$= \sum_{p=-\infty}^{+\infty} \alpha(p)\,\varphi_{x,y}(n-p) .$$

Let us prove (ii).

$$\varphi_{U(\beta)x,y} = \tilde{\varphi}_{y,U(\beta)x} = (\varphi_{y,x} * \beta)^{\sim} = \varphi_{x,y} * \tilde{\beta} .$$

Let us prove (iii) and (iv). We have

$$\varpi_{x,U(\alpha)y} = \frac{1}{2\pi}\mathcal{F}^\star\,\varphi_{x,U(\alpha)y} = \frac{1}{2\pi}\mathcal{F}^\star\,\varphi_{x,y} \,.\, \mathcal{F}^\star\alpha = \varpi_{x,y} \,.\, \mathcal{F}^\star\alpha$$

$$\varpi_{U(\beta)x,y} = \frac{1}{2\pi}\mathcal{F}^\star\,\varphi_{U(\beta)x,y} = \frac{1}{2\pi}\mathcal{F}^\star\,\varphi_{x,y} \,.\, \mathcal{F}^\star\tilde{\beta} = \varpi_{x,y} \,.\, \overline{\mathcal{F}^\star\beta} .$$

4. STATIONARY RANDOM PROCESSES OF THE SECOND ORDER

In this section, we shall study random processes described by a sequence $n \to X(n)$ of scalar- or vector-valued random variables.

Such a process is said to be ***stationary*** if the probability distribution of the variables

$$X(n_1 - p), \ldots, X(n_k - p)$$

is identical to that of the variables $X(n_1), \ldots, X(n_k)$ for all values of the instants $n_1, \ldots, n_k$ (where k is arbitrary) and p.

Let H denote the space of the scalar random variables Z such that

$$\mathbf{E}(|Z|^2) < +\infty$$

and let $\overset{\circ}{H}$ denote the hyperplane of H constituted by the centered variables. A process

$$n \to X(n) = \begin{bmatrix} X^1(n) \\ \vdots \\ X^m(n) \end{bmatrix}$$

is said to be ***stationary of second order*** if

$$\begin{cases} X^i(n) \in H & (\forall n, i) \\ \mathbf{E}(X(n)) \quad \text{is independent of } n & \\ \mathbf{v}(X^i(n_1 - p), X^j(n_2 - p)) = \mathbf{v}(X^i(n_1), X^j(n_2)) & (\forall i, j, n_1, n_2, p). \end{cases}$$

Such a process is said to be *centered* if $\mathbf{E}(X(n)) = 0$. The process $n \to X(n) - \mathbf{E}(X(n))$ is centered.

Theorem 10. *A necessary and sufficient condition for a centered process*

$$n \to X(n) = \begin{bmatrix} X^1(n) \\ \vdots \\ X^m(n) \end{bmatrix}$$

to be stationary of second order is that there exist a unitary operator U on H such that

$$X^i(n - 1) = UX^i(n) \qquad (\forall i),$$

that is, a unitary representation $p \to U^p$ of $\mathbf{Z}$ in H such that

$$X^i(x - p) = U^p X^i(n).$$

Proof. The proof is analogous to that given for continuous processes. To prove the sufficiency, note that, if U exists,

$$\begin{aligned}\mathbf{v}(X^i(n_1 - p), X^j(n_2 - p)) &= \langle X^i(n_1 - p), X^j(n_2 - p) \rangle \\ &= \langle U^p X^i(n_1), U^p X^j(n_2) \rangle \\ &= \langle X^i(n_1), X^j(n_2) \rangle \\ &= \mathbf{v}(X^i(n_1), X^j(n_2)) .\end{aligned}$$

To prove the necessity, take the set $\mathfrak{N}_X$ of elements of $\mathring{H}$ of the form

$$\xi = \sum_{ih} a_i^h X^i(t_h)$$

and denote its closure by H_X. Consider the set Γ of pairs $\begin{bmatrix} \xi \\ \eta \end{bmatrix}$ such that ξ and η have representations of the form

$$\xi = \sum_{ih} a_i^h X^i(t_h)$$

$$\eta = \sum_{ih} a_i^h X^i(t_h - 1) .$$

One can verify that Γ is a vector subspace of $H_X \times H_X$ and that, if $\begin{bmatrix} \xi \\ \eta \end{bmatrix}$ and $\begin{bmatrix} \xi' \\ \eta' \end{bmatrix}$ belong to Γ, then

$$\langle \xi, \xi' \rangle = \langle \eta, \eta' \rangle ,$$

from which we conclude that, for every $\xi \in \mathfrak{N}_X$, there exists exactly one $\eta \in \mathfrak{N}_X$ such that $\begin{bmatrix} \xi \\ \eta \end{bmatrix} \in \Gamma$. Therefore, there exists a unitary operator U defined on H_X such that $U\xi = \eta$, and we have

$$X^i(n - 1) = UX^i(n) .$$

We can extend U to the entire space H by setting, for example, $U\xi = \xi$ for $\xi \in H_X^\perp$. Thus, we obtain a unitary operator in H that satisfies the conditions of the theorem.

We can also give a theorem analogous to Theorem 7A of Chapter 7. Let $t \to \begin{bmatrix} X(n) \\ Y(n) \end{bmatrix}$ denote a random process that is stationary of second order and centered.

The sequence $\varphi_{X,Y}$ defined by

$$\begin{aligned}\varphi_{X,Y}(p) = \mathbf{v}(X(n - p), Y(n)) &= \langle X(n - p), Y(n) \rangle \\ &= \langle U^p X(n), Y(n) \rangle\end{aligned}$$

is called an *intercorrelation function.*

In accordance with section 3,

$$\varphi_{X,Y} = \mathcal{F} \varpi_{X,Y} ,$$

where $\varpi_{X,Y}$ is a measure on **T**.

In particular, $\varphi_{X,X} = \mathcal{F}\varpi_{X,X}$, where $\varpi_{X,X}$ is a positive measure on **T**.

The measures $\varpi_{X,Y}$ and $\varpi_{Y,X}$ are called *spectral mutual power distributions* of X and Y. The measure $\varpi_{X,X}$ is called the *spectral power distribution* of X. We have the relations

$$\varphi_{Y,X} = \tilde{\varphi}_{X,Y}, \qquad \varpi_{Y,X} = \overline{\varpi_{X,Y}}.$$

Theorem 11. *Let $X: n \to X(n)$ denote a centered stationary random process of the second order. Suppose that $\boldsymbol{\alpha} \in l^1$. The process Z defined by*

$$Z(n) = \sum_{p=-\infty}^{+\infty} \boldsymbol{\alpha}(p) X(n-p) = U(\boldsymbol{\alpha}) X(n)$$

and the process

$$n \to \begin{bmatrix} X(n) \\ Z(n) \end{bmatrix}$$

are centered processes stationary of second order.

We set $Z = \boldsymbol{\alpha} * X$.

Proof. We have

$$U(\boldsymbol{\alpha}) X(n) = \sum_{p=-\infty}^{+\infty} \boldsymbol{\alpha}(p) U^p X(n) = \sum_{p=-\infty}^{+\infty} \boldsymbol{\alpha}(p) X(n-p).$$

On the other hand,

$$UZ(n) = UU(\boldsymbol{\alpha}) X(n) = U(\boldsymbol{\alpha}) UX(n) = U(\boldsymbol{\alpha}) X(n-1) = Z(n-1).$$

Consequently, the process Z and the process

$$n \to \begin{bmatrix} X(n) \\ Z(n) \end{bmatrix}$$

are processes stationary of second order.

Properties of convolutions. One can show, just as in the case of continuous processes, that convolutions of sequences have the following properties:

$$(\boldsymbol{\alpha} * \boldsymbol{\beta}) * X = \boldsymbol{\alpha} * (\boldsymbol{\beta} * X)$$

$$(\boldsymbol{\delta}_p * X)(n) = X(n-p)$$

$$\varphi_{\boldsymbol{\alpha} * X, Y} = \tilde{\boldsymbol{\alpha}} * \varphi_{X,Y}$$

$$\varphi_{X, \boldsymbol{\beta} * Y} = \boldsymbol{\beta} * \varphi_{X,Y}$$

$$\varpi_{\alpha * X, Y} = \overline{\mathcal{F}^{\star} \alpha} \,.\, \varpi_{X,Y}$$

$$\varpi_{X, \beta * Y} = \mathcal{F}^{\star} \beta \,.\, \varpi_{X,Y} \,.$$

In particular,

$$\varpi_{\alpha * X, \alpha * X} = |\mathcal{F}^{\star} \alpha|^2 \,.\, \varpi_{X,X} \,.$$

5. DISCRETE SERVO SYSTEMS

A stationary linear system is defined by a relation of the form

$$\mathbf{s} = \mathbf{f} * \mathbf{e} \,,$$

between the input $\mathbf{e}$ and the output $\mathbf{s}$. The sequence $\mathbf{f}$ will be called the *impulse response* (this is the value of $\mathbf{s}$ for $\mathbf{e} = \delta$). For physical reasons, we shall assume that it has positive support.

One can also define a stationary linear system by

(1) the response $\mathbf{f} * \mathcal{Y}^{\star}$ with unit step $\mathcal{Y}^{\star}$.

(2) the transfer function $F(z)$ representing the Laplace transform of the sequence $\mathbf{f}$: it establishes the following relation between the Laplace transforms $E(z)$ of $\mathbf{e}$ and $S(z)$ of $\mathbf{s}$:

$$S(z) = F(z)\, E(z)$$

(if e has positive support and has a Laplace transform).

(3) The harmonic response $F(e^{i\omega})$. It is defined if F is defined for $|z| > \rho$, where $\rho < 1$, and also for systems governed by a difference equation.

Example. If

$$\mathbf{s}_n - 4\,\mathbf{s}_{n-1} + 7\,\mathbf{s}_{n-2} = \mathbf{e}_n \,,$$

we obtain

$$\left(1 - \frac{4}{z} + \frac{7}{z^2}\right) S(z) = E(z) \,,$$

so that

$$S(z) = F(z)\, E(z) \,,$$

where

$$F(z) = \frac{1}{1 - \dfrac{4}{z} + \dfrac{7}{z^2}} = \frac{z^2}{z^2 - 4z + 7} \,.$$

We have

$$\frac{z}{z^2 - 4z + 7} = \frac{\frac{1}{2} - \frac{i}{\sqrt{3}}}{z - (2 + i\sqrt{3})} + \frac{\frac{1}{2} + \frac{i}{\sqrt{3}}}{z - (2 - i\sqrt{3})}$$

and

$$F(z) = \left(\frac{1}{2} - \frac{i}{\sqrt{3}}\right) \frac{z}{z - (2 + i\sqrt{3})} + \left(\frac{1}{2} + \frac{i}{\sqrt{3}}\right) \frac{z}{z - (2 - i\sqrt{3})}.$$

Consequently,

$$\mathbf{f}(n) = \left(\frac{1}{2} - \frac{i}{\sqrt{3}}\right) (2 + i\sqrt{3})^n + \left(\frac{1}{2} + \frac{i}{\sqrt{3}}\right) (2 - i\sqrt{3})^n.$$

Let us set $2 + i\sqrt{3} = \sqrt{7}\, e^{i\alpha}$. We then obtain

$$\begin{aligned} \mathbf{f}(n) &= \left(\frac{1}{2} - \frac{i}{\sqrt{3}}\right) 7^{n/2} e^{in\alpha} + \left(\frac{1}{2} + \frac{i}{\sqrt{3}}\right) 7^{n/2} e^{-in\alpha} \\ &= 7^{n/2} \left(\cos n\alpha + \frac{2}{\sqrt{3}} \sin n\alpha\right). \end{aligned}$$

The calculation of formulas for the transfer functions of discrete servo systems are the same as in the continuous case. Suppose that $\mathbf{f}$ has positive support. We can, in general, define a servo system with unit feedback by

$$(\mathbf{e} - \mathbf{s}) * \mathbf{f} = \mathbf{s},$$

that is,

$$(\mathbf{f} + \boldsymbol{\delta}) * \mathbf{s} = \mathbf{e} * \mathbf{f}.$$

Indeed, we know (Chapter 10, Theorem 13) that, if $\mathbf{f}(0) \neq -1$, then $\mathbf{f} + \boldsymbol{\delta}$ has an inverse in d'_+, which we denote by $\mathbf{g}_1$ and the relation between $\mathbf{e}$ and $\mathbf{s}$ can be written

$$\mathbf{s} = \mathbf{g} * \mathbf{e}, \quad \text{with } \mathbf{g} = \mathbf{g}_1 * \mathbf{f}.$$

If $\mathbf{g}$ has a Laplace transform G, this transform is equal to

$$G(z) = \frac{F(z)}{1 + F(z)}$$

and constitutes the transfer function in closed loop. In the case in which the feedback system has transfer function $H(z)$, we similarly find that the transfer function in closed loop is

$$G(z) = \frac{F(z)}{1 + H(z)\, F(z)}.$$

We shall say that a system with impulse response $\mathbf{f}$ is *stable* if

$$\lim_{n \to +\infty} \mathbf{f}(n) = 0,$$

that it is *quasistable* if the sequence $\mathbf{f}$ is bounded, and that it is *unstable* if the sequence $\mathbf{f}$ is not bounded.

If $F = \mathfrak{L}\mathbf{f}$ is meromorphic in the exterior of a disk of radius $\rho < 1$, a necessary and sufficient condition for stability is that F be holomorphic for $|z| \geqslant 1$.

In particular, if the relationship between $\mathbf{e}$ and $\mathbf{s}$ is given by

$$a_0\, \mathbf{s}(n-p) + a_1\, \mathbf{s}(n-p+1) + \cdots + a_p\, \mathbf{s}(n) = \mathbf{e}(n)\,,$$

we have

$$S(z) = F(z) \,.\, E(z)\,,$$

where

$$F(z) = \frac{z^p}{a_0 + a_1 z + \cdots + a_p z^p}\,.$$

The system will be stable if all the zeros of the denominator are of absolute value less than 1; it will be quasistable if all these zeros are of absolute value not exceeding 1; it will be unstable otherwise.

In the case of discrete servo systems, we can use a criterion analogous to Nyquist's criterion.

6. SAMPLING. BLOCKING

Theorem 12. *Let f denote a function that vanishes for $t < 0$, that is continuous for $t > 0$, and that approaches a limit $f(+0)$ as t approaches 0 from above. Let F denote the Laplace transform of f and let ξ_0 denote the abscissa of integrability.*

I. *If $F(p) = O(1/p^2)$ as $p \to +\infty$ (which implies that $f(+0) = 0$), the Laplace transform of $f^\star$ has abscissa of integrability not exceeding ξ_0 and*

$$(\mathfrak{L}f^\star)(p) = \sum_{k=-\infty}^{+\infty} F(p + k \,.\, 2\,\mathrm{i}\,\pi) \qquad \textit{for} \quad \mathfrak{R}(p) > \xi_0\,.$$

II. *If $F(p) = A/p + O(1/p^2)$ for $p \to +\infty$, the Laplace transform of $f^\star(p)$ has abscissa of integrability not exceeding ξ_0 and*

$$(\mathfrak{L}f^\star)(p) = \lim_{M \to +\infty} \sum_{k=-M}^{+M} F(p + k \,.\, 2\,\mathrm{i}\,\pi) + \frac{f(+0)}{2}\,.$$

Proof. I. Suppose that $p = \xi + \mathrm{i}\,\omega$, where $\xi > \xi_0$. Let Φ_ξ denote the function defined by

$$\Phi_\xi(\omega) = F(\xi + \mathrm{i}\,\omega)\,.$$

We have

$$\Phi_\xi \in \mathbf{L}^1, \Phi_\xi = \mathcal{F}^\star\{\mathrm{e}^{-\xi t} f(t)\}$$

and

$$\sum_{k=-\infty}^{+\infty} F(p + k\,.\,2\,\mathrm{i}\,\pi) = \sum_{k=-\infty}^{+\infty} \Phi_\xi(\omega + k\,.\,2\,\pi) = (\Delta * \Phi_\xi)(\omega)$$

$$\mathcal{F}(\Delta * \Phi_\xi) = \mathcal{F}\Delta\,.\,\mathcal{F}\Phi_\xi = 2\,\pi\left(\sum_{n=-\infty}^{+\infty}\delta_n\right).\,\mathrm{e}^{-\xi \cdot} f$$

$$= 2\,\pi \sum_{n=0}^{+\infty} \mathrm{e}^{-n\xi} f(n)\,\delta_n = 2\,\pi\,\mathrm{e}^{-\xi\cdot} f^\star\,.$$

Since $\Delta * \Phi_\xi$ is a bounded periodic function, the sequence $\{\mathrm{e}^{-\xi n} f^\star\}$ is bounded for every $\xi > \xi_0$. Thus, $\mathrm{e}^{-\xi t} f^\star$ is a bounded measure for $\xi > \xi_0$. Thus, the Laplace transform of $f^\star$ has abscissa of convergence not exceeding ξ_0, and

$$\sum_{k=-\infty}^{+\infty} F(\xi + \mathrm{i}\,\omega + k\,.\,2\,\mathrm{i}\,\pi) = \mathcal{F}^\star\{\mathrm{e}^{-\xi\cdot} f^\star\}(\omega)$$

$$= (\mathcal{L}f^\star)(\xi + \mathrm{i}\,\omega)\,.$$

II. Let us write

$$\sum_{k=-\infty}^{+\infty}{}' \quad \text{for} \quad \lim_{M\to+\infty} \sum_{k=-M}^{+M}.$$

We set

$$f_1(t) = \begin{cases} f(t) - f(+0) & \text{for } t \geqslant 0 \\ 0 & \text{for } t < 0. \end{cases}$$

We have

$$F_1(p) = (\mathcal{L}f_1)(p) = F(p) - \frac{f(+0)}{p}$$

$$f^\star = f_1^\star + f(+0)\,\mathcal{Y}^\star$$

$$(\mathcal{L}\mathcal{Y}^\star)(p) = \sum_{k=0}^{+\infty} \mathrm{e}^{-np} = \frac{1}{1 - \mathrm{e}^{-p}}\,,$$

so that

$$(\mathcal{L}f^\star)(p) = (\mathcal{L}f_1^\star)(p) + \frac{f(+0)}{1 - \mathrm{e}^{-p}}$$

$$= \sum_{k=-\infty}^{+\infty} F_1(p + k\,.\,2\,\mathrm{i}\,\pi) + \frac{f(+0)}{1 - \mathrm{e}^{-p}}$$

$$= \sum_{k=-\infty}^{+\infty}{}' \left[F(p + k\,.\,2\,\mathrm{i}\,\pi) - \frac{f(+0)}{p + k\,.\,2\,\mathrm{i}\,\pi}\right] + \frac{f(+0)}{1 - \mathrm{e}^{-p}}\,.$$

Since the quantity $\sum_{k=-\infty}^{+\infty}{}' \frac{1}{p+k\,.\,2\,i\,\pi}$ exists, so does the quantity

$$\sum_{k=-\infty}^{+\infty}{}' F(p+k\,.\,2\,i\,\pi)\,,$$

and we may write

$$(\mathfrak{L}f^{\star})(p) = \sum_{k=-\infty}^{+\infty}{}' F(p+k\,.\,2\,i\,\pi)$$

$$(\alpha) \qquad + f(+0)\left[\frac{1}{1-e^{-p}} - \sum_{k=-\infty}^{+\infty}{}' \frac{1}{p+k\,.\,2\,i\,\pi}\right].$$

In view of the classical formula

$$\cot z = \sum_{k=-\infty}^{+\infty}{}' \frac{1}{z+k\pi}\,,$$

we easily obtain

$$(\beta) \qquad \sum_{k=-\infty}^{+\infty}{}' \frac{1}{p+k\,.\,2\,i\,\pi} = \frac{1}{2}\coth\frac{p}{2}\,.$$

Also,

$$(\gamma) \qquad \frac{1}{1-e^{-p}} = \frac{1}{2}\coth\frac{p}{2} + \frac{1}{2}\,.$$

If we substitute (β) and (γ) into (α), we obtain the desired formula.

Remark. If the period of sampling is T, we find, by setting

$$f^{\star} = \sum_{n=0}^{+\infty} f(nT)\,\delta_{nT}\,,$$

that

$$(\mathfrak{L}f^{\star})(p) = \frac{1}{T}\sum_{k=-\infty}^{+\infty}{}' F\left(p + k\cdot\frac{2\,i\pi}{T}\right) + \frac{f(+0)}{2}\,.$$

In particular, we have

$$(\mathfrak{L}f^{\star})(i\,\omega) = \frac{1}{T}\sum_{k=-\infty}^{+\infty}{}' F\left(i\,\omega + k\cdot\frac{2\,i\,\pi}{T}\right) + \frac{f(+0)}{2}\,.$$

This formula enables us to construct the Nyquist locus of $f^{\star}$ point by point. We note that the mapping $\omega \to (\mathfrak{L}f^{\star})(i\,\omega)$ is periodic with period $2\,\pi$.

The operation that assigns to every sequence **a** the function f defined by

$$f(t) = \mathbf{a}_n \quad \text{for} \quad t \in [nT, (n+1)T]$$

is called *blocking.* A system performing this operation is called a *blocking unit.* By definition,

$$f = \underline{\mathbf{a}} * \alpha \,,$$

where

$$\underline{\mathbf{a}} = \sum_{n=-\infty}^{+\infty} \mathbf{a}(n)\, \delta_{nT}$$

and

$$\alpha(t) = \begin{cases} 1 & \text{for} \quad t \in [0, 1[\,, \\ 0 & \text{for} \quad t \notin [0, 1[\,. \end{cases}$$

Let us suppose that $\mathbf{a}$ has positive support. Let

$$A(p) = \sum_{n=0}^{+\infty} \mathbf{a}(n)\, \mathrm{e}^{-pn}$$

denote the Laplace transform of $\underline{\mathbf{a}}$. Since

$$\alpha = \mathcal{Y} - \delta_T * \mathcal{Y},$$

we have

$$(\mathcal{L}\alpha)(p) = \frac{1}{p} - \frac{\mathrm{e}^{-pT}}{p} = \frac{1 - \mathrm{e}^{-pT}}{p}\,.$$

Therefore, the Laplace transform of f is

$$(\mathcal{L}f)(p) = A(p)\, \frac{1 - \mathrm{e}^{-pT}}{p}\,.$$

Sampling of continuous random processes. Let $t \to X(t)$ (where $t \in \mathbf{R}$) denote a centered stationary random process of second order. The process $n \to X(n)$ (for $n \in \mathbf{Z}$) is a discrete stationary second-order process which we shall denote by $X^\star$. Thus, we set

$$X^\star(n) = X(n)$$

for $n \in \mathbf{Z}$. Then,

$$\varphi_{X^\star, X^\star}(p) = \mathbf{v}\,(X^\star(n-p), X^\star(p)) = \varphi_{X,X}(p)$$

and, consequently,

$$\varphi_{X^\star, X^\star} = \varphi^\star_{X,X}\,.$$

From this, we conclude that $\varpi_{X^\star,X^\star} = \mathring{\varpi}_{X.X}$ and, hence, $\dot{\varpi}_{X^\star,X^\star} = \Delta * \varpi_{X,X}$. In particular, if the density of $\varpi_{X.X}$ is $\Phi_{X,X}$, then the density of $\omega_{X^\star,X^\star}$ will be $\Phi_{X^\star,X^\star} = \mathring{\Phi}_{X,X}$ (on **T**). The evolute of **R** of this spectral density is

$$\dot{\Phi}_{X^\star,X^\star} = \Delta * \Phi_{X,X} .$$

If $\Phi_{X,X}$ is continuous, we have

$$\Phi_{X^\star,X^\star}(\omega) = \sum_{k=-\infty}^{+\infty} \Phi_{X,X}(\omega + k \,.\, 2\pi) .$$

12

Convex Sets

1. DEFINITIONS AND ELEMENTARY PROPERTIES

All the vector spaces considered in the present chapter are real.

Definition. For a given vector space E, a *convex combination* of the elements $x_1, \ldots, x_k$ of E is defined as any element x of the form

$$x = \sum_{i=1}^{k} \lambda_i x_i , \qquad \text{with} \qquad \sum_{i=1}^{k} \lambda_i = 1, \ \lambda_i \geqslant 0 \, (\forall i) .$$

We denote by Λ_k the set of elements

$$\lambda = (\lambda_1, \ldots, \lambda_k) \quad \text{in } \mathbf{R}^k$$

such that

$$\sum_{i=1}^{k} \lambda_i = 1 \qquad \text{and} \qquad \lambda_i \geqslant 0 \qquad (\forall i) .$$

We denote by

$[x_1, x_2]$ the set of all $x = \lambda_1 x_1 + \lambda_2 x_2$ with $\lambda \in \Lambda_2$, ,
$]x_1, x_2[$ the set of all $x = \lambda_1 x_1 + \lambda_2 x_2$ with $\lambda \in \Lambda_2$, $\lambda_1 \neq 0$, $\lambda_2 \neq 0$,
$]x_1, x_2]$ the set of all $x = \lambda_1 x_1 + \lambda_2 x_2$ with $\lambda \in \Lambda_2$, $\lambda_2 \neq 0$,
$[x_1, x_2[$ the set of all $x = \lambda_1 x_1 + \lambda_2 x_2$ with $\lambda \in \Lambda_2$, $\lambda_1 \neq 0$.

The first of these four cases is called a *segment* with *end points* x_1 and x_2.

A subset A of a vector space E is said to be *convex* if the condition $x_1, x_2 \in A$ implies $[x_1, x_2] \subset A$.

Examples. All balls (no matter what their centers) of a normed vector space are convex sets.

The set Λ_k is convex in $\mathbf{R}^k$.

Theorem 1. *Let A denote a convex subset of a vector space E. For every $x_1, \ldots, x_k \in A$, every convex combination of $x_1, \ldots, x_k$ belongs to A.*

Proof. Since the property mentioned in the conclusion in the theorem holds for $k = 1$ and $k = 2$, we prove the theorem by induction. Let us suppose the theorem true for $k - 1$. Then,

$$\sum_{i=1}^{k} \lambda_i x_i = \sum_{i=1}^{k-1} \lambda_i x_i + \lambda_k x_k = l \sum_{i=1}^{k-1} \frac{\lambda_i}{l} x_i + \lambda_k x_k, \quad \text{where} \quad l = \sum_{i=1}^{k-1} \lambda_i .$$

Since

$$\sum_{i=1}^{k-1} \frac{\lambda_i}{l} = 1,$$

we have

$$\sum_{i=1}^{k-1} \frac{\lambda_i}{l} x_i \in A.$$

Since $l + \lambda_k = 1$, we have

$$\sum_{i=1}^{k} \lambda_i x_i \in A .$$

Theorem 2. *Let E and F denote two vector spaces and let f denote an affine mapping of E into F. Then,*

(a) *The image under f of every convex subset of E is a convex subset of F;*
(b) *The inverse image under f of every convex subset B of F is a convex subset of E.*

Proof. (a) follows from the fact that the image under f of the segment $[x_1, x_2]$ is the segment $[f(x_1), f(x_2)]$.

(b) Suppose that x_1 and x_2 are members of $f^{-1}(B)$. Let us set $y_1 = f(x_1)$ and $y_2 = f(x_2)$. Then, y_1 and y_2 belong to B, so that $[y_1, y_2] \subset B$ and

$$f([x_1, x_2]) = [y_1, y_2], \quad \text{so that} \quad [x_1, x_2] \subset \overset{-1}{f}[y_1, y_2] \subset \overset{-1}{f}(B).$$

Theorem 3. *The intersection of every collection of convex subsets of a vector space E is convex.*

Proof. The result follows immediately from the definition of convex sets.

Theorem 4. *The product of convex subsets A_i of vector spaces E_i is convex in the product space of the spaces E_i.*

Proof. This follows immediately from the definitions.

Theorem 5. *Let A and B denote two convex subsets of a vector space E and let α and β denote two real numbers. The set $\alpha A + \beta B$ (consisting of all elements of the form $\alpha x + \beta y$, where $x \in A$ and $y \in B$) is convex.*

Proof. The set $\alpha A + \beta B$ is the image under the linear mapping $x, y \to \alpha x + \beta y$ of the convex subset $A \times B$ of E^2.

Definition. The convex hull of a subset A of a vector space E is defined as the set $\mathcal{K}(A)$ constituting the intersection of all convex subsets of E that contain A.

By virtue of Theorem 3, $\mathcal{K}(A)$ is convex. It is the "smallest" convex set that contains A.

Theorem 6. *The convex hull of a subset A of E is the set of the convex combinations of elements of A, that is, the set of elements of the form*

$$x = \sum_{i=1}^{k} \lambda_i x_i, \quad \text{with} \begin{cases} \lambda \in \Lambda_k, x_i \in A & (\forall i) \\ k \text{ arbitrary.} \end{cases}$$

Proof. Let A' denote the set of convex combinations of elements of A. One can show directly that A' is convex. Also, in accordance with Theorem 1, $A' \subset \mathcal{K}(A)$. Therefore, $A' = \mathcal{K}(A)$.

Theorem 7 (Carathéodory). *Let A denote a subset of an n-dimensional vector space E. The convex hull $\mathcal{K}(A)$ of A is the set of elements of the form*

$$x = \sum_{i=1}^{n+1} \lambda_i x_i, \quad \text{with} \quad \lambda \in \Lambda_{n+1}, \quad x_i \in A \quad (\forall i).$$

Proof. Every $x \in \mathcal{K}(A)$ is of the form

$$x = \sum_{i=1}^{k} \lambda_i x_i, \quad \text{with} \quad \lambda \in \Lambda_k, x_i \in A \qquad (\forall i).$$

Let us show that, if $k > n + 1$, we can find, among all the representations of this form, that one for which one of the numbers $\lambda_i = 0$. If $k > n + 1$, the elements $x_1 - x_k, \ldots, x_{k-1} - x_k$ are linearly dependent. Therefore, there exist $\mu_1, \ldots, \mu_{k-1} \in \mathbf{R}$, not all zero, such that

$$\mu_1(x_1 - x_k) + \cdots + \mu_{k-1}(x_1 - x_{k-1}) = 0 .$$

Let us set

$$\mu_k = -\sum_{i=1}^{k-1} \mu_i .$$

Thus, we have

$$\sum_{i=1}^{k} \mu_i x_i = 0 , \qquad \sum_{i=1}^{k} \mu_i = 0 .$$

Therefore, we can write, for every $t \in \mathbf{R}$,

$$x = \sum_{i=1}^{k} (\lambda_i - t\mu_i) x_i , \quad \text{with} \quad \sum_{i=1}^{k} (\lambda_i - t\mu_i) = 1 .$$

Let I denote the set of indices i such that $\mu_i > 0$ (this set I is not empty). Let i_0 denote a member of I such that

$$\frac{\lambda_{i_0}}{\mu_{i_0}} = \min_{i \in I} \frac{\lambda_i}{\mu_i} = t_0 .$$

We have

$$x = \sum_{i=1}^{k} (\lambda_i - t_0 \mu_i) x_i , \quad \text{where} \quad \begin{cases} \lambda - t_0 \mu \in \Lambda_k \\ \lambda_{i_0} - t_0 \mu_{i_0} = 0 . \end{cases}$$

Consequently, x is a convex combination of $k - 1$ points of E.

By repetition of this procedure, we can write x successively as a convex combination of $k - 1, k - 2, \ldots, n + 1$ points of E, which completes the proof of the theorem.

2. CONVEX SETS IN TOPOLOGICAL VECTOR SPACES

Theorem 8. *In a topological vector space E, the closure of a convex set A is a convex set.*

Proof. We shall confine ourselves to the case in which E is a metric space. Let x and y denote members of $\bar{A}$. There exists a sequence $\{x_n\}$ and a sequence $\{y_n\}$ with values in A such that

$$\lim_{n \to \infty} x_n = x \quad \text{and} \quad \lim_{n \to \infty} y_n = y .$$

Let λ denote a number in the interval $[0, 1]$. Then,

$$\lambda x_n + (1 - \lambda) y_n \in A ,$$

and, consequently,

$$\lambda x + (1 - \lambda)\, y = \lim_{n \to \infty} \left(\lambda x_n + (1 - \lambda)\, y_n\right) \in \bar{A}\,.$$

Definition. Let A denote a subset of a topological vector space E. The intersection $\overline{\mathcal{K}}(A)$ of the closed convex subsets of E that contain A is called the *closed convex hull* of A.

The set $\overline{\mathcal{K}}(A)$ is closed and convex and it is contained in every closed subset of E that contains A. Let $\overline{\mathcal{K}(A)}$ denote the closure of $\mathcal{K}(A)$. Then, $\mathcal{K}(A) \subset \overline{\mathcal{K}}(A)$, so that $\overline{\mathcal{K}(A)} \subset \overline{\mathcal{K}}(A)$ and, from Theorem 8, $\overline{\mathcal{K}}(A) \subset \overline{\mathcal{K}(A)}$, so that $\overline{\mathcal{K}}(A) = \overline{\mathcal{K}(A)}$. Thus, the closed convex hull of A is the closure of the convex hull of A.*

Theorem 9. *The convex hull $\mathcal{K}(A)$ of a compact subset A of a finite-dimensional vector space is compact.***

Proof. By virtue of Carathéodory's theorem (Theorem 7), every $x \in \mathcal{K}(A)$ is of the form

$$x = \sum_{i=1}^{k+1} \lambda_i x_i\,, \qquad \text{with} \quad \lambda \in \Lambda_{k+1}, x_i \in A \qquad (\forall i)\,.$$

In other words, $\mathcal{K}(A)$ is the image of $\Lambda_{k+1} \times A^{k+1}$ under the mapping

$$\lambda, x_1, \ldots, x_{k+1} \to \sum_{i=1}^{k+1} \lambda_i x_i\,.$$

Since Λ_{k+1} and A are compact, this image is compact.

Theorem 10. *Let A denote a convex subset of a topological vector space E, let x_0 denote an interior point of A, and let x_1 denote a point of $\bar{A}$. Then, every point in $[x_0, x_1[$ is an interior point of A.*

Proof. Let y denote a member of $[x_0, x_1[$. Let h denote the homothetic transformation with center y that maps x_0 into x_1. The homothetic ratio λ of this transformation is negative. Let V denote an open neighborhood of x_0 that is contained in A. Then,

---

* On the other hand, it is not necessarily the convex hull of the closure of A.

** This theorem does not hold for infinite-dimensional vector spaces. If E is a Banach space, we have the following theorem:

The closed convex hull $\overline{\mathcal{K}}(A)$ of a compact subset of the space is compact.

Note also that, even in the finite-dimensional case, the convex hull of a closed set is not necessarily closed. For example, if A is the union of the two subsets of $\mathbf{R}^2$ defined respectively by the relations $\{xy \geqslant 1, x \geqslant 0, y \geqslant 0\}$ and $\{xy \leqslant -1, x \geqslant 0, y \leqslant 0\}$, then A is closed and its convex envelope is the open half-plane $x > 0$.

$h(V)$ is an open neighborhood of x_1 that has a nonempty intersection with A. Therefore, there exists a $z \in V$ such that $h(z) \in A$ and we have

$$h(z) - y = \lambda(z - y) = \lambda(z - h(z)) + \lambda(h(z) - y),$$

so that

$$y - h(z) = \frac{\lambda}{\lambda - 1}(z - h(z)).$$

Therefore, y is the image of z under the homothetic transformation of g with center $h(z)$ and homothetic ratio $t = \lambda/\lambda - 1$. This means that $t \in]0, 1[$. Therefore, $g(V)$ is an open set contained in A.* Since $y \in g(V)$, it follows that y is an interior point of A.

Corollaries. I. *The interior of a convex set is a convex set (as can be shown by applying the preceding theorem, taking $x_1 \in \mathring{A}$).*

II. *If $\mathring{A} \neq \varnothing$, then $\overline{A} = \overline{\mathring{A}}$ and $\mathring{A} = \mathring{\overline{A}}$.*

Proof of II. The preceding theorem implies that, if $\mathring{A}$ is not empty, every point $x_1 \in \overline{A}$ belongs to the closure of $\mathring{A}$, which means that $\overline{A} = \overline{\mathring{A}}$. Let us show that $\mathring{A} = \mathring{\overline{A}}$. We have trivially $\mathring{A} \subset \mathring{\overline{A}}$. Suppose that $x_0 \in \mathring{\overline{A}}$. By making a translation, we can assume that $x_0 = 0$. Let B denote a symmetric open neighborhood of x_0 that is contained in $\overline{A}$. Since $\overline{A} = \overline{\mathring{A}}$, there exists a $y \in \mathring{A} \cap B$, and since $-y \in \overline{A}$, we conclude that $0 \in \mathring{A}$. We recall that any subset M of a vector space E that can be obtained by translation from a vector subspace V (that is, any subset M of the form $M = a + V$, where $a \in E$ and V is a vector subspace of E) is called an *affine variety*. If V is a hyperplane, then M is called an *affine hyperplane*. The dimension of M is defined as the dimension of V.

Let A denote any nonempty subset of a vector space. Then, the intersection of all the affine varieties containing A is an affine variety, called the *affine variety generated by* A. It is the set of points of the form

$$x = \sum_{i=1}^{k} \lambda_i x_i, \quad \text{where} \quad \sum_{i=1}^{k} \lambda_i = 1, \quad k \text{ arbitrary.}$$

*Note that, if A is convex and g is a homothetic mapping with center $y \in A$ and homothetic ratio $t \in [0, 1]$, we have $g(A) \subset A$. This is true because, for every $x \in A$,

$$g(x) = y + t(x - y) = tx + (1 - t)y$$

is a convex combination of x and y.

If A is a convex set, the dimension of A is defined as the dimension of the affine variety generated by A and is denoted by dim (A).

Theorem 11. *Let A denote a convex subset of an n-dimensional vector space E. A necessary and sufficient condition for $\mathring{A}$ to be nonempty is that* dim $(A) = n$.

Proof. Suppose that dim $(A) = n$. By translation, we may assume that $0 \in A$. The affine variety generated by A is then identical to the vector subspace generated by A. Since this subspace is of dimension n, there exist n independent vectors $x_1, \ldots, x_n$ in A, and A contains the set of elements of the form

$$x = \sum_{i=1}^{n} \lambda_i x_i, \quad \text{where} \quad \lambda_i \geqslant 0, \ \sum_{i=1}^{n} \lambda_i \leqslant 1 .$$

The mapping that assigns to $\lambda = [\lambda_1, \ldots, \lambda_n] \in \mathbf{R}^n$ the point $x = \sum_{i=1}^{n} \lambda_i x_i$ is a homeomorphism of $\mathbf{R}^n$ onto E.

The interior of the set

$$\left\{ \lambda_i \geqslant 0, \ \sum_{i=1}^{n} \lambda_i \leqslant 1 \right\}$$

is the set

$$\left\{ \lambda_i > 0, \ \sum_{i=1}^{n} \lambda_i < 1 \right\}$$

Therefore, the set consisting of the elements

$$x = \sum_{i=1}^{n} \lambda_i x_i, \quad \text{where} \quad \left\{ \lambda_i > 0, \sum_{i=1}^{n} \lambda_i < 1 \right\}$$

is an open subset of A. Therefore, $\mathring{A} \neq \varnothing$.

If dim $(A) < n$, we have trivially $\mathring{A} = \varnothing$.

Definition. Let E denote a finite-dimensional vector space, let A denote a nonempty convex subset of E, and let M denote the affine variety generated by A. A point x belonging to A is called an internal point of A if, with respect to M, x is interior to A. The set of internal points is called the inside of A; it is the interior of A with respect to M.

Note that inside (A) is never empty and that, for a closed subset A, we have $A = \overline{\text{inside}\ (A)}$ and that, for any convex subset A, we have $A \subset \overline{\text{inside}\ (A)}$.

Example. We say that $k + 1$ points of a vector space E are *affinely independent* if the dimension of the affine variety that they generate is k. The convex hull Σ of $k+1$ affinely independent points $a_1, \ldots, a_{k+1}$ is called a *k-dimensional simplex.* Necessarily,

$$k = \dim(\Sigma) \leqslant \dim(E).$$

The inside of Σ is thus the set of points of the form

$$x = \sum_{i=1}^{k+1} \lambda_i a_i, \quad \text{where} \quad \sum_{i=1}^{k+1} \lambda_i = 1 \quad \text{and} \quad \lambda_i > 0.$$

3. SEPARATION OF CONVEX SETS

In the remainder of the chapter, we shall confine ourselves to finite-dimensional vector spaces.* The word "hyperplane" is to be understood in the sense of an affine hyperplane.

Let H denote a hyperplane and suppose that the equation that it represents is $u(x) = \alpha$. The sets defined by the inequalities $u(x) \geqslant \alpha$ and $u(x) \leqslant \alpha$ are called the *closed half-spaces determined by H.* The sets defined by the inequalities $u(x) > \alpha$ and $u(x) < \alpha$ are called the *open half-spaces determined by H.*

We note first of all that, if H is a hyperplane and A a convex set such that $H \cap A = \varnothing$, then A is situated in one of the open hyperplanes determined by H. This is true because, if there exist two points x_1 and x_2 in A such that $u(x_1) < \alpha$ and $u(x_2) > \alpha$, there exists a convex combination of x_1 and x_2 belonging to H.

Definition. Let A and B denote two nonempty subsets of a vector space E and let H denote a hyperplane** contained in E. The sets A and B are said to be *separated* (resp. *strictly separated*) by H if A is contained in one of the closed (resp. open) half-spaces determined by H and if B is contained in the other. In other words, A and B are separated by a hyperplane H whose equation is $u(x) = \alpha$ if

$$\begin{cases} u(x) \geqslant \alpha & \text{for } x \in A, \\ u(x) \leqslant \alpha & \text{for } x \in B, \end{cases} \quad \text{or} \quad \begin{cases} u(x) \leqslant \alpha & \text{for } x \in A, \\ u(x) \geqslant \alpha & \text{for } x \in B. \end{cases}$$

Similarly, A and B are strictly separated by H if

$$\begin{cases} u(x) > \alpha & \text{for } x \in A, \\ u(x) < \alpha & \text{for } x \in B, \end{cases} \quad \text{or} \quad \begin{cases} u(x) < \alpha & \text{for } x \in A. \\ u(x) > \alpha & \text{for } x \in B. \end{cases}$$

From time to time, we shall say "separated in the broad sense" to mean "separated."

---

* We shall, however, in passing make certain notes regarding infinite-dimensional topological vector spaces.

**In the infinite-dimensional case, we need to replace "hyperplane" with "closed hyperplane."

Theorem 12. *Let E denote a finite-dimensional vector space, let A denote an open convex set, and let M denote an affine variety disjoint from A. Then, there exists a hyperplane passing through M and disjoint from A.*

Proof. (a) Suppose that dim $(E) = 2$. Let us show that if $0 \notin A$, there exists a straight line D passing through 0 that is disjoint from A. Let C denote the cone with vertex at 0 generated by A. Then,

$$C = \bigcup_{k>0} kA .$$

Since A is open, the set kA is open for every $k > 0$ and, hence, C is open. Furthermore, $0 \notin C$. Finally, C is convex since y_1 and y_2 belong to C, where

$$y_1 = k_1 x_1 \quad (k_1 > 0, x_1 \in A) \qquad \text{and} \qquad y_2 = k_2 x_2 \quad (k_2 > 0, x_2 \in A) .$$

For $\lambda \in \Lambda_2$, we have

$$\begin{aligned} \lambda_1 y_1 + \lambda_2 y_2 &= \lambda_1 k_1 x_1 + \lambda_2 k_2 x_2 \\ &= \mu_1 x_1 + \mu_2 x_2 \qquad (\text{where} \quad \mu_1 = \lambda_1 k_1, \mu_2 = \lambda_2 k_2) \\ &= (\mu_1 + \mu_2) \left(\frac{\mu_1}{\mu_1 + \mu_2} x_1 + \frac{\mu_2}{\mu_1 + \mu_2} x_2 \right) , \end{aligned}$$

and, hence,

$$\lambda_1 y_1 + \lambda_2 y_2 \in C.$$

Let us look at the complement G of $\{0\}$. The set G is not convex. Therefore $C \neq G$. Since C is connected, it has a boundary point x different from 0. Let D_x denote the line generated by x and 0. We set

$$D_x = D_x^+ \cup \{0\} \cup D_x^- ,$$

where D_x^+ is the set of all elements kx for $k > 0$ and where D_x^- is the set of all elements kx where $k < 0$. Since $x \notin C$, we have $D_x^+ \cap C = \varnothing$.

Furthermore, $-x \notin C$ because, otherwise, we should have $0 \in C$ (by Theorem 10).

Therefore, $D_x \cap C = \varnothing$ and $D_x \cap A = \varnothing$.

(b) Let us suppose that dim $(E) \geqslant 2$. Let us show that if $0 \notin A$, there exists a straight line D passing through 0 that is disjoint from A. Let Q denote a plane passing through 0. If $Q \cap A = \varnothing$, then every line in Q that passes through 0 satisfies the condition. If $Q \cap A \neq \varnothing$, then Q contains a line that is disjoint from $Q \cap A$. This line also satisfies the condition.

(c) Let A denote a convex open set and M an affine variety disjoint from A. We may assume that $0 \in M$. Let p denote the maximum dimension of the vector subspaces passing through M that are disjoint from A. We need to show that $p = \dim(E) - 1$. Let N denote a p-dimensional vector subspace passing through M and disjoint from A. Let P denote a complementary subspace of N. Let us project N and A parallel to N onto P. Then, N is projected into 0 and A is projected onto an open subset A' of Q to which 0 does not belong. If $p < \dim(E) - 1$, we have $\dim(P) \geqslant 2$. Thus, there exists in P a straight line D that passes through 0 and that is disjoint from A'. The subspaces N and D generate a vector subspace of dimension $p + 1$ that is disjoint from A. This contradicts the definition of p.

Thus, the proof of the theorem is complete.

Corollary. *Let A denote a convex set that does not contain 0. There exists a linear form u such that $u(x) \geqslant 0$ for every $x \in A$.*

Proof. If $\dim(A) < \dim(E)$, then A is contained in an affine hyperplane H whose equation is $u(x) = \alpha$. We may assume that $\alpha \geqslant 0$. (In the opposite case, we should set $u' = -u$ and write the equation for H in the form $u'(x) = -\alpha$.) Thus, $u(x) \geqslant 0$ for $x \in A$. Furthermore, if $\dim(A) = \dim(E)$, there exists a hyperplane H passing through 0 that is disjoint from $\mathring{A}$. In other words, there exists a linear form u such that $u(x) > 0$ for every $x \in \mathring{A}$. From this conclude that $u(x) \geqslant 0$ for every $x \in \overline{\mathring{A}} = \bar{A}$.

Theorem 13. *Let A and B denote two disjoint convex sets. There exists a hyperplane separating them.**

Proof. The set $A - B$ (consisting of elements $z = x - y$ for $x \in A$ and $y \in B$) is a convex set to which 0 does not belong. By virtue of the above corollary, there exists a linear form u such that $u(z) \geqslant 0$ for every $z \in A - B$, that is, such that $u(x) \geqslant u(y)$ for every $x \in A$ and $y \in B$. We set

$$\alpha = \inf_{x \in A} u(x).$$

We then have

$$u(x) \geqslant \alpha \quad \text{for} \quad x \in A \qquad \text{and} \qquad u(x) \leqslant \alpha \quad \text{for} \quad x \in B.$$

Thus, the hyperplane H whose equation is $u(x) = \alpha$ separates A and B.

Corollary. *Let A and B denote two convex sets with no common internal point. There exists a hyperplane separating these points.*

---

*This theorem is false for infinite-dimensional spaces.

Proof. The insides of A and B are disjoint convex sets. Therefore, there exists a hyperplane H separating them. Since A and B are contained in the closures of their insides, the hyperplane H separates A and B.

Theorem 14. *Let A denote a compact convex set and let B denote a closed convex set. If A and B are disjoint, there exists a hyperplane separating them.**

Proof. Let $d: x, y \to d(x, y)$ denote a Euclidean distance in E. We define

$$\delta = \inf_{\begin{cases} x \in A \\ x \in B \end{cases}} d(x, y).$$

For every x, the function $y \to d(x, y)$ attains its greatest lower bound $d(x, B)$. The function $x \to d(x, B)$ is continuous and, since A is compact, it attains its greatest lower bound. Therefore, there exists $x_0 \in A$ and $y_0 \in B$ such that $\delta = d(x_0, y_0)$. Since A and B are disjoint, we have $\delta > 0$. Let us consider the set A_1 consisting of those $z \in E$ such that $d(z, A) \leqslant \delta/3$ and the set B_1 consisting of those $z \in E$ such that $d(z, B) \leqslant \delta/3$.

The set A_1 is convex since it is the set of those $z = x + u$, where $x \in A$ and u belong to the closed ball of radius $\delta/3$ with center at 0.

Similarly, B_1 is convex. One can verify immediately that A_1 and B_1 are disjoint. Therefore, there exists a hyperplane H separating A_1 and B_1. This hyperplane strictly separates A and B.

Remark. Let us give an example of two closed convex sets that cannot be strictly separated by a hyperplane: In $\mathbf{R}^2$, let A denote the line $x = 0$ and let B denote the set $\{ xy \geqslant 1, x, y \geqslant 0 \}$.

4. SUPPORT HYPERPLANES AND EXTREMAL POINTS

Definition. Let A denote a convex subset of a finite-dimensional vector space. A hyperplane H is called a *support hyperplane* of A if H contains at least one point of A and if A is contained in one of the closed half-spaces determined by H.

Remark. Every hyperplane containing A is a support hyperplane.

Theorem 15. *Let A denote a convex subset of a vector space E. Through every point of A passes a support hyperplane H.*

---

*This theorem is valid for certain infinite-dimensional spaces, for example, Banach and Fréchet spaces.

Proof. Let x_0 denote a boundary point of A. If dim $(A) <$ dim (E), we need only take for H a hyperplane containing A. If dim $(A) =$ dim (E), there exists a hyperplane passing through x_0 that is disjoint from $\mathring{A}$. Since $\bar{A} = \bar{\mathring{A}}$, it follows that H is a support hyperplane of A.

Theorem 16. *Every nonempty compact set has either one or two support hyperplanes parallel to a given hyperplane.*

Proof. Suppose that $u(x) = \alpha$ is the equation of the given hyperplane H. If we set

$$m = \min_{x \in A} u(x)\,, \qquad M = \max_{x \in A} u(x)\,,$$

the hyperplanes $u(x) = m$ and $u(x) = M$ are the support hyperplanes of A that are parallel to H. (These may coincide if A is contained in a hyperplane parallel to H.)

Definition. A point x belonging to a convex set A is called an extreme point of A if x is not a convex combination of two points in A and different from x.

Remark. Every extreme point of A belongs to the boundary of A.

Example. Consider a closed ball of radius ρ with a Euclidean metric. Every boundary point is an extreme point.

Theorem 17. *Let A denote a compact convex subset of a vector space E. Every support hyperplane contains at least one extreme point.*

Proof. We shall show this by induction on the dimension of A. If dim $(A) = 0$, the set A is reduced to a point x_0. Every support hyperplane contains x_0, which is thus an extreme point. Let us suppose the theorem is true for dim $A < p$. Then, on the basis of Theorem 16, every compact convex subset of E of dimension of less than p has extreme points.

Now suppose that dim $(A) = p$. Let H denote a support hyperplane of A. We have dim $(A \cap H) \leqslant$ dim A. If dim $(A \cap H) < p$, then $A \cap H$ has an extreme point which is also an extreme point of A. If dim $(A \cap H) = p$ (that is, if $A \subset H$), let us consider in H a support hyperplane H_0 of A. We have

$$\dim(H_0) = \dim(H) - 1.$$

The set $A \cap H_0$ has an extreme point, which is also an extreme point of A.

5. FACETS. ASYMPTOTIC CONES

Notation. Let A denote a convex subset of a finite-dimensional vector space E and let x denote a point belonging to A. We denote by G_x (or by G_x^A) the set of all x and all straight lines D passing through x such that x belongs to the interior (relative to D) of $D \cap A$.

Let h denote an element of E. A necessary and sufficient condition that the straight line of direction h and through x belong to G_x is that there exist a positive number k such that $[x - kh, x + kh] \subset A$.

One can easily verify that G_x is convex. Specifically, let h_1 and h_2 denote members of E such that the lines with direction h_1 and h_2 that pass through x belong to G_x. There exist positive numbers k_1 and k_2 such that

$$[x - k_1 h_1, x + k_1 h_1] \subset A \quad \text{and} \quad [x - k_2 h_2, x + k_2 h_2] \subset A .$$

Define $k = \min(k_1, k_2)$. Then,

$$[x - kh_1, x + kh_1] \subset A \quad \text{and} \quad [x - kh_2, x + kh_2] \subset A .$$

Define $\lambda = (\lambda_1, \lambda_2) \in \Lambda_2$. Then,

$$x - k(\lambda_1 h_1 + \lambda_2 h_2) \in A \quad \text{and} \quad x + k(\lambda_1 h_1 + \lambda_2 h_2) \in A$$

and, consequently,

$$[x - k(\lambda_1 h_1 + \lambda_2 h_2), x + k(\lambda_1 h_1 + \lambda_2 h_2)] \subset A .$$

Thus, the line with direction $\lambda_1 h_1 + \lambda_2 h_2$ that passes through x belongs to G_x. Thus, G_x is convex. Since G_x is a symmetric cone (with vertex x), it is an affine variety.

Definition. The intersection of A with the affine variety G_x (in the above notation) is called a *facet* of x in A and is denoted by F_x (or by F_x^A).

By construction, F_x is a convex set. A necessary and sufficient condition for y to belong to F_x is that y be equal to x or that x be a convex combination of y and a point $a \in A$ and different from x. The following theorem gives the essential properties of facets.

Theorem 18. *Let A denote a nonempty convex subset of a finite-dimensional vector space and let F_x denote the facet of a point $x \in A$. Then,*

(1) *x is internal to F_x and* $\dim(F_x) = \dim(G_x)$.
(2) *If x is internal to A, then $F_x = A$ and conversely.*
(3) *If x is an extreme point of A, then $F_x = \{x\}$ and conversely.*
(4) *F_x is the greatest convex set contained in A to which x is an internal point.*

(5) *If* $y \in F_x$, *the facet* F_y *of* y *in* A *coincides with the facet of* y *in* F_x *(and, in particular,* $F_y \subset F_x$*). A necessary and sufficient condition for* F_y *to be equal to* F_x *is that* y *be internal to* F_x. *A necessary and sufficient condition for* y *to be extreme with respect to* A *is that it be extreme with respect to* F_x.

Proof. (1) Every line passing through x and contained in G_x intersects A along a segment with end-points different from x and of which x is an internal point. Consequently, x belongs to the interior of F_x with respect to G_x. Therefore, x is internal to F_x, and $\dim(F_x) = \dim(G_x)$.

(2) If x is internal to A, then G_x is the affine variety generated by A and, hence, $F_x = A$. Conversely, if $F_x = A$, then x is internal to A and hence to F_x.

(3) If x is an extreme point of A, then $G_x = \{x\}$ and, consequently, $F_x = \{x\}$. Conversely, if $F_x = \{x\}$, we have $G_x = \{x\}$. If x is the convex combination of two points x_1 and x_2 distinct from x, the line passing through x_1, x_2, and x belongs to G_x, which is contradictory. Therefore, x is extreme.

(4) Let B denote a convex set contained in A and containing x as an internal point. Let F_x^A and F_x^B denote the facets of x in A and B, respectively. Obviously, $F_x^B \subset F_x^A$. Since x is internal to B, we have $F_x^B = B$ so that $B \subset F_x^A$.

(5) Suppose that $y \in F_x$ and $z \in F_y$. Let us show that $z \in F_x$. We may assume that $y \neq x$ and $z \neq y$ because, otherwise, the conclusion is trivial. Thus, there exists an $a \in A$ such that $x = \lambda_1 a + \lambda_2 y$, where λ_1 and λ_2 are positive and $\lambda_1 + \lambda_2 = 1$. Similarly, there exists a $b \in A$ such that $y = \mu_1 b + \mu_2 z$, where μ_1 and μ_2 are positive and $\mu_1 + \mu_2 = 1$. We then have

$$x = \lambda_1 a + \lambda_2(\mu_1 b + \mu_2 z) = (\lambda_1 + \lambda_2 \mu_1)\frac{\lambda_1 a + \lambda_2 \mu_1 b}{\lambda_1 + \lambda_2 \mu_1} + \lambda_2 \mu_2 z .$$

Since $(\lambda_1 + \lambda_2 \mu_1) + \lambda_2 \mu_2 = 1$, we see that x is a convex combination with nonzero coefficients of z and of the point

$$\frac{\lambda_1 a + \lambda_2 \mu_1 b}{\lambda_1 + \lambda_2 \mu_1}$$

which belongs to A. Therefore, $z \in F_x$.

Furthermore, F_y is the largest convex set contained in A that contains y as an internal point. Since we know that $F_y \subset F_x$, we see that F_y is also the largest convex set contained in F_x that contains y as an internal point. In other words, F_y coincides with the facet of y in F_x.

If $F_y = F_x$, then y is internal to F_x, since it is internal to F_y. Conversely, if y is internal to F_x, then the facet of y in F_x is identical to F_x. The same is true of the facet of y in A.

For y to be extreme with respect to A, it is necessary and sufficient that the facet F_y with respect to A be reduced to $\{y\}$, that is, that the facet of y with respect to F_x be reduced to $\{y\}$, in short, that y be extreme with respect to F_x. The following theorem gives the essential relationships between facets and support hyperplanes:

Theorem 19. *Let H denote a support hyperplane of A and let x denote a member of $H \cap A$. Then $F_x \subset H \cap A$ and $F_x = H \cap A$ if and only if x is internal to $H \cap A$.*

Proof. We have $G_x \subset H$, so that $F_x \subset H \cap A$. If $F_x = H \cap A$, then x is internal to $H \cap A$ since it is internal to F_x. Conversely, if x is internal to $H \cap A$, we have $F_x \supset H \cap A$, so that, finally, $F_x = H \cap A$.

Theorem 20 (Krein-Milman). *Every compact convex set A is the convex hull of its extreme points.**

Proof. We prove this theorem by induction on the dimension of A. Suppose the theorem true if dim $(A) \leqslant p - 1$. Let us suppose that dim $(A) = p$. Let x denote a member of A. If $F_x \neq A$, then x belongs to the convex hull of the extreme points of F_x and *a fortiori* to the convex hull of the set of the extreme points of A. Let us suppose now that $F_x = A$. Let y denote an extreme point of A. The line passing through x and y intersects A along a segment of which y is an end-point. Let z denote the other end-point. Then, $F_z \neq A$. Consequently, z belongs to the convex hull of the set of extreme points of F_z and *a fortiori* to the convex hull of the set of extreme points of A. Since x is a convex combination of y and z, the theorem is proven.

Proposition 1. *Let A denote a closed convex set and let x denote a member of A. The sets of elements h such that $x + th \in A$ for every nonnegative t is a closed convex cone independent of x.*

Definition. This closed convex cone is called the *asymptotic cone* of A.

Proof. One can immediately verify that the set in question is a convex cone. For given nonnegative t, the set of all h such that $x + th \in A$ is a closed set. The cone in question is the intersection of these closed sets. Therefore, it is closed. Let y denote a member of A. Suppose that $x + th \in A$ for every nonnegative t. The set A is convex and closed, and it contains the point y and the ray with origin x and direction h. Therefore, A contains the closed convex hull of the set formed by y and this ray. Now, this closed convex hull contains the ray with origin y and direction h. Therefore, $y + th \in A$ for every nonnegative t. Thus, the cone in question is independent of x.

*This theorem is false for infinite-dimensional spaces. For certain infinite-dimensional spaces, for example, Banach and Fréchet spaces, we have the following more restrictive statement: *Every compact convex space is the closed convex hull of its extreme* points.

Proposition 2. *Let A denote a closed convex set. A necessary and sufficient condition for A to be bounded is that its asymptotic cone be reduced to 0.*

Proof. The condition is trivially necessary. Let us show that it is sufficient. Let us suppose that A is not bounded. Equip the space with a Euclidean norm and let x_0 denote a member of A and let $\{x_n\}$ denote a sequence of elements of A such that

$$\lim_{n \to +\infty} \| x_n \| = +\infty.$$

Set $x_n = x_0 + t_n h_n$, where $\| h_n \| = 1$, $t_n \geqslant 0$, and $\lim_{n \to +\infty} t_n = +\infty$. By virtue of the compactness of the unit sphere, we may assume that the sequence $\{h_n\}$ is convergent. Let us now set $h = \lim_{n \to +\infty} h_n$. Let t denote a nonnegative number. For sufficiently large n, we have $t_n \geqslant t$, from which we conclude that $x_0 + th_n \in A$. By taking the limit as $n \to +\infty$, we see that $x_0 + th \in A$. Thus, h belongs to the asymptotic cone of A. Since $\| h \| = 1$, this asymptotic cone is not reduced to 0.

Theorem 21. *Let A denote a closed convex set that does not contain any straight line. Then, every $x \in A$ is of the form $x = x_1 + x_2$, where x_1 belongs to the convex hull of the set of extreme points of A and x_2 belongs to the asymptotic cone of A. Conversely, every point of this form belongs to A.*

Proof. For the first part, we use induction on the dimension of A. If $\dim(A) = 0$, the conclusion of the theorem is obvious. Let us suppose it true for $\dim(A) \leqslant p - 1$. Let us suppose that $\dim(A) = p$. If A is bounded, the result follows from the Krein-Milman theorem. If A is not bounded, let h denote a nonzero vector belonging to the asymptotic cone. Let us consider the straight line passing through x containing h. This line intersects A along a ray issuing from y. Since y belongs to the boundary of A, we have $\dim F_y < \dim(A)$. Also, $y = y_1 + y_2$, where y belongs to the convex hull of the set of extreme points of F_y (and, hence, to the convex hull of the extreme points of A) and y_2 belongs to the asymptotic cone of F_y (and, consequently, to the asymptotic cone of A). Finally, $x + y = th$, where $t \geqslant 0$. Let us set $x_1 = y_1$ and $x_2 = y_2 + th$. Then, the decomposition $x = x_1 + x_2$ is of the desired form. This completes the proof of the first part of the theorem, and the second part is trivial.

6. CONVEX FUNCTIONS

Definitions. A real function f defined on a convex set A is said to be *convex* if

$$f(\lambda_1 x_1 + \lambda_2 x_2) \leqslant \lambda_1 f(x_1) + \lambda_2 f(x_2), \quad \forall x_1, x_2 \in A \text{ and } \lambda \in \Lambda_2 .$$

It is said to be *concave* if the opposite inequality holds, that is, if $-f$ is convex.

Remark. An affine function is both convex and concave.

Theorem 22. *A necessary and sufficient condition for a function f defined on a convex set A to be convex is that the set $\overline{T}_f$ of the ordered pairs*

$$\begin{bmatrix} x \\ y \end{bmatrix} \in A \times \mathbf{R}$$

such that $y \geqslant f(x)$ be a convex set.

Proof of the necessity. Suppose that f is convex and that

$$\begin{bmatrix} x_1 \\ y_1 \end{bmatrix}, \begin{bmatrix} x_2 \\ y_2 \end{bmatrix} \in A \times \mathbf{R}, \qquad \text{where} \qquad y_1 \geqslant f(x_1),\ y_2 \geqslant f(x_2) \qquad \text{and} \qquad \lambda \in \Lambda_2 .$$

Then,

$$\lambda_1 \begin{bmatrix} x_1 \\ y_1 \end{bmatrix} + \lambda_2 \begin{bmatrix} x_2 \\ y_2 \end{bmatrix} = \begin{bmatrix} \lambda_1 x_1 + \lambda_2 x_2 \\ \lambda_1 y_1 + \lambda_2 y_2 \end{bmatrix} \in \overline{T}_f ,$$

since

$$\lambda_1 y_1 + \lambda_2 y_2 \geqslant \lambda_1 f(x_1) + \lambda_2 f(x_2) \geqslant f(\lambda_1 x_1 + \lambda_2 x_2) .$$

Proof of the sufficiency. Suppose that the set $y \geqslant f(x)$ is convex. Let x_1 and x_2 denote members of A and let λ denote a member of Λ_2. We have

$$\begin{bmatrix} x_1 \\ f(x_1) \end{bmatrix} \in \overline{T}_f , \qquad \begin{bmatrix} x_2 \\ f(x_2) \end{bmatrix} \in \overline{T}_f .$$

Consequently,

$$\begin{bmatrix} \lambda_1 x_1 + \lambda_2 x_2 \\ \lambda_1 f(x_1) + \lambda_2 f(x_2) \end{bmatrix} \in \overline{T}_f ,$$

that is,

$$\lambda_1 f(x_1) + \lambda_2 f(x_2) \geqslant f(\lambda_1 x_1 + \lambda_2 x_2) .$$

Consequently, f is convex.

From this theorem, one can easily deduce the following corollaries:

Corollary 1. *If f is convex, then*

$$f\left(\sum_{i=1}^{k} \lambda_i x_i\right) \leqslant \sum_{i=1}^{k} \lambda_i f(x_i) \qquad (\forall \lambda \in \Lambda_k,\ k \text{ arbitrary}).$$

Corollary 2. *Let $\{f_i\}$ denote a family of convex functions defined on a convex set. Then, the function $\sup_i f_i$ is convex.*

13

Programming Problems

1. STATEMENT OF THE PROBLEM AND GENERAL NOTATIONS

Any problem of finding the minimum (or maximum) of a function f defined on a subset of $\mathbf{R}^m$ by means of equations or inequalities is called a *programming problem* or simply a *program.*

We shall give the general formulation of such a problem by introducing "general notations" to which we shall refer in the theorems that provide the fundamental result of the theory.

Let f denote a continuously differentiable function defined on an open subset Ω of $\mathbf{R}^m$. Suppose that

$\varphi^1, \ldots, \varphi^j, \ldots, \varphi^q$, are q continuously differentiable functions defined on Ω,
$\psi^1, \ldots, \psi^k, \ldots, \psi^r$, are r continuously differentiable functions defined on Ω,
$\tilde{\alpha}^1, \ldots, \tilde{\alpha}^j, \ldots, \tilde{\alpha}^q$, are q real numbers,
$\tilde{\beta}^1, \ldots, \tilde{\beta}^k, \ldots, \tilde{\beta}^r$, are r real numbers.

A *permissible set* is defined as the set Δ consisting of the points $x \in \mathbf{R}^m$ and satisfying the conditions

$$\begin{cases} x \in \Omega\,, \\ \varphi^j(x) = \tilde{\alpha}^j \qquad (j = 1, \ldots, q)\,, \\ \psi^k(x) \geqslant \tilde{\beta}^k \qquad (k = 1, \ldots, r)\,. \end{cases}$$

We shall say that f has, at a point $\tilde{x} \in \Delta$,

a *global minimum* if $f(x) \geqslant f(\tilde{x})$ for every $x \in \Delta$,

a *global strict minimum* if $f(x) > f(\tilde{x})$ for every $x \in \Delta$ distinct from $\tilde{x}$, *a local minimum* if there exists a neighborhood A of $\tilde{x}$ in $\mathbf{R}^m$ such that $f(x) \geqslant f(\tilde{x})$ for every $x \in \Delta \cap A$,

a local strict minimum if there exists a neighborhood A of $\tilde{x}$ in $\mathbf{R}^m$ such that $f(x) > f(\tilde{x})$ for every $x \in \Delta \cap A$ distinct from $\tilde{x}$.

We obtain analogous definitions by replacing "minimum" with "maximum" and the symbols $\geqslant$ and $>$ with $\leqslant$ and $<$.

Let us set

$$\alpha^j = \varphi^j(x), \qquad \beta^k = \psi^k(x),$$

$$\alpha = \begin{bmatrix} \alpha^1 \\ \vdots \\ \alpha^q \end{bmatrix}, \quad \beta = \begin{bmatrix} \beta^1 \\ \vdots \\ \beta^r \end{bmatrix}, \quad \tilde{\alpha} = \begin{bmatrix} \tilde{\alpha}^1 \\ \vdots \\ \tilde{\alpha}^q \end{bmatrix}, \quad \tilde{\beta} = \begin{bmatrix} \tilde{\beta}^1 \\ \vdots \\ \tilde{\beta}^r \end{bmatrix},$$

$$\alpha = \varphi(x), \qquad \beta = \psi(x).$$

Thus, the set Δ is the intersection of Ω with the set defined by

$$\varphi(x) = \tilde{\alpha}, \qquad \psi(x) \geqslant \tilde{\beta}.$$

Note that φ and ψ are continuously differentiable mappings of $\mathbf{R}^m$ into $\mathbf{R}^q$ and $\mathbf{R}^r$, respectively.

The equations $\varphi^j(x) = \tilde{\alpha}^j$ will be called *relations*. The inequalities $\psi^k(x) \geqslant \tilde{\beta}^k$ will be called *constraints*. The function f will be called the *objective*.

We say that a constraint $\psi^k(x) \geqslant \tilde{\beta}^k$ is *saturated* at a point $\tilde{x}$ if $\psi^k(\tilde{x}) = \tilde{\beta}^k$. We denote by $K(\tilde{x})$ the set of all k such that the constraint $\psi^k(x) \geqslant \tilde{\beta}^k$ is saturated at $\tilde{x}$.

Definition. Let $\tilde{x}$ denote a given point in $\tilde{x} \in \Delta$. Relations and constraints are said to be *regular* at $\tilde{x}$ if the derivatives

$$(\varphi^j)'(\tilde{x}) \quad (j = 1, \ldots, q), \qquad (\psi^k)'(\tilde{x}) \quad (k \in K(\tilde{x}))$$

are linearly independent.

A programming problem is said to be *linear* if the functions φ^j (for $j = 1, \ldots, q$), ψ^k (for $k = 1, \ldots, r$), and f are linear.

Remarks. I. The constraint $\psi^k(x) \geqslant \tilde{\beta}^k$ can be replaced by the set of conditions $\psi^k(x) - \delta^k = \tilde{\beta}^k$, where $\delta^k \geqslant 0$. The variable δ^k is called the slack variable.

II. Every equation $\varphi^j(x) = \tilde{\alpha}^j$ can be replaced by the set consisting of the two inequalities $\varphi^j(x) \geqslant \tilde{\alpha}^j$, $\varphi^j(x) \leqslant \tilde{\alpha}^j$. However, this replacement can interfere with the application of the theorems.

2. FIRST-ORDER CONDITIONS

A *first-order condition* for a minimum is defined as a condition that can be expressed with the aid of the first derivatives of the functions f, φ^j and ψ^k. Without supplementary hypotheses, first-order conditions are simply necessary conditions. First, let us establish the Farkas-Minkowski theorem, which is based on the following lemma:

Lemma 1. *In a finite-dimensional vector space, let x_0 denote a point and let C denote a closed convex cone. If $x_0 \notin C$, there exists a linear form U such that*

$$U(x_0) < 0,$$
$$U(x) \geqslant 0 \quad \textit{for} \quad x \in C.$$

Proof. In accordance with Theorem 14 of Chapter 12, there exists a linear form U and a real number γ such that

$$U(x_0) < \gamma$$

and

$$U(x) > \gamma \quad \text{for} \quad x \in C.$$

The second inequality implies, when we set $x = 0$, that $\gamma < 0$ and hence that $U(x_0) < 0$. Also, for $x \in C$, we have

$$U(tx) = tU(x) > \gamma \qquad (\forall t \geqslant 0)$$

and, hence, $U(x) > \gamma/t$. If we now let t approach $+\infty$, we see that

$$U(x) \geqslant 0.$$

Theorem 1 (Farkas-Minkowski). *Let $V^1, ..., V^p, W$ denote $p+1$ linear forms on a finite-dimensional vector space E. A necessary and sufficient condition for the implication*

$$V^i(x) \geqslant 0 \quad (i = 1, ..., p) \quad \Rightarrow \quad W(x) \geqslant 0,$$

to be valid is that there exist nonzero numbers $v_1, ..., v_p$ such that

$$W = \sum_{i=1}^{p} v_i V^i.$$

Proof. (*a*) That the condition is sufficient is trivial.

(*b*) Let us show that the condition is necessary. The equation

$$W = \sum_{i=1}^{p} v_i V^i, \quad \text{with} \quad v_i \geqslant 0,$$

means that W belongs to the convex cone (with vertex) C that is generated in the dual E' of E by $V^1, ..., V^p$ (this cone is closed). Let us suppose that this is not the case. In accordance with Lemma 1, there exists an element $x \in E$ such that $W(x) < 0$ and $V(x) \geqslant 0$ ($\forall V \in C$). In particular, we have $V^i(x) \geqslant 0$ ($\forall i = 1, ..., p$), which is ruled out by the hypothesis.

Remark. If the implication $V^i(x) \geqslant 0\ (\forall i) \Rightarrow W(x) \geqslant 0$ is assumed only for $\| x \| \leqslant \varepsilon$ (where $\varepsilon > 0$), we still have

$$W = \sum_{i=1}^{p} v_i V^i , \quad \text{with} \quad v_i \geqslant 0 .$$

This is true since $\dfrac{\| \varepsilon x \|}{\| x \|} = \varepsilon$ and hence

$$V^i(x) \geqslant 0 , (\forall i) \quad \Rightarrow \quad V^i \left(\frac{\varepsilon x}{\| x \|} \right) \geqslant 0 , (\forall i) \quad \Rightarrow \quad W \left(\frac{\varepsilon x}{\| x \|} \right) \geqslant 0$$

$$\Rightarrow \quad W(x) \geqslant 0 .$$

Corollary. *Let* $U^1, \ldots, U^q, V^1, \ldots, V^r, W$, *denote* $q + r + 1$ *linear forms defined on a finite-dimensional vector space* E. *A necessary and sufficient condition for this implication*

$$\left\{ \begin{array}{ll} U^j(x) = 0 & (\forall j) \\ V^k(x) \geqslant 0 & (\forall k) \end{array} \right\} \Rightarrow W(x) \geqslant 0 ,$$

to be valid is that there exist numbers $\mu_1, \ldots, \mu_q, v_1, \ldots, v_r$ *such that*

$$\left\{ \begin{array}{l} v_k \geqslant 0 \qquad (\forall k) \\ W = \sum_{j=1}^{q} \mu_j\ U^j + \sum_{k=1}^{r} v_k V^k . \end{array} \right.$$

Proof. (*a*) That the condition is sufficient is trivial.

(*b*) Let us show that the condition is necessary. The implication may be written

$$\left\{ \begin{array}{ll} U^j(x) \geqslant 0 & (\forall j) \\ (- U^j)(x) \geqslant 0 & (\forall j) \\ V^k(x) \geqslant 0 & (\forall k) \end{array} \right\} \Rightarrow W(x) \geqslant 0 .$$

Thus, there exist nonnegative numbers

$$\mu_j^+ \ (j = 1, \ldots, q), \qquad \mu_j^- \ (j = 1, \ldots, q), \qquad v_k \ (k = 1, \ldots, r)$$

such that

$$W = \sum_{j=1}^{q} \mu_j^+ U^j + \sum_{j=1}^{q} \mu_j^- (- U^j) + \sum_{k=1}^{r} v_k V^k .$$

Let us set

$$\mu_j = \mu_j^+ - \mu_j^- .$$

We then obtain

$$W = \sum_{j=1}^{q} \mu_j U^j + \sum_{k=1}^{r} v_k V^k$$

where $v_k \geqslant 0$ for every k, which completes the proof.

The following theorem gives a necessary and sufficient condition for optimality in linear programming:

Theorem 2. *The notations are the "general notations." Suppose also that the functions f, φ^j (for $j = 1, ..., q$) and ψ^k (for $k = 1, ..., r$) are linear.*

Then, a necessary and sufficient condition for f to have global minimum over Δ at a point $\tilde{x} \in \Delta$ is that there exist real numbers μ_j (for $j = 1, ..., q$) and ν_k (for $k = 1, ..., r$) such that

(a) $\nu_k \geqslant 0 \quad (\forall k), \qquad \nu_k = 0 \quad$ *for* $\; k \notin K(\tilde{x})$

(b) $f = \sum_{j=1}^{q} \mu_j \varphi^j + \sum_{k=1}^{r} \nu_k \psi^k .$

Proof. Let us denote the set $K(\tilde{x})$ simply by K. To prove the necessity of the condition, let h denote a member of $\mathbf{R}^m$ such that

$$\begin{cases} \varphi^j(h) = 0 & (\forall j = 1, ..., q) \\ \psi^k(h) \geqslant 0 & (\forall k \in K) . \end{cases}$$

There exists a positive number ε such that $\| h \| \leqslant \varepsilon \Rightarrow \tilde{x} + h \in \Delta$. We then have $f(\tilde{x} + h) \geqslant f(\tilde{x})$, that is, $f(h) \geqslant 0$. Therefore, there exist numbers μ_j (for $j = 1, ..., q$) and ν_k (for $k \in K$) such that

$$\begin{cases} f = \sum_{j=1}^{q} \mu_j \varphi^j + \sum_{k \in K} \nu_k \psi^k \\ \nu_k \geqslant 0 . \end{cases}$$

If we set $\nu_k = 0$ for $k \notin K$, we obtain conditions (a) and (b).

Let us now prove the sufficiency. If (a) and (b) are satisfied, then $f(h) \geqslant 0$ for every h such that

$$\begin{cases} \varphi^j(h) = 0 & (\forall j = 1, ..., q) \\ \psi^k(h) \geqslant 0 & (\forall k \in K) . \end{cases}$$

Therefore, f has a global minimum at $\tilde{x}$ over the set

$$\left\{ \begin{array}{ll} \varphi^j(x) = \tilde{\alpha}^j & (j = 1, ..., q) \\ \psi^k(x) \geqslant \tilde{\beta}^k & (k \in K) \end{array} \right\}$$

and *a fortiori* over Δ.

With the aim of establishing a necessary condition for optimality in nonlinear programming, we first prove the following lemma:

Lemma 2. *The notations are still the general notations. If the relations and constraints are regular at a point $\tilde{x} \in \Delta$, then, for every vector h satisfying the conditions*

$$\begin{cases} (\varphi^j)'(\tilde{x})(h) = 0 & (\forall j = 1, ..., q) \\ (\psi^k)'(\tilde{x})(h) \geqslant 0 & (\forall k \in K(\tilde{x})) , \end{cases}$$

there exists a differentiable arc $t \to \xi(t)$ *defined on an interval* $[0, t_0]$, *where* $t_0 > 0$, *such that*

$$\begin{cases} \xi(0) = \tilde{x} \\ \xi'(0) = h \\ \xi(t) \in \Delta . \end{cases}$$

Proof. One can always assume that $K(\tilde{x}) = \{1, ..., r'\}$, that $\tilde{\alpha}^j = 0$ (for $j = 1, ..., q$), and that $\tilde{\beta}^k = 0$ (for $k = 1, ..., r'$). Let us set

$$\omega^j = \varphi^j \quad (\text{for } j = 1, ..., q),$$
$$\omega^{q+k} = \psi^k \quad (\text{for } k = 1, ..., r'),$$

and, for $q + r' < i \leqslant m$, let us choose continuously differentiable functions ω^i such that the derivatives $(\omega^i)'(\tilde{x})$ are linearly independent for $i = 1, ..., m$. The functions ω^i constitute local coordinates of $\tilde{x}$ in $\mathbf{R}^m$. In other words, there exists a neighborhood A of $\tilde{x}$ such that, on A, the mapping

$$x \to X = \omega(x) = \begin{bmatrix} \omega^1(x) \\ \vdots \\ \omega^m(x) \end{bmatrix}$$

is a differentiable homeomorphism, with a differentiable inverse, of A onto a neighborhood of 0 in $\mathbf{R}^m$.

Let us set $H = \omega'(\tilde{x})(h)$. Then,

$$H^i = 0 \quad \text{for} \quad i = 1, ..., q$$
$$H^i \geqslant 0 \quad \text{for} \quad i = q + 1, ..., q + r' .$$

Let us set

$$\Xi(t) = tH .$$

Then,

$$\Xi(0) = 0 , \qquad \Xi'(0) = H ,$$

Also, there exists a $t_0 > 0$ such that $\Xi(t) \in \omega(A)$ for $t \in [0, t_0]$. We then have $\Xi(t) \in \omega(\Delta)$. If we now set

$$\xi(t) = \overset{-1}{\omega}(\Xi(t)) \qquad (\text{for } t \in [0, t_0]),$$

we have

$$\begin{cases} \xi(t) \in \Delta \\ \xi(0) = \tilde{x} \\ \xi'(0) = h . \end{cases}$$

Theorem 3. *The notations are the general notations (see section 1). Let us suppose that the relations and constraints are regular at a point* $\tilde{x} \in \Delta$. *Then if f has a local minimum at* $\tilde{x}$, *there exist real numbers* μ_j *(for* $j = 1, \ldots, q$*) and* ν_k *(for* $k = 1, \ldots, r$*) such that*

(a) $\nu_k \geqslant 0 \quad (\forall k = 1, \ldots, r), \quad \nu_k = 0 \;\; \textit{for} \;\; k \notin K(\tilde{x})$,

(b) $$f'(\tilde{x}) = \sum_{j=1}^{q} \mu_j (\varphi^j)'(\tilde{x}) + \sum_{k=1}^{r} \nu_k (\psi^k)'(\tilde{x}).$$

Condition (a) can be written

(a$_1$) $$\nu_k \geqslant 0 \quad (\forall k = 1, \ldots, r), \; \nu_k[\psi^k(\tilde{x}) - \tilde{\beta}^k] = 0,$$

or, by setting $\nu = [\nu_1, \ldots, \nu_r]$,

(a$_2$) $$\nu \geqslant 0 \qquad \nu[\psi(\tilde{x}) - \tilde{\beta}] = 0.$$

Condition (b) can also be written, if we set $\mu = [\mu_1, \ldots, \mu_q]$,

(b$_1$) $$f'(\tilde{x}) = \mu\varphi'(\tilde{x}) + \nu\varphi'(\tilde{x}),$$

or

(b$_2$) $$dz = \sum_{j=1}^{q} \mu_j \, d\alpha^j + \sum_{k=1}^{r} \nu_k \, d\beta^k \quad \text{at} \quad \tilde{x},$$

or

(b$_3$) $$dz = \mu \, d\alpha + \nu \, d\beta \text{ at } \tilde{x},$$

and, finally, if we set

$$\Phi = f - \sum_{j=1}^{q} \mu_j \varphi^j - \sum_{k=1}^{r} \nu_k \psi^k = f - \mu\varphi - \nu\psi,$$

it can be written

(b$_4$) $$\Phi'(\tilde{x}) = 0.$$

Proof. Let h denote a member of $\mathbf{R}^m$ such that

$$\begin{cases} (\varphi^j)'(\tilde{x})(h) = 0 & (j = 1, \ldots, q), \\ (\psi^k)'(\tilde{x})(h) \geqslant 0 & (k \in K(\tilde{x})). \end{cases}$$

Let us show that $f'(\tilde{x})(h) \geqslant 0$. There exists (from Lemma 2) a differentiable arc $t \to \xi(t)$ defined for $t \in [0, t_0]$, where $t_0 > 0$, such that

$$\begin{cases} \xi(t) \in \Delta \\ \xi(0) = \tilde{x} \\ \xi'(0) = h. \end{cases}$$

If we now set $F(t) = f(\xi(t))$, we have $F(t) \geqslant F(0)$ for every $t \in [0, t_0]$ and, consequently,

$$F'(0) = f'(\tilde{x})(h) \geqslant 0 .$$

According to Theorem 1, there exist numbers μ_j (for $j = 1, \ldots, q$) and v_k (for $k \in K(\tilde{x})$) such that

$$f'(\tilde{x}) = \sum_{j=1}^{q} \mu_j(\varphi^j)'(\tilde{x}) + \sum_{k \in K(\tilde{x})} v_k(\psi^k)'(\tilde{x})$$

where $v_k \geqslant 0$. If we set $v_k = 0$ for $k \neq K(\tilde{x})$, we obtain conditions (a) and (b).

Remark. The hypothesis of regularity of the relations and constraints can be weakened. For example, it can be replaced with the hypothesis that the linear forms $(\varphi^j)'(\tilde{x})$ are independent and that there exists a vector h such that

$$(\varphi^j)'(\tilde{x})(h) = 0 \qquad (j = l, \ldots, q)$$
$$(\psi^k)'(\tilde{x})(h) > 0 \qquad (k \in K(\tilde{x}))$$

(cf. Abadie, *Problèmes d'optimisation*, Institut Blaise Pascal, 1965).

Definition. The constants μ^j and v^k introduced in Theorems 2 and 3 are called *multipliers.* The constants μ^j are called *Lagrangian multipliers.* The constants v^k are generally called *Kuhn-Tucker multipliers* because of the essential contribution of these two authors to nonlinear programming.

Let us recall that, if $r = 0$, that is, if the problem reduces to the classical problem of constrained minima, we say that f is *stationary* at a point $\tilde{x}$ in Δ at which the relations are regular if condition (b) is satisfied. This condition then reduces to

$$f'(\tilde{x}) = \sum_{j=1}^{q} \mu_j(\varphi^j)'(\tilde{x}) .$$

The corresponding value of f is called a *stationary value.*

3. SUFFICIENT CONDITIONS IN NONLINEAR PROGRAMMING

We shall establish for a global minimum a sequence of sufficient conditions that are based on convexity hypotheses (Theorem 4) and then sufficient conditions for a local minimum that are expressed in terms of second derivatives (Theorem 5). Finally, Theorem 6 will enable us to study the variations in the minimum attained by a function when we give infinitesimal increments to the parameters $\tilde{\alpha}^j$ and $\tilde{\beta}^k$ defining the relations and the constraints.

It is essential for the interpretation of multipliers. The following proposition gives an important relation between the concept of support hyperplane and that of a tangent hyperplane as it is introduced in classical differential calculus.

Proposition 1. *Let g denote a convex continuous function that is differentiable at a point $\tilde{x}$ of $\mathbf{R}^m$. Suppose that $g'(\tilde{x}) \neq 0$. Then, the subset A of $\mathbf{R}^m$ defined by the inequality $g(x) \leqslant g(\tilde{x})$ is convex and it admits at the point $\tilde{x}$ a unique support hyperplane whose equation is*

$$g'(\tilde{x})(x - \tilde{x}) = 0$$

(tangent hyperplane at $\tilde{x}$ to the hypersurface $g(x) = g(\tilde{x})$). The set A is contained in the half-space $g'(\tilde{x})(x - \tilde{x}) \leqslant 0$.

Proof. We may assume that $\tilde{x} = 0$ and $g(\tilde{x}) = 0$. Let A denote the convex set defined by the inequality $g(x) \leqslant 0$. Let us show that 0 is a boundary point of A. Let h denote a member of $\mathbf{R}^m$ such that $g'(0)(h) \neq 0$. We have

$$(\alpha) \qquad g(th) = tg'(0)(h) + o(t).$$

Thus, there exist points arbitrarily close to 0 such that $g(x) > 0$.

Therefore, A has at least one support hyperplane H at $\tilde{x}$. If H contains a vector h such that $g'(0)(h) \neq 0$, then H, from (α), contains points x such that $g(x) < 0$, which is impossible since such a point belongs to the interior of A, which is disjoint from H. Therefore, the equation for the hyperplane H is $g'(0)(x) = 0$. On the other hand, if $g'(0)(h) < 0$, there exist positive values of t such that $g(th) < 0$. Consequently, the closed half-space in which A is contained is the half-space defined by the equation $g'(0)(x) \leqslant 0$.

Proposition 2. *Let g denote a convex continuous function that is at a point $\tilde{x}$ in $\mathbf{R}^m$. If $g'(\tilde{x}) = 0$, then g has a minimum at $\tilde{x}$.*

Proof. Suppose that $\tilde{x} = 0$ and $g(\tilde{x}) = 0$. Let z denote a member of $\mathbf{R}$. Let A denote the subset of $\mathbf{R}^{m+1}$ defined by the inequality $g(x) - z \leqslant 0$. According to Proposition 1, the set A has a support hyperplane at 0 the equation for which is $z = 0$, and A lies in the half-space $z \geqslant 0$. Therefore, the inequality $g(x) - z \leqslant 0$ implies the inequality $z \geqslant 0$. If we set $z = g(x)$, we conclude that $g(x) \geqslant 0$ for every $x \in \mathbf{R}^m$, which completes the proof.

Theorem 4. *The notations are the general notations. Suppose also that the fuction f is convex, that the functions φ^j are linear, and that the functions ψ^k are concave. Suppose that there exist a point $\tilde{x} \in \Delta$ and numbers μ_j (for $j = 1, \ldots, q$) and ν_k (for $k = 1, \ldots, r$) such that*

$$\text{(a)} \qquad \nu_k \geqslant 0, \quad \nu_k = 0 \quad \textit{for } k \notin K(\tilde{x}),$$

$$\text{(b)} \qquad f'(\tilde{x}) = \sum_{j=1}^{q} \mu_j(\varphi^j)'(\tilde{x}) + \sum_{k=1}^{r} \nu_k(\psi^k)'(\tilde{x}).$$

Then, f has a global minimum over Δ at $\tilde{x}$.

Proof. For purposes of simplicity, let us suppose that $\tilde{\alpha}^j = 0\ (\forall j)$ and $\tilde{\beta}_k = 0\ (\forall k)$. Define

$$\Phi = f - \sum_{j=1}^{q} \mu_j \varphi^j - \sum_{k=1}^{r} \nu_k \psi^k .$$

This function Φ is convex and $\Phi'(\tilde{x}) = 0$. Consequently, Φ has a global minimum at $\tilde{x}$. We have $\Phi(\tilde{x}) = f(\tilde{x})$. Also $\Phi(x) \leqslant f(x)$ for all $x \in \Delta$. Therefore, f has a minimum over Δ at $\tilde{x}$. Henceforth, we shall assume that the functions f, φ^j (for $j = 1, \ldots, q$) and ψ^k (for $k = 1 \ldots, r$) are twice continuously differentiable.

Theorem 5. *The notations are the general ones. Let us suppose that, at a point $\tilde{x} \in \Delta$,*

(i) *the relations and constraints are regular,*

(ii) *there exist numbers μ_j (for $j = 1, \ldots, q$) and ν_k (for $k = 1, \ldots, r$) satisfying the conditions (a) and (b) of Theorem 3,*

(iii) $$\Phi''(\tilde{x})(h)(h) > 0 \quad \textit{for all} \quad h \in \mathfrak{M}, \neq 0$$

where $\mathfrak{M}$ is the vector subspace consisting of vectors h such that

$$(\varphi^j)'(\tilde{x})(h) = 0 \quad \textit{for} \quad j = 1, \ldots, q$$
$$(\psi^k)'(\tilde{x})(h) = 0 \quad \textit{for} \quad k \in K'$$

where K' denotes the set of k such that $\nu_k \neq 0$ $(K' \subset K(\tilde{x}))$. Then, f has a strict local minimum over Δ at $\tilde{x}$.

Proof. We may, for simplicity, assume that $\tilde{\alpha}^j = 0$ and $\tilde{\beta}^k = 0$ $(\forall j, k)$.

If f does not have a strict local minimum over Δ at $\tilde{x}$, there exist a sequence $\{x_n\}$ of points $\mathbf{R}^m$ such that

$$\begin{cases} x_n \in \Delta,\ x_n \neq 0 \\ \lim\limits_{n \to +\infty} x_n = \tilde{x} \\ f(x_n) \leqslant f(\tilde{x}) . \end{cases}$$

Because of the compactness of the unit sphere, we may also assume that

$$\frac{x_n - \tilde{x}}{\| x_n - \tilde{x} \|}$$

has a limit h as $n \to +\infty$. Finally, we have

$$\begin{cases} \varphi^j(x_n) = 0 & (\forall j = 1, \ldots, q), \\ \psi^k(x_n) \geqslant 0 & (\forall k = 1, \ldots, r), \\ \lim\limits_{n \to +\infty} x_n = \tilde{x}, & \\ \lim\limits_{n \to +\infty} \dfrac{x_n - \tilde{x}}{\| x_n - \tilde{x} \|} = h \neq 0, & \\ f(x_n) \leqslant f(\tilde{x}). & \end{cases}$$

By virtue of condition (*b*) of Theorem 3, we have

$$f'(\tilde{x})(h) = \sum_{j=1}^{q} \mu_j(\varphi^j)'(\tilde{x})(h) + \sum_{k \in K'} \nu_k(\psi^k)'(\tilde{x})(h),$$

which reduces to

$$f'(\tilde{x})(h) = \sum_{k \in K'} \nu_k(\psi^k)'(\tilde{x})(h).$$

Also, the inequality

$$\psi^k(x_n) \geqslant 0 \qquad (\forall k = 1, \ldots, r)$$

implies $\psi^k(x_n) \geqslant \psi^k(\tilde{x})$ for $k \in K(\tilde{x})$. From this we get $(\psi^k)'(\tilde{x})(h) \geqslant 0$ for $k \in K(\tilde{x})$ and *a fortiori* for $k \in K'$.

(*a*) Let us suppose that the derivatives $(\psi^k)'(\tilde{x})(h)$ do not all vanish for $k \in K'$. Since $\nu_k > 0$ for $k \in K'$, we then have

$$f'(\tilde{x})(h) > 0.$$

Now, we have the finite Taylor expansion

$$f(x_n) = f(\tilde{x}) + f'(\tilde{x})(h) \cdot \| x_n - \tilde{x} \| + o(\| x_n - \tilde{x} \|).$$

Therefore, $f(x_n) > f(\tilde{x})$ for sufficiently large n, which is contradictory.

(*b*) Let us suppose now that the derivatives $(\psi^k)'(\tilde{x})(h)$ all vanish for $k \in K'$. In other words, let us suppose that $h \in \mathfrak{M}$. Since $\Phi'(\tilde{x}) = 0$, we can write the finite expansion

$$\Phi(x_n) = \Phi(\tilde{x}) + \Phi''(\tilde{x})(h)(h) \cdot \| x_n - \tilde{x} \|^2 + o(\| x_n - \tilde{x} \|^2).$$

Therefore, for n sufficiently great,

$$\Phi(x_n) > \Phi(\tilde{x}).$$

Now, we have $\Phi(x) \leqslant f(x)$ for every $x \in \Delta$ and $\Phi(\tilde{x}) = f(\tilde{x})$. From this we have $f(x_n) > f(\tilde{x})$ for sufficiently large n, which is contradictory. This completes the proof of the theorem.

Condition (iii) of the above theorem leads us to ask the following question: Under what conditions is the restriction of a quadratic form $Q(x)$ to a vector subspace positive definite? To study this problem, let us first give a few definitions and properties regarding quadratic forms. Let E denote a Euclidean vector space of finite dimension m and let Q denote a quadratic form on E. Then, there exists exactly one self-adjoint operator A on E such that

$$Q(x) = \langle x, Ax \rangle .$$

The eigenvalues of A are called the *modes* of Q. If $\lambda_1, \ldots, \lambda_m$ are the modes of Q and if S is an orthonormal basis for E, then

$$Q(x) = \sum_{i=1}^{m} \lambda_i (x_S^i)^2 .$$

A necessary and sufficient condition for Q to be positive-definite is that the modes of Q be positive.

In the following two lemmas, $\mathbf{R}^m$ is equipped with its natural Euclidean structure.

Lemma 3. *In the space $\mathbf{R}^m$, the modes of the quadratic form $Q(x) = \bar{x}Ax$ (where $A = \bar{A}$) are the stationary values of $Q(x)$ on the stationary values of $Q(x)$ on the sphere $\bar{x}x = 1$.*

Proof. For Q to be stationary on the sphere $\bar{x}x = 1$ at a point x, it is necessary and sufficient that there exist a $\lambda \in \mathbf{R}$ such that

$$Q'(x)\,\mathrm{d}x = \lambda\, \mathrm{d}(\bar{x}x)$$

that is,

$$2\,\bar{x}A\,\mathrm{d}x = 2\,\lambda\bar{x}\,\mathrm{d}x$$

or

$$\bar{x}A = \lambda\bar{x}$$

or, finally,

$$Ax = \lambda x .$$

Then,

$$\bar{x}Ax = \bar{x}(\lambda x) = \lambda\bar{x}x = \lambda ,$$

which completes the proof.

Lemma 4. *Let $Q(x) = \bar{x}Ax$ (where $A = \bar{A}$) denote a quadratic form on $\mathbf{R}^m$ and let B denote a linear mapping of rank p of $\mathbf{R}^m$ into $\mathbf{R}^p$ (that is, B is a $p \times m$ matrix the rows of which are linearly independent). The modes of the restriction of $Q(x)$ to the subspace* ker (B) *are the solutions of the equation*

$$\det \left[\begin{array}{c|c} \lambda - A & \bar{B} \\ \hline B & 0 \end{array} \right] = 0 .$$

Proof. Let us set

$$C(\lambda) = \left[\begin{array}{c|c} \lambda - A & \overline{B} \\ \hline B & 0 \end{array}\right].$$

According to Lemma 3, the desired modes are stationary values of Q on the variety $\{\overline{x}x = 1,\ Bx = 0\}$. For Q to be stationary at a point x of that variety, it is necessary and sufficient that there exist a $\lambda \in \mathbf{R}$ and a $\mu = [\mu_1, \ldots, \mu_p] \in (\mathbf{R}^p)'$ such that

$$Q'(x)\,\mathrm{d}x = \lambda\,\mathrm{d}(\overline{x}x) + 2\,\mu B\,\mathrm{d}x\,,$$

that is,

$$(\alpha) \qquad \overline{x}A = \lambda\overline{x} + \mu B\,,$$

or

$$Ax = \lambda x + \overline{B}\overline{\mu}\,.$$

Let us suppose that this condition is satisfied. Then,

$$C(\lambda)\left[\frac{x}{\overline{\mu}}\right] = \left[\begin{array}{c|c} (\lambda - A) & \overline{B} \\ \hline B & 0 \end{array}\right]\left[\frac{x}{\overline{\mu}}\right] = 0\,.$$

Therefore, $\det C(\lambda) = 0$ and

$$Q(x) = \overline{x}Ax = \overline{x}(\lambda x + \overline{B}\overline{\mu}) = \lambda\,.$$

Thus, every stationary value of Q on the variety $\{\overline{x}x = 1,\ Bx = 0\}$ is a solution of the equation $\det C(\lambda) = 0$.

Conversely, let us suppose that $\det C(\lambda) = 0$. Since this equation might conceivably have complex roots, let us think of $C(\lambda)$ as operating in $\mathbf{C}^m$. For every matrix M, we denote by $\overline{M}$ the matrix $\overline{M}$ defined by $(\overline{M})^i_j = \overline{M^j_i}$. There exist $x \in \mathbf{C}^m$ and $\overline{\mu} \in \mathbf{C}^p$ such that

$$\left[\frac{x}{\overline{\mu}}\right] \neq 0 \quad \text{and} \quad C(\lambda)\left[\frac{x}{\overline{\mu}}\right] = 0$$

and this implies

$$(\beta) \qquad \begin{cases} (\lambda - A)\,x + \overline{B}\,\overline{\mu} = 0 \\ Bx = 0\,. \end{cases}$$

If x were equal to 0, we would have $\overline{B}\overline{\mu} = 0$, which would imply $\overline{\mu} = 0$ (since the hypothesis that rank $(B) = p$ implies ker $(\overline{B}) = 0$), which is excluded. Therefore, $x \neq 0$. Let us multiply the left-hand side of (β) on the left by $\overline{x}$. We get

$$\overline{x}(\lambda - A)\,x + \overline{x}\,\overline{B}\,\overline{\mu} = 0\,,$$

that is,

$$\lambda\overline{x}x = \overline{x}Ax\,.$$

Since the quantities $\bar{x}x$ and $\bar{x}Ax$ are real (we recall that $A = \bar{A}$), it follows that λ is real. Therefore, we may assume that x and μ are real. Also we can choose x in such a way that $\bar{x}x = 1$. Then, x belongs to the variety $\{\bar{x}x = 1, Bx = 0\}$. Condition (α) is satisfied and Q is therefore stationary at x on that variety. Furthermore,

$$Q(x) = \bar{x}Ax = \bar{x}(\lambda x + \bar{B}\,\bar{\mu}) = \lambda\,.$$

Thus, λ is a stationary value of Q on the variety $\{\bar{x}x = 1, Bx = 0\}$, which completes the proof.

Corollary. *If the restriction of Q to the subspace* ker (B) *is positive-definite, we have*

$$\det \left[\begin{array}{c|c} A & \bar{B} \\ \hline B & 0 \end{array}\right] \neq 0\,.$$

* * *

We shall now let the parameters $\tilde{\alpha}$ and $\tilde{\beta}$ vary. We denote by $\Delta(\tilde{\alpha}, \tilde{\beta})$ the permitted set. For every $x \in \Delta(\tilde{\alpha}, \tilde{\beta})$, we denote by $K(x, \tilde{\beta})$ the set defined by

$$k \in K(x, \tilde{\beta}) \quad \Leftrightarrow \quad \psi^k(x) = \tilde{\beta}^k\,.$$

Theorem 6. *Suppose that, for particular values $\tilde{\alpha}_0$ and $\tilde{\beta}_0$ of $\tilde{\alpha}$ and $\tilde{\beta}$, there exists a point $\tilde{x}_0 \in \Delta(\tilde{\alpha}_0, \tilde{\beta}_0)$ such that the following conditions are satisfied:*

(i) *the relations and constraints are regular at $\tilde{x}_0$;*
(ii) *there exist numbers μ_j^0 (for $j = 1, \ldots, q$) and ν_k^0 (for $k = 1, \ldots, r$) satisfying the conditions (a) and (b) of Theorem 3;*
(iii) $\nu_k^0 \neq 0$ *for every* $k \in K(\tilde{x}_0, \tilde{\beta}_0)$;
(iv) *Condition* (iii) *of Theorem 5 is satisfied.*

Then, there exist open neighborhoods of A, U and V of $\tilde{x}_0$, $\tilde{\alpha}_0$ and $\tilde{\beta}_0$ such that, for $\tilde{\alpha} \in U$ and $\tilde{\beta} \in V$,

(*a*) *the restriction of f to $A \cap \Delta(\tilde{\alpha}, \tilde{\beta})$ has a unique strict minimum at a point*

$$\tilde{x} = \theta(\tilde{\alpha}, \tilde{\beta})\,,$$

(*b*) *the function θ is continuously differentiable,*
(*c*) *Properties* (i), (ii), (iii) *and* (iv) *of the present theorem are satisfied at $\tilde{x}$,*
(*d*) *the set $K(\tilde{x}, \tilde{\beta})$ is fixed,*
(*e*) *the multipliers μ_j and ν_k are continuously differentiable functions of $\tilde{\alpha}$ and $\tilde{\beta}$,*
(*f*) *the value of $\tilde{z} = f(\theta(\tilde{\alpha}, \tilde{\beta}))$ of the minimum of f is a continuously differentiable function of $\tilde{\alpha}$ and $\tilde{\beta}$, and*

$$\mathrm{d}\tilde{z} = \sum_{j=1}^{q} \mu_j\,\mathrm{d}\tilde{\alpha}^j + \sum_{k=1}^{r} \nu_k\,\mathrm{d}\tilde{\beta}^k\,,$$

that is,

$$d\tilde{z} = \mu \, d\tilde{\alpha} + \nu \, d\tilde{\beta} .$$

Proof. We may assume that A, U, and V are chosen so that, for $\tilde{x} \in A$, $\tilde{\alpha} \in U$ and $\tilde{\beta} \in V$, we have $\psi^k(x) > \tilde{\beta}^k$ or $k \notin K(\tilde{x}_0, \tilde{\beta}_0)$. Therefore, by a change of notation, we may assume that $K(\tilde{x}_0, \tilde{\beta}_0) = \{1, ..., r\}$. Let us also assume that A is chosen so that the derivatives $(\varphi^j)'(\tilde{x})$ and $(\psi^k)'(\tilde{x})$ are independent for every $x \in A$. Conditions (a) and (b) then imply

$$(\mathrm{S}) \qquad \begin{cases} (1) \ f'(\tilde{x}) - \mu\varphi'(\tilde{x}) - \nu\psi'(\tilde{x}) = 0 , \\ (2) \ \varphi(\tilde{x}) = \tilde{\alpha} , \\ (3) \ \nu_k [\psi^k(\tilde{x}) - \tilde{\beta}^k] = 0 \qquad (k = 1, ..., r) . \end{cases}$$

Let us differentiate this system at the point $\tilde{x}_0$. We obtain

$$(\mathrm{S}')_1 \qquad \begin{cases} (1') \ f''(\tilde{x}_0) \, d\tilde{x} - \overset{0}{\mu} \varphi''(\tilde{x}_0) \, d\tilde{x} - d\mu \, . \, \varphi'(\tilde{x}_0) - \\ \qquad\qquad - \overset{0}{\nu} \psi''(\tilde{x}_0) \, d\tilde{x} - d\nu \, . \, \psi'(\tilde{x}_0) = 0 , \\ (2') \ \varphi'(\tilde{x}_0) \, d\tilde{x} = d\tilde{\alpha} , \\ (3') \ \overset{0}{\nu}_k [(\psi^k)'(\tilde{x}_0) \, d\tilde{x} - d\tilde{\beta}^k] + [\psi^k(\tilde{x}_0) - \tilde{\beta}_0^k] \, d\nu_k = 0 \\ \qquad\qquad (\forall k = 1, ..., r) , \end{cases}$$

that is,

$$(\mathrm{S}')_2 \qquad \begin{cases} (1') \ \Phi''(\tilde{x}_0) \, d\tilde{x} - d\mu \, . \, \varphi'(\tilde{x}_0) - d\nu \, . \, \psi'(\tilde{x}_0) = 0 , \\ (2') \ \varphi'(\tilde{x}_0) \, d\tilde{x} = d\tilde{\alpha} , \\ (3') \ (\psi^k)'(\tilde{x}_0) \, d\tilde{x} - d\tilde{\beta}^k = 0 \qquad (\forall k = 1, ..., r) , \end{cases}$$

or (as we can obtain by transposing the left-hand side of (1'))

$$(\mathrm{S}')_3 \qquad \begin{cases} (1') \ \overline{\Phi''(x_0) \, d\tilde{x}} - \overline{\varphi'(x_0)} \, d\bar{\mu} - \overline{\psi'(x_0)} \, d\bar{\nu} = 0 , \\ (2') \ \varphi'(\tilde{x}_0) \, d\tilde{x} = d\tilde{\alpha} , \\ (3') \ \psi'(\tilde{x}_0) \, d\tilde{x} = d\tilde{\beta} . \end{cases}$$

These relations are of the form

$$(\mathrm{S}')_4 \qquad \mathcal{A} \begin{bmatrix} d\tilde{x} \\ \hline d\bar{\mu} \\ \hline d\bar{\nu} \end{bmatrix} = \begin{bmatrix} 0 \\ \hline d\tilde{\alpha} \\ \hline d\tilde{\beta} \end{bmatrix} ,$$

where

$$\mathcal{A} = \left[\begin{array}{c|c|c} H & -\overline{\varphi'(\tilde{x}_0)} & -\overline{\psi'(\tilde{x}_0)} \\ \hline \varphi'(\tilde{x}_0) & 0 & 0 \\ \hline \psi'(\tilde{x}_0) & 0 & 0 \end{array} \right]$$

H being the symmetric matrix defined by

$$\Phi''(\tilde{x})(h_1)(h_2) = \bar{h}_1 H h_2 ,$$

that is, having coefficients

$$H_i^j = \Phi''_{x^i x^j}(\tilde{x}) .$$

We then have

$$\overline{\Phi''(x_0)\, d\tilde{x}} = H\, d\tilde{x} .$$

Now, $\mathcal{A}$ is nonsingular (by virtue of the corollary to Lemma 4). Therefore, there exist open neighborhoods A, U, and V of $\tilde{x}_0, \tilde{\alpha}_0$, and $\tilde{\beta}_0$ satisfying the conditions stated at the beginning of the proof and such that, for $x \in A, \tilde{\alpha} \in U$, and $\tilde{\beta} \in V$, the system (S) has a unique solution in $\tilde{x}$, μ, and ν, that is, it is equivalent to a set of equations of the form

$$\begin{aligned} \tilde{x} &= \theta(\tilde{\alpha}, \tilde{\beta}) \\ \mu &= M(\tilde{\alpha}, \tilde{\beta}) \\ \nu &= N(\tilde{\alpha}, \tilde{\beta}) \end{aligned}$$

where the functions θ, L, and M are continuously differentiable.

By restricting A, U, and V if necessary, we may assume that $\nu_k > 0$ ($\forall k = 1, \ldots, r$) (since $\nu_k^0 > 0$). Therefore, $\psi^k(\tilde{x}) = \tilde{\beta}^k$ ($\forall k$), which shows, in particular, that $\tilde{x} \in \Delta(\tilde{\alpha}, \tilde{\beta}$ and that no constraint that is saturated at $\tilde{x}_0$ ceases to be saturated as $\tilde{\alpha}$ and $\tilde{\beta}$ vary in U and V.

Still restricting A, U, and V, if necessary, we can suppose that Condition (iii) of Theorem 5 is satisfied at every point $\tilde{x}$. Therefore, f has a unique strict minimum at $\tilde{x}$ on $A \cap \Delta(\tilde{\alpha}, \tilde{\beta})$. Let $\tilde{z} = f(\tilde{x})$ denote this minimum.

If we multiply both sides of $(S')_4$ by $[0 \mid \mu^0 \mid \nu^0]$, we obtain

$$[\mu^0\, \varphi'(\tilde{x}_0) + \nu^0\, \psi'(\tilde{x}_0)]\, d\tilde{x} = \mu^0\, d\tilde{\alpha} + \nu^0\, d\tilde{\beta} ,$$

that is,

$$f'(\tilde{x}_0)\, d\tilde{x} = \mu^0\, d\tilde{\alpha} + \nu^0\, d\tilde{\beta} ,$$

or, finally,

$$d\tilde{z} = \mu^0\, d\tilde{\alpha} + \nu^0\, d\tilde{\beta} .$$

The relation $d\tilde{z} = \mu\, d\tilde{\alpha} + \nu\, d\tilde{\beta}$ that we have just proven at $\tilde{x}_0$ remains valid at every point $\tilde{x} = \theta(\tilde{\alpha}, \tilde{\beta})$ since the hypotheses of the present theorem remain valid at such a point. This completes the proof.

APPENDIX

A Saddle-Point Theorem. A function of two variables $x, y \to F(x, y)$ defined for $x \in A$ and $y \in B$ is said to have a *saddle point* at $(\tilde{x}, \tilde{y})$ if

$$F(x, \tilde{y}) \leqslant F(\tilde{x}, \tilde{y}) \leqslant F(\tilde{x}, y) \qquad (\forall x \in A, y \in B)$$

or if the opposite inequalities hold.

Because of the form of condition (α) in the theorem that we are about to present, that theorem is often called the "saddle-point theorem." It gives a new form for the optimality conditions in "convex programming" (that is, nonlinear programming with suitable convexity conditions). Here, we mention that it is often cited in the literature. For reasons of convenience, we shall depart from the usual notations.

The saddle-point theorem (Kuhn-Tucker). *Let f denote a convex function defined on $\mathbf{R}^n$ into $\mathbf{R}$ and let g^i (for $i = 1, \ldots, p$) and h^j (for $j = 1, \ldots, q$) denote concave functions defined on $\mathbf{R}^n$ into $\mathbf{R}$. Let Δ denote the set of points $x \in \mathbf{R}^n$ such that*

$$g^i(x) \geqslant 0 \ (i = 1, \ldots, \ p), \quad h^j(x) \geqslant 0 \ (j = 1, \ldots, q).$$

(1) *Suppose that the functions f, g^i, and h^j are continuously differentiable, that $\tilde{x} \in \Delta$, that the constraints are regular at $\tilde{x}$, and that f has a minimum over Δ at the point $\tilde{x}$. Then, there exists a $\tilde{\lambda} = [\tilde{\lambda}_1, \ldots, \tilde{\lambda}_p] \geqslant 0$ such that the function*

$$\Phi(x, \lambda) = f(x) - \sum_{i=1}^{p} \lambda_i g^i(x) = f(x) - \lambda g(x)$$

satisfies the following condition:

$$(\alpha)\begin{cases} \Phi(\tilde{x}, \lambda) \leqslant \Phi(\tilde{x}, \tilde{\lambda}) \leqslant \Phi(x, \tilde{\lambda}) \\ \text{for every } x \text{ such that } h^j(x) \geqslant 0 \ (\text{for } j = 1, \ldots, q) \\ \text{and for every } \lambda \geqslant 0. \end{cases}$$

(2) *Conversely, if $h^j(\tilde{x}) \geqslant 0$ (for $j = 1, \ldots, q$) and if condition (α) is satisfied, then $\tilde{x} \in \Delta$ and f has a minimum over Δ at the point $\tilde{x}$.*

Proof. (1) For f to have a minimum over Δ at $\tilde{x}$, it is necessary that there exist a $\tilde{\lambda} = [\tilde{\lambda}_1, \ldots, \tilde{\lambda}_p]$ and a $\tilde{\mu} = [\tilde{\mu}_1, \ldots, \tilde{\mu}_q]$ such that

$$f'(\tilde{x}) = \sum_i \tilde{\lambda}_i (g^i)'(\tilde{x}) + \sum_j \tilde{\mu}_j (h^j)'(\tilde{x})$$

$$\tilde{\lambda} \geqslant 0, \quad \tilde{\mu} \geqslant 0, \quad \tilde{\lambda}_i = 0 \ \text{ if } \ g^i(\tilde{x}) > 0, \quad \tilde{\mu}_j = 0 \ \text{ if } \ h^j(\tilde{x}) > 0 \,.$$

Thus,

$$\Phi'_x(\tilde{x}, \tilde{\lambda}) = \sum \tilde{\mu}_j (h^j)'(\tilde{x})\,.$$

In accordance with Theorem 4, the mapping $x \to \Phi(x, \tilde{\lambda})$ has a global minimum at $\tilde{x}$ over the set of all x such that $h^j(x) \geqslant 0$, so that

$$\Phi(\tilde{x}, \tilde{\lambda}) \leqslant \Phi(x, \tilde{\lambda}), \qquad \forall x \text{ such that } h^j(x) \geqslant 0 \qquad (j = 1, \ldots, q)\,.$$

Also,

$$\Phi(\tilde{x}, \tilde{\lambda}) - \Phi(\tilde{x}, \lambda) = \sum_i (\lambda_i - \tilde{\lambda}_i)\, g^i(\tilde{x}) = (\lambda - \tilde{\lambda})\, g(\tilde{x}),$$

with $g(\tilde{x}) \geqslant 0$ and $\tilde{\lambda}\, g(\tilde{x}) = 0$, so that

$$\Phi(\tilde{x}, \tilde{\lambda}) - \Phi(\tilde{x}, \lambda) \geqslant 0 \qquad \text{for} \qquad \lambda \geqslant 0.$$

Thus, condition (α) is satisfied.

(2) Let us assume that condition (α) is satisfied. Consider the condition

$$\Phi(\tilde{x}, \tilde{\lambda}) - \Phi(\tilde{x}, \lambda) = (\lambda - \tilde{\lambda})\, g(\tilde{x}) \geqslant 0 \qquad \forall \lambda \geqslant 0.$$

If we let λ_i approach $+\infty$, keeping the other components of λ equal to 0, we obtain

$$g^i(\tilde{x}) \geqslant 0.$$

Consequently, $g(\tilde{x}) \geqslant 0$ and $\tilde{x} \in \Delta$. Also, if we set λ equal to 0, we obtain $\tilde{\lambda}\, g(\tilde{x}) \leqslant 0$. Since $\tilde{\lambda} \geqslant 0$, this means that

$$\tilde{\lambda}\, g(\tilde{x}) = 0.$$

Thus, for every $x \in \Delta$,

$$f(x) \geqslant \Phi(x, \tilde{\lambda}) \geqslant \Phi(\tilde{x}, \tilde{\lambda}) = f(\tilde{x}),$$

which completes the proof.

Remarks. I. This theorem is most frequently applied in the following two cases:

(a) $0 = b$

(b) $q = n$ and the inequalities $h^j(x) \geqslant 0$ are of the form $x^j \geqslant 0$ (for $j = 1, \ldots, n$).

Then, condition (α) is written

$$\Phi(\tilde{x}, \lambda) \leqslant \Phi(\tilde{x}, \tilde{\lambda}) \leqslant \Phi(x, \tilde{\lambda})$$

for every $x \geqslant 0$ and $\lambda \geqslant 0$.

II. The second part of the saddle-point theorem makes no differentiability assumptions on the functions f, g^i, and h^j. One can also make analogous statements in the first part without assuming differentiability (cf., for example, [49], Theorem 7-1-1).

14

Linear Programming

1. PRELIMINARIES

We shall now present certain properties of sets defined by a system of finitely many linear equations and inequalities. We resume the "general notations" of the preceding chapter.

We denote by Δ the set of functions defined in $\mathbf{R}^m$ by

$$\begin{cases} \varphi^j(x) = \tilde{\alpha}^j & (j = 1, \ldots, q), \\ \psi^k(x) \geqslant \tilde{\beta}^k & (k = 1, \ldots, r), \end{cases}$$

where φ^j and ψ^k are linear functions.

For such a set, the extreme points are often called *vertices*.

Proposition 1. *Let $\tilde{x}$ denote a member of Δ and let $K(\tilde{x})$ denote the set of indices k such that $\psi^k(\tilde{x}) = \tilde{\beta}^k$. The affine variety $G_{\tilde{x}}$ generated by the facet $F_{\tilde{x}}$ of $\tilde{x}$ is defined by the equations*

$$\begin{cases} \varphi^j(x) = \tilde{\alpha}^j & (j = 1, \ldots, q), \\ \psi^k(\tilde{x}) = \tilde{\beta}^k & (k \in K(\tilde{x})). \end{cases}$$

Proof. We may assume that $\tilde{x} = 0$ and $\tilde{\alpha}^j = 0$ (for $j = 1, \ldots, q$) and hence that $\tilde{\beta}^k = 0$ for $k \in K(\tilde{x})$. Let D denote a line passing through 0 that is generated by a nonzero vector h.

If $\varphi^j(h) \neq 0$, then $D \cap \Delta = \{0\}$, so that $D \notin G_{\tilde{x}}$.

If $\psi^k(h) > 0$ for $k \in K(\tilde{x})$, then $\psi^k(th) < 0$ for $t < 0$ and hence $D \notin G_{\tilde{x}}$.

If $\psi^k(h) < 0$ for $k \in K(\tilde{x})$, we have, similarly, $D \notin G_{\tilde{x}}$.

Thus, if $D \in G_{\tilde{x}}$, then $\varphi^j(h) = 0$ (for $j = 1, ..., q$) and $\psi^k(h) = 0$ (for $k \in K(\tilde{x})$).

Conversely, if these conditions are satisfied, then

$$\left.\begin{array}{ll} \psi^j(th) = 0 & (j = 1, ..., q) \\ \psi^k(th) = 0 & (k \in K(\tilde{x})) \end{array}\right\} \text{ for every } t \in \mathbf{R}$$

and

$$\psi^k(th) \geqslant \tilde{\beta}^k \qquad k \notin K(\tilde{x})$$

for sufficiently small t. Therefore, $\tilde{x}$ is internal to $\Delta \cap D$ and $D \in G_{\tilde{x}}$.

Corollaries. I. *A necessary and sufficient condition for a point* $\tilde{x} \in \Delta$ *to be extreme is that the equations*

$$\left\{\begin{array}{ll} \varphi^j(x) = 0 & (j = 1, ..., q) \\ \psi^k(x) = 0 & (k \in K(\tilde{x})) \end{array}\right\}$$

have only the zero solution.

II. *If $\tilde{x}$ is an extreme point of Δ, there exist at least $m - q$ saturated constraints at $\tilde{x}$.*

III. *Δ has only finitely many extreme points.*

IV. *The convex hull of extreme points is compact.*

Proposition 2. *Let Δ_0 denote the convex hull of the set of extreme points of Δ and let f denote a linear form. If Δ contains no straight line, then either*

f is not bounded below (resp. above)

or

$$\min_{x \in \Delta} f(x) = \min_{x \in \Delta_0} f(x)$$

$$\left(resp. \max_{x \in \Delta} f(x) = \max_{x \in \Delta_0} f(x)\right).$$

Let C_Δ denote the asymptotic cone of Δ. Then, in the first case,

there exists an $h \in C_\Delta$ such that

$$f(h) < 0 \; (resp. \; f(h) > 0)$$

and, in the second case,

$$f(h) \geqslant 0 \; (resp. \; f(h) \leqslant 0) \; \textit{for every } h \in C_\Delta.$$

Proof. Since Δ_0 is compact, f is bounded on Δ_0 and it attains its greatest lower bound on Δ_0. If there exists an $h \in C_\Delta$ such that $f(h) < 0$, then

$$x + th \in \Delta \quad \text{and} \quad f(x + th) = f(x) + tf(h)$$

for $x \in \Delta$ and $t \geqslant 0$. From this we conclude that

$$\lim_{t \to +\infty} f(x + th) = -\infty .$$

Consequently, f is not bounded below on Δ.

Let us assume now that $f(h) \geqslant 0$ for every $h \in C_\Delta$. Suppose that $x \in \Delta$. There exists an $x_1 \in \Delta_0$ and an $h \in C_\Delta$ such that $x = x_1 + h$. Then, $f(x) = f(x_1) + f(h)$ and, consequently, $f(x) \geqslant f(x_1)$ and

$$\inf_{x \in \Delta} f(x) \geqslant \min_{x \in \Delta_0} f(x) .$$

Since the inverse inequality is trivial, we see that both sides are equal, which proves that f attains its greatest lower bound on Δ and that this greatest lower bound is equal to the minimum of f on Δ_0.

Corollary. *If Δ contains no straight line and if f is bounded below on Δ, there exists an extreme point at which f attains its minimum.*

Remark. The asymptotic cone of the set Δ is defined by the relations

$$\varphi^j(h) = 0 \quad (j = 1, \ldots, q)$$

$$\psi^k(h) \geqslant 0 \quad (k = 1, \ldots, r) .$$

2. THE SIMPLEX METHOD

We shall now explain a classical method of solving linear programs, namely, the so-called *simplex method*. This method has several variations from the point of view of its application. Numerous more elaborate methods have been derived from it. One can find an exposition of these methods in specialized texts.

Linear programs are often presented in one of the following particular forms:

Program I. *Maximize the linear form ux over the set*

$$\begin{cases} ax \leqslant b \\ x \geqslant 0 \end{cases}$$

where $x \in \mathbf{R}^n$, $b \in \mathbf{R}^p$, $a \in L(\mathbf{R}^n, \mathbf{R}^p)$, *and* $u \in (\mathbf{R}^n)'$.

Program II. *Maximize the linear form UX over the set*

$$\begin{cases} AX = b \\ X \geqslant 0 \end{cases}$$

where $X \in \mathbf{R}^{n+p}$, $b \in \mathbf{R}^p$, $A \in L(\mathbf{R}^{n+p}, \mathbf{R}^p)$, *and* $U \in (\mathbf{R}^{n+p})'$.

Form I can be converted to Form II as follows: Let us introduce the slack variables $x'^j = b^j - a^j x$. In other words, let us set

$$x' = \begin{bmatrix} x'^1 \\ \vdots \\ x'^p \end{bmatrix} = b - ax\,.$$

Then, the relation $ax \leqslant b$ is written $x' \geqslant 0$.

Let us form the matrix $A = [a \mid \mathbf{1}]$ and the $(n+p)$-tuple $X = \left[\frac{x}{x'}\right]$. By definition,

$$\begin{cases} AX = b \\ X \geqslant 0\,. \end{cases}$$

Let us set $U = [u \mid 0] \in (\mathbf{R}^{n+p})'$. We have $UX = ux$. Consequently, the problem is converted to Form II.

Example. The program

$$\begin{cases} 10\,x^1 + 6\,x^2 + 3\,x^3 \leqslant 100 \\ 2\,x^1 + 3\,x^2 + 7\,x^3 \leqslant 100 \\ x^1, x^2, x^3 \geqslant 0 \\ \text{MAX}\; x^1 + x^2 + x^3 \end{cases}$$

can be put in the form

$$\begin{cases} 10\,X^1 + 6\,X^2 + 3\,X^3 + X^4 = 100 \\ 2\,X^1 + 3\,X^2 + 7\,X^3 + X^5 = 100 \\ X^1, X^2, X^3, X^4, X^5 \geqslant 0 \\ \text{MAX}\; X^1 + X^2 + X^3\,. \end{cases}$$

We note that the permitted set Δ contains no straight line either in Form II or Form I. Therefore, Proposition 2 is applicable.

Let us give some conditions for optimality for programs of Forms I and II. For this, it will be sufficient to apply Theorem 2 of Chapter 13.

An optimality condition for program I. A necessary and sufficient condition for an admissible point $\tilde{x}$ to be optimal is that there exist a $v \in (\mathbf{R}^p)'$ and a $v \in (\mathbf{R}^p)'$ such that

$$\begin{cases} u = va + w\,, \\ v \geqslant 0\,, \\ w \leqslant 0\,, \\ v(b - a\tilde{x}) = 0\,, \\ w\tilde{x} = 0\,, \end{cases}$$

in other words, that there exist a $w \in (\mathbf{R}^n)'$ such that

$$\begin{cases} u - va \leqslant 0, \\ v \geqslant 0, \\ v(b - a\tilde{x}) = 0, \\ (u - va)\,\tilde{x} = 0. \end{cases}$$

An optimality condition for program II. A necessary and sufficient condition for an admissible point $\tilde{X}$ to be optimal is that there exist a $v \in (\mathbf{R}^p)'$ and a $w \in (\mathbf{R}^{n+p})'$ such that

$$\begin{cases} U = vA + w, \\ w \leqslant 0, \\ w\tilde{X} = 0, \end{cases}$$

in other words, that there exist a $v \in (\mathbf{R}^p)'$ such that

$$\begin{cases} U - vA \leqslant 0, \\ (U - vA)\,\tilde{X} = 0. \end{cases}$$

Consider a partition of the matrix A of the following form:

$$A = [A_\gamma \mid A_{\bar{\gamma}}]$$

where the matrix A_γ is nonsingular. Let us set $X = \begin{bmatrix} X^\gamma \\ \hline X^{\bar{\gamma}} \end{bmatrix}$. Then, the equation $AX = b$ may be written

$$A_\gamma X^\gamma + A_{\bar{\gamma}} X^{\bar{\gamma}} = b.$$

A particular solution of this equation is

$$\tilde{X} = \begin{bmatrix} \tilde{X}^\gamma \\ \hline \tilde{X}^{\bar{\gamma}} \end{bmatrix} = \begin{bmatrix} \overset{-1}{A_\gamma} b \\ \hline 0 \end{bmatrix}.$$

If $\overset{-1}{A_\gamma} b \geqslant 0$, this solution is admissible; that is, it belongs to Δ. Also, it is a vertex of the permitted set since the system

$$\begin{cases} AX = 0 \\ X^{\bar{\gamma}} = 0 \end{cases} \quad \text{implies } A_\gamma X^\gamma = 0$$

and therefore has only the zero solution.

Conversely, let $\tilde{X}$ denote a vertex of Δ and let δ denote the support of $\tilde{X}$ (the set of indices i such that $\tilde{X}^i \neq 0$). We know that δ contains at most p indices. The system

$$\begin{Bmatrix} AX = 0 \\ X^{\bar{\delta}} = 0 \end{Bmatrix}$$

has only the zero solution. Therefore, the columns of A_δ are independent. Thus, there is a set γ of p indices such that $\delta \subset \gamma$ and A_γ is nonsingular. Since $\tilde{X}^\gamma = 0$, we have

$$\tilde{X} = \left[\frac{\tilde{X}^\gamma}{\tilde{X}^{\bar{\gamma}}}\right] = \left[\frac{\overset{-1}{A_\gamma}\, b}{0}\right].$$

Note now that, if K is a nonsingular matrix, the condition $AX = b$ is equivalent to the condition $KAX = Kb$. Therefore, program II can be put in the form

$$II' \begin{cases} KAX = Kb \\ X \geqslant 0 \\ \text{MAX } UX \end{cases} \quad \text{that is,} \quad \begin{cases} A'X = b' \\ X \geqslant 0 \\ \text{MAX } UX \end{cases} \quad \text{with} \quad \begin{cases} A' = KA \\ b' = Kb\,. \end{cases}$$

We shall say that the program is in *simplicial form* if

$$A = [A_\gamma \mid A_{\bar{\gamma}}] \quad \text{with} \quad A_\gamma = \mathbf{1} \quad \text{and} \quad b \geqslant 0\,.$$

Associated with every simplicial form is the vertex

$$\tilde{X} = \left[\frac{\tilde{X}^\gamma}{\tilde{X}^{\bar{\gamma}}}\right] = \left[\frac{b}{0}\right].$$

Conversely, let us again start with Form II. Suppose that $A = [A_\gamma \mid A_{\bar{\gamma}}]$, where A_γ is nonsingular, $\overset{-1}{A_\gamma}\, b \geqslant 0$, and

$$\tilde{X} = \left[\frac{\tilde{X}^\gamma}{\tilde{X}^{\bar{\gamma}}}\right] = \left[\frac{\overset{-1}{A_\gamma}\, b}{0}\right]$$

is the corresponding summit. By taking $K = \overset{-1}{A_\gamma}$, we obtain the simplicial form

$$II' \begin{cases} A'X \leqslant b' \\ X \geqslant 0 \\ \text{MAX } UX \end{cases} \quad \text{with} \quad \begin{cases} A' = [A'_\gamma \mid A'_{\bar{\gamma}}] = [\mathbf{1} \mid \overset{-1}{A_\gamma}\, A_{\bar{\gamma}}] \\ b' = \overset{-1}{A_\gamma}\, b\,. \end{cases}$$

An admissible solution is always the vertex $\tilde{X} = \left[\frac{b'}{0}\right]$.

A simplicial form is said to be *degenerate* if b has zero components.

We shall now suppose that we have put the program in a nondegenerate simplicial form. Note that if program II has been derived from program I by introduction of slack variables and if $b \gg 0$, it will be in nondegenerate simplicial form from the very beginning of the calculation.

Let us now see whether the corresponding $\tilde{X}$ is optimal. We have

$$\tilde{X} = \begin{bmatrix} \tilde{X}^{\gamma} \\ \hline \tilde{X}^{\bar{\gamma}} \end{bmatrix} = \begin{bmatrix} b \\ \hline 0 \end{bmatrix}.$$

Let us apply the general optimality condition. A necessary and sufficient condition for $\tilde{X}$ to be optimal is that there exist a $v \in (\mathbf{R}^p)'$ such that

$$\begin{cases} U - vA \leqslant 0, \\ (U - vA)\tilde{X} = 0. \end{cases}$$

Similarly, the second condition can be written $(U - vA)_{\gamma} = 0$, that is,

$$U_{\gamma} - v = 0.$$

A necessary and sufficient condition for optimality is therefore

$$\boxed{U - U_{\gamma} A \leqslant 0}\ .$$

Moreover, it can be reduced to $U_{\bar{\gamma}} - U_{\gamma} A_{\bar{\gamma}} \leqslant 0$. We can find this condition again directly. The equation $AX = b$ may be written

$$X^{\gamma} + A_{\bar{\gamma}} X^{\bar{\gamma}} = b,$$

so that

$$X^{\gamma} = b - A_{\bar{\gamma}} X^{\bar{\gamma}}.$$

We then have

$$\begin{aligned} UX &= U_{\gamma} X^{\gamma} + U_{\bar{\gamma}} X^{\bar{\gamma}} \\ &= U_{\gamma}(b - A_{\bar{\gamma}} X^{\bar{\gamma}}) + U_{\bar{\gamma}} X^{\bar{\gamma}} \\ &= U_{\gamma} b + (U_{\bar{\gamma}} - U_{\gamma} A_{\bar{\gamma}}) X^{\bar{\gamma}}. \end{aligned}$$

Now, since we have assumed that $b \gg 0$, the solution $\tilde{X}$ remains admissible if $X^{\bar{\gamma}}$ is a sufficiently small nonnegative number.

If $U_{\bar{\gamma}} - U_{\gamma} A_{\bar{\gamma}} \leqslant 0$, we have $UX \leqslant U_{\gamma} b$, where equality holds when $X = \tilde{X}$. Consequently, $\tilde{X}$ is optimal. One can also see that a strict maximum will exist at $\tilde{X}$ if and only if

$$U_{\bar{\gamma}} - U_{\gamma} A_{\bar{\gamma}} \lll 0.$$

On the other hand, if the inequality $U_{\gamma} - U_{\gamma} A_{\bar{\gamma}} \leqslant 0$ does not hold, then there exists an index $e \in \bar{\gamma}$ such that

$$U_e - U_{\gamma} A_e > 0.$$

If we take $X^i = 0$ for $i \in \bar{\gamma}$ different from e and $X^e > 0$ so small that again $X^{\gamma} \geqslant 0$, then $UX > U_{\gamma} b$ and $\tilde{X}$ is not optimal.

We again find the condition for optimality that we stated earlier. Furthermore, we obtain a procedure for improving the solution $\widetilde{X}$. Specifically, consider an index* $e \in \bar{\gamma}$ (called the input index) such that

$$U_e - U_\gamma A_e > 0 .$$

Let us take $X^i = 0$ for $i \in \bar{\gamma}$ different from e. We obtain

$$X^\gamma = b - A_e X^e$$

so that

$$UX = U_\gamma b + (U_e - U_\gamma A_e) X^e .$$

Thus, it will be desirable for us to take X^e as great as possible while we keep $X^\gamma \geqslant 0$, that is,

$$X^{\gamma(j)} = b^j - X^e A_e^j \geqslant 0 \quad \text{for every} \quad j = 1, ..., p .$$

If $A_e \leqslant 0$, this condition is satisfied for every $X^e \geqslant 0$. Hence, UX is not bounded above on Δ. For every j such that $A_e^j > 0$, we need to have $X^e \leqslant b^j/A_e^j$. Let σ denote a value of j for which b_j/A_e^j assumes the smallest positive value. Let us take

$$X^e = \frac{b^\sigma}{A_e^\sigma} .$$

Then, we have $X^{\gamma(\sigma)} = 0$. The index $s = \gamma(\sigma)$ is called the *output* index. Let γ' denote the set of indices obtained from γ by replacing s with e. We obtain a new solution

$$\widetilde{X}' = \left[\begin{array}{c} \widetilde{X}^{\gamma'} \\ \hline X^{\bar{\gamma}'} \end{array}\right], \quad \text{with} \quad \widetilde{X}^{\bar{\gamma}'} = 0 .$$

We can associate with this solution a new simplicial form

$$\left\{\begin{array}{l} A' X \leqslant b' \\ X \geqslant 0 \\ \text{MAX } UX \end{array}\right.$$

where $A' = KA$, $b' = Kb$, $K = \overset{-1}{A}_{\gamma'}$ and $A' = [\mathbf{1} \mid A_{\gamma'}]$.

Obviously,

$$\widetilde{X}^{\gamma'} = b' .$$

---

*In practice, we can choose the index e that maximizes $U_e - U_\gamma A_e$. But there is nothing imperative about this choice.

If the index s chosen is the only one for which b^i/A_e^i assumes its smallest positive value, we shall have $b' \gg 0$, and the new simplicial form is nondegenerate. In the opposite case, we obtain a degenerate simplicial form. Thus, for us to be able to use the simplex method, we must have a modification of which we shall not speak here. We assume only that we obtain a new nondegenerate simplicial form. We apply the optimality test to it. If $\tilde{X}'$ is optimal, the process is terminated. If not, either UX is not bounded above or $\tilde{X}'$ can be improved. Thus, we obtain a sequence of vertices of Δ, the calculating operations terminating either when we have obtained the optimal summit or when we have shown that UX is not bounded above on Δ.

Let us now see how these calculations are pursued. The matrix $A_{\gamma'}$ differs from the matrix 1 by a single column. Therefore, we need the following rule:

Rule. *For the matrix*

$$M = \left[\begin{array}{c|c|c} & a^1 & \\ \mathbf{1} & \vdots & 0 \\ & a^{j-1} & \\ \hline 0 & a^j & 0 \\ \hline & a^{j+1} & \\ 0 & \vdots & \mathbf{1} \\ & a^p & \end{array}\right] = [\varepsilon_1 \mid \dots \mid \varepsilon_{j-1} \mid a \mid \varepsilon_{j+1} \mid \dots \mid \varepsilon_p].$$

jth column

If $a_j \neq 0$, then M is nonsingular and

$$\overset{-1}{M} = \left[\begin{array}{c|c|c} & -a^1/a^j & \\ \mathbf{1} & \vdots & 0 \\ & -a^{j-1}/a^j & \\ \hline 0 & 1/a^j & 0 \\ \hline & -a^{j+1}/a^j & \\ 0 & \vdots & \mathbf{1} \\ & -a^p/a^j & \end{array}\right] = \left[\begin{array}{c|c|c|c|c|c|c} & & & -a^1/a^j & & & \\ & & & \vdots & & & \\ & & & -a^{j-1}/a^j & & & \\ \varepsilon_1 & \dots & \varepsilon_{j-1} & 1/a^j & \varepsilon_{j+1} & \dots & \varepsilon_p \\ & & & -a^{j+1}/a^j & & & \\ & & & \vdots & & & \\ & & & -a^p/a^j & & & \end{array}\right].$$

To prove this rule, it suffices to show that $\overset{-1}{M}M = \mathbf{1}$. We leave this to the reader.

Let us now represent the passage from a simplicial form to the following form: The calculations can be presented schematically as follows (pp. 317 and 318): The columns of A (and of U) are distributed in two groups, one group consisting of those whose index belongs to γ and the other consisting of those whose index belongs to $\bar{\gamma}$. In practice, these two groups are jumbled together. To

$$
\begin{array}{llccccccccccl}
\text{Column indices:} & & & \begin{matrix} i=\gamma(j) \\ i\neq s\end{matrix} & & s=\gamma(\sigma) & & \begin{matrix} i'\in\bar{\gamma} \\ i'\neq e\end{matrix} & & e & & & \\
& U = & [\;\cdots & U_i & \cdots & U_s & \cdots & U_{i'} & \cdots & U_e & \cdots\;] & & \\
& A = & [\;\cdots & \varepsilon_j & \cdots & \varepsilon_\sigma & \cdots & A_{i'} & \cdots & A_e & \cdots\;] & [\,b\,] & \\
[U_\gamma] & U_\gamma A = & [\;\cdots & U_i & \cdots & U_s & \cdots & U_\gamma A_{i'} & \cdots & U_\gamma A_e & \cdots\;] & [U_\gamma B] & = U\tilde{X} \\
[K] & KA = A' = & [\;\cdots & \varepsilon_j & \cdots & A'_s & \cdots & A'_{i'} & \cdots & \varepsilon_\sigma & \cdots\;] & [\,b'\,] &
\end{array}
$$

with

$$
K = \overset{-1}{A_{\gamma'}} = \left[\begin{array}{c|c|c|c|c|c|c}
 & & & -A_e^1/A_e & & & \\
 & & & \vdots & & & \\
 & & & -A_e^{\sigma-1}/A_e & & & \\
\varepsilon_1 & \cdots & \varepsilon_{\sigma-1} & 1/A_e & \varepsilon_{\sigma+1} & \cdots & \varepsilon_p \\
 & & & -A_e^{\sigma+1}/A_e & & & \\
 & & & \vdots & & & \\
 & & & -A_e^p/A_e & & &
\end{array}\right]
$$

Each component of A' is the product of the row of K that appears at its left and the column of A that appears above it.

Example. Let us treat the following very simple example by the simplex method:

$$
\begin{cases}
10\,X^1 + 6\,X^2 + 3\,X^3 + X^4 = 100 \\
\;\;2\,X^1 + 3\,X^2 + 7\,X^3 + X^5 = 100 \\
X^1, X^2, X^3, X^4, X^5 \geqslant 0 \\
\text{MAX}\, X^1 + X^2 + X^3\,.
\end{cases}
$$

The calculations are presented in the following form:

$$U = \begin{bmatrix} 1 & 1 & 1 & 0 & 0 \end{bmatrix}$$

$$\gamma = \{4.5\} \quad \tilde{X} = \begin{bmatrix} 0 \\ 0 \\ 0 \\ 100 \\ 100 \end{bmatrix} \qquad A = \begin{bmatrix} 10 & 6 & 3 & 1 & 0 \\ 2 & 3 & 7 & 0 & 1 \end{bmatrix} \quad \begin{bmatrix} 100 \\ 100 \end{bmatrix} = b$$

$$e = 1, \sigma = 1, s = 4 \qquad U_\gamma = \begin{bmatrix} 0 & 0 \end{bmatrix} \qquad U_\gamma A = \begin{bmatrix} 0 & 0 & 0 & 0 & 0 \end{bmatrix} \quad \begin{bmatrix} 0 \end{bmatrix} = U\tilde{X}$$

$$\gamma' = \{1.5\} \quad \tilde{X}' = \begin{bmatrix} 10 \\ 0 \\ 0 \\ 0 \\ 80 \end{bmatrix} \qquad K = \begin{bmatrix} 0.1 & 0 \\ -0.2 & 1 \end{bmatrix} \qquad A' = \begin{bmatrix} 1 & 0.6 & 0.3 & 0.1 & 0 \\ 0 & 1.8 & 6.4 & -0.2 & 1 \end{bmatrix} \quad \begin{bmatrix} 10 \\ 80 \end{bmatrix} = b'$$

$$e' = 3, \sigma' = 2, s' = 5 \qquad U_{\gamma'} = \begin{bmatrix} 1 & 0 \end{bmatrix} \qquad U_{\gamma'} A' = \begin{bmatrix} 1 & 0.6 & 0.3 & 0.1 & 0 \end{bmatrix} \quad \begin{bmatrix} 10 \end{bmatrix} = U\tilde{X}'$$

$$\gamma'' = \{1.3\} \quad \tilde{X}'' = \begin{bmatrix} 6.24 \\ 0 \\ 12.50 \\ 0 \\ 0 \end{bmatrix} \qquad K' = \begin{bmatrix} 1 & -0.047 \\ 0 & 0.156 \end{bmatrix} \qquad A'' = \begin{bmatrix} 1 & 0.515\,5 & 0 & 0.109\,4 & -0.047 \\ 0 & 0.282\,0 & 1 & -0.031\,5 & 0.156 \end{bmatrix} \quad \begin{bmatrix} 6.24 \\ 12.50 \end{bmatrix} = b''$$

$$e'' = 2, \sigma'' = 1, s'' = 1 \qquad U_{\gamma''} = \begin{bmatrix} 1 & 1 \end{bmatrix} \qquad U_{\gamma''} A'' = \begin{bmatrix} 1 & 0.797\,5 & 1 & 0.077\,9 & 0.109 \end{bmatrix} \quad \begin{bmatrix} 18.74 \end{bmatrix} = U\tilde{X}''$$

$$\gamma''' = \{2.3\} \quad \tilde{X}''' = \begin{bmatrix} 0 \\ 12.11 \\ 9.09 \\ 0 \\ 0 \end{bmatrix} \qquad K'' = \begin{bmatrix} 1.94 & 0 \\ -0.547 & 1 \end{bmatrix} \qquad A''' = \begin{bmatrix} 1.940 & 1 & 0 & 0.212\,5 & -0.091\,2 \\ 0.547 & 0 & 1 & -0.091\,2 & 0.181\,6 \end{bmatrix} \quad \begin{bmatrix} 12.11 \\ 9.09 \end{bmatrix} = b'''$$

$$\tilde{X}''' \text{ is optimal} \qquad U_{\gamma'''} = \begin{bmatrix} 1 & 1 \end{bmatrix} \qquad U_{\gamma'''} A''' = \begin{bmatrix} 1.393 & 1 & 1 & 0.121\,3 & 0.090\,4 \end{bmatrix} \quad \begin{bmatrix} 21.20 \end{bmatrix} = U\tilde{X}'''$$

separate them, we would need to proceed to recopy the schemes at each stage in the procedure, which would create a considerable risk of error. In the group of columns with index belonging to γ, we need to single out the column of index s and, in the other group, the column with index e.

Remarks: I. At the first stage of the calculation, we have $U_e - U_\gamma A_e = 1$ for every $e \in \bar{\gamma}$. The rule of "taking $e \in \bar{\gamma}$ such that $U_e - U_\gamma A_e$ is maximized" does not enable us to separate the different possible values of e.

We can choose e so as to maximize X^e. This leads to taking $e = 2$. The reader can verify that the optimal vertex is then attained in two stages instead of three.

II. Consider the program put in the following two equivalent forms:

$$\begin{cases} AX = b \\ X \geqslant 0 \\ \text{MAX } UX \end{cases} \quad \text{and} \quad \begin{cases} (KA)X = Kb \\ X \geqslant 0 \\ \text{MAX } UX \end{cases}$$

where K is a nonsingular matrix.

The condition for optimality of $\tilde{X}$ may be written as follows: For the first form, there exists a v such that

$$\begin{cases} U - vA \leqslant 0\,, \\ (U - vA)\tilde{X} = 0\,. \end{cases}$$

For the second form, there exists a v' such that

$$\begin{cases} U - v'(KA) \leqslant 0\,, \\ (U - v'KA)\tilde{X} = 0\,. \end{cases}$$

Then, the equation $v = v'K$ associates with every form v satisfying the first condition, a form v' satisfying the second condition and vice versa.

Let us suppose that the two forms are nondegenerate simplicial forms with

$$A = [A_\gamma \mid A_{\bar{\gamma}}]\,, \quad A_\gamma = \mathbf{1}\,,$$

and

$$A' = [A_{\gamma'} \mid A_{\bar{\gamma}'}]\,, \quad A_{\gamma'} = \mathbf{1}\,.$$

We know that $v' = U_{\gamma'}$. We have $vA = v'A'$, so that

$$v = vA_\gamma = (vA)_\gamma = (v'A')_\gamma = (U_{\gamma'}A')_\gamma\,.$$

Let us apply this result to the example treated. Recalling the difference in notation, we have

$$v = (U_{\gamma'''}A''')_{\gamma},$$

that is,

$$v = [0.12, \quad 0.09].$$

3. DUALITY

Theorem 1 (known as the duality theorem). *Suppose that* $a \in L(\mathbf{R}^n, \mathbf{R}^p)$, $x \in \mathbf{R}^n$, $b \in \mathbf{R}^p$, $u \in (\mathbf{R}^n)'$ *and* $v \in (\mathbf{R}^p)'$. *Consider the two (so-called dual) programs*

$$(\mathrm{I}) \begin{cases} ax \leqslant b \\ x \geqslant 0 \\ \text{MAX } ux \end{cases} \qquad (\mathrm{I}^*) \begin{cases} va \geqslant u \\ v \geqslant 0 \,. \\ \text{MIN } vb \end{cases}$$

The following assertions hold:

(1) *If* x *and* v *are two admissible solutions of* (I) *and* (I*), *then*

$$ux \leqslant vb \,.$$

(2) *If* (I) *and* (I*) *have admissible solutions, they have optimal solutions.*

(3) *If* (I) *has an optimal solution, so does* (I*), *and conversely.*

(4) *A necessary and sufficient condition for two admissible solutions* $\tilde{x}$ *and* $\tilde{v}$ *of* (I) *and* (I*), *respectively, to be optimal is that*

$$u\tilde{x} = \tilde{v}b \,.$$

(5) *If* $\tilde{x}$ *is an optimal solution of* (I), *the optimal solutions of* (I*) *are the admissible solutions* $\tilde{v}$ *satisfying the equations*

$$(u - \tilde{v}a)\,\tilde{x} = 0 \,,$$

$$\tilde{v}(b - a\tilde{x}) = 0 \,.$$

(6) *If* $\tilde{v}$ *is an optimal solution of* (I*), *the optimal solutions of* (I) *are the admissible solutions* $\tilde{x}$ *satisfying the equations*

$$(u - \tilde{v}a)\,\tilde{x} = 0 \,,$$

$$\tilde{v}(b - a\tilde{x}) = 0 \,.$$

Proof. (1) Let x and v denote two admissible solutions of (I) and (I*). Then,

$$(\alpha) \qquad ux \leqslant va\,x \leqslant vb \,.$$

This is true because the inequalities $u \leqslant va$ and $x \geqslant 0$ imply $ux \leqslant vax$, whereas the inequalities $ax \leqslant b$ and $v \geqslant 0$ imply $vax \leqslant vb$.

(2) If (I) and (I*) have admissible solutions, the set of values of ux for admissible x is bounded above. Therefore, ux attains a maximum on the permitted set of (I). Similarly, the set of values vb for admissible v is bounded below, so that vb attains a minimum on the permitted set of (I*).

(3) Let $\tilde{x}$ denote an optimal solution of (I). There exists a $\tilde{v} \geqslant 0$ such that

$$\begin{cases} u - \tilde{v}a \leqslant 0\,, \\ \tilde{v}(b - a\tilde{x}) = 0\,, \\ (u - \tilde{v}a)\,\tilde{x} = 0\,. \end{cases}$$

Thus, $\tilde{v}$ is an admissible solution of (I*) and the two conditional inequalities (α) become equalities. Therefore, $u\tilde{x} = \tilde{v}b$. Consequently, $\tilde{v}$ is an optimal solution of (I*). We also see that, if (I) and (I*) have admissible solutions, then

$$(\beta) \qquad \max_{\left\{\substack{ax \leqslant b \\ x \geqslant 0}\right.} ux = \min_{\left\{\substack{va \geqslant u \\ v \geqslant 0}\right.} vb\,.$$

Furthermore, if (I*) has an optimal solution $\tilde{v}$, there exist two multipliers $\tilde{x} \in \mathbf{R}^n$ and $\tilde{y} \in \mathbf{R}^p$ such that

$$\tilde{x} \geqslant 0\,, \quad \tilde{y} \geqslant 0\,, \quad b = a\tilde{x} + \tilde{y}\,, \quad (u - \tilde{v}a)\,\tilde{x} = 0\,, \quad \tilde{v}\tilde{y} = 0\,.$$

From this, we conclude that

$$a\tilde{x} \leqslant b$$

$$\tilde{v}(b - a\tilde{x}) = 0\,.$$

Consequently, $\tilde{x}$ is an optimal solution of (I).

(4) This follows immediately from equation (β).

(5) Let $\tilde{x}$ denote an optimal solution of (I). A necessary and sufficient condition for an admissible solution $\tilde{v}$ of (I*) to be optimal is that $u\tilde{x} = \tilde{v}b$. In accordance with (α), this condition may be written

$$(u - \tilde{v}a)\,\tilde{x} = 0\,, \qquad \tilde{v}(b - a\tilde{x}) = 0\,.$$

(6) The proof is analogous to that of (5).

Corollary. *There exist three possible situations for programs* (I) *and* (I*):

(1) *The two programs have no admissible solutions.*
(2) *One of the programs has admissible solutions but has no optimal solution, whereas the other program has no admissible solution.*
(3) *Both programs have optimal solutions.*

Remark. Suppose that $\tilde{v}$ is an optimal solution of (I*). From (6) of the preceding theorem, every optimal solution $\tilde{x}$ of (I) must satisfy the relations

$$\tilde{x}^i = 0\,, \quad \text{if} \quad u_i - \tilde{v}a_i \neq 0\,,$$
$$a^j\tilde{x} = b^j\,, \quad \text{if} \quad \tilde{v}_j \neq 0\,.$$

In general, these conditions are sufficient to determine $\tilde{x}$. Specifically, in the program (I*), there exist

n constraints of the form $u_i - \tilde{v}a_i \neq 0$,
p constraints of the form $v_j \geqslant 0$.

Let p_1 and p_2 denote the number of constraints of these two types that are saturated at $\tilde{v}$. Let us assume (which is the general case) that $p_1 + p_2 = p$. To determine $\tilde{x}$, we have

$n - p_1$ equations of the form $\tilde{x}^i = 0$,
$p - p_2$ equations of the form $a^j\,\tilde{x} = b^j$,

which, in general, enable us to determine $\tilde{x}$.

Example. Consider the program

$$\text{(I)}\begin{cases} 10\,x^1 + 6\,x^2 + 3\,x^3 \leqslant 100 \\ 2\,x^1 + 3\,x^2 + 7\,x^3 \leqslant 100 \\ x^1, x^2, x^3 \geqslant 0 \\ \text{MAX}\; x^1 + x^2 + x^3\,. \end{cases}$$

The dual program may be written

$$\text{(I*)}\begin{cases} 10\,v_1 + 2\,v_2 \geqslant 1 \\ 6\,v_1 + 3\,v_2 \geqslant 1 \\ 3\,v_1 + 7\,v_2 \geqslant 1 \\ v_1, v_2 \geqslant 0 \\ \text{MIN}\; 100\,v_1 + 100\,v_2\,. \end{cases}$$

One can solve it graphically (see Fig. 1). We find that the optimal solution is $\tilde{v}_1 = 0.12$, $\tilde{v}_2 = 0.09$.

Let us find the optimal solutions $\tilde{x}$ of program (I). We have

$$\tilde{x}^1 = 0$$

(since the constraint $10\,v_1 + 2\,v_2 \geqslant 1$ is not saturated),

$$10\,\tilde{x}^1 + 6\,\tilde{x}^2 + 3\,\tilde{x}^3 = 100$$

(since the constraint $v_1 \geqslant 0$ is not saturated), and

$$2\,\tilde{x}^1 + 3\,\tilde{x}^2 + 7\,\tilde{x}^3 = 100$$

(since the constraint $v_2 \geqslant 0$ is not saturated). From this we get the system

$$6\,\tilde{x}^2 + 3\,\tilde{x}^3 = 100\,,$$

$$3\,\tilde{x}^2 + 7\,\tilde{x}^3 = 100\,,$$

which yields $\tilde{x}^2 = 12.1$, $\tilde{x}^3 = 9.1$.

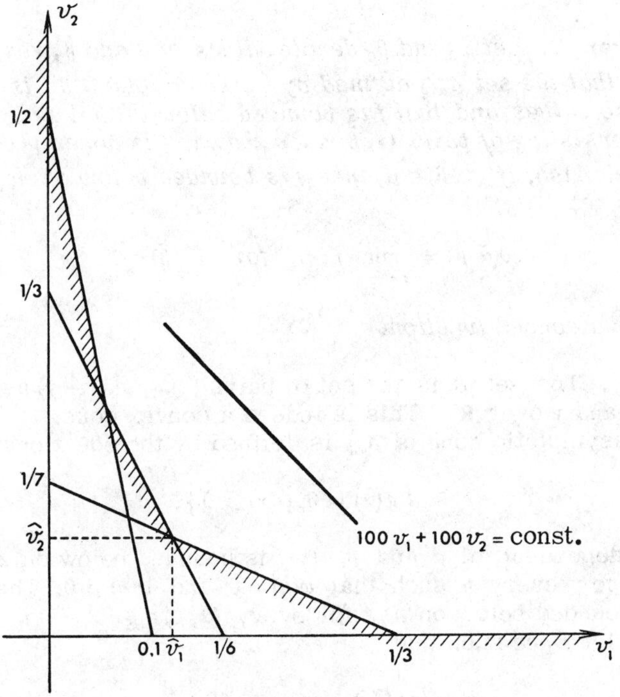

FIG. 1.

When we introduce new variables, the probram in question yields the program used as an example of the simplex method in the preceding section. One can verify that the solutions found agree with each other.

4. PARAMETRIZATION

Consider a linear program of the form (in the "general notations" of Chapter 13)

$$\begin{cases} \varphi^j(x) = \tilde{\alpha}^j & (j = 1, \ldots, q) \\ \psi^k(x) \geqslant \tilde{\beta}^k & (k = 1, \ldots, r) \\ \text{MIN}\, f(x)\,. \end{cases}$$

The study of the variation of the optimal point (if it is unique) and of the minimum value of the objective as $\tilde{\alpha}$ and $\tilde{\beta}$ vary is called parametrization. This problem has already been solved in the case of nonlinear programming (see Chapter 13, Theorem 6) but it provided only local results. In the case of linear programming, the results can be improved so that they take a global structure. For Theorems 2 and 3, the notations are the general notations of Chapter 13.

Theorem 2. *Let $\tilde{\alpha}_0$ and $\tilde{\beta}_0$ denote values of $\tilde{\alpha}$ and $\tilde{\beta}$, respectively. Suppose that the set $\Delta_{\tilde{\alpha},\tilde{\beta}}$ defined by $\{\varphi(x) = \tilde{\alpha},\ \psi(x) \geqslant \tilde{\beta}\}$ is nonempty for these values and that f is bounded below in that set. Then, the set B consisting of pairs $(\tilde{\alpha}, \tilde{\beta})$ such that $\Delta_{\tilde{\alpha},\tilde{\beta}}$ is nonempty is a convex cone. Also, if $(\tilde{\alpha}, \tilde{\beta}) \in B$, then f is bounded below on $\Delta_{\tilde{\alpha},\tilde{\beta}}$.*

If we set

$$\theta(\tilde{\alpha}, \tilde{\beta}) = \min_{x \in \Delta_{\tilde{\alpha},\tilde{\beta}}} f(x) \quad \textit{for} \quad (\tilde{\alpha}, \tilde{\beta}) \in B\,,$$

then θ is a convex function.

Proof. The set B is the set of pairs $(\varphi(x), \psi(x) - y)$ as x ranges over $\mathbf{R}^m$ and y over $\mathbf{R}^r_+$. This is indeed a convex cone.

The asymptotic cone of $\Delta_{\alpha,\beta}$ is defined by the equations

$$\{\varphi(x) = 0,\ \psi(x) \geqslant 0\}.$$

It is independent of $\tilde{\alpha}$ and $\tilde{\beta}$. If f is bounded below on $\Delta_{\tilde{\alpha}_0,\tilde{\beta}_0}$, then $f(h) \geqslant 0$ for every h such that $\varphi(h) = 0$ and $\psi(h) \geqslant 0$. Therefore, f will be bounded below on $\Delta_{\tilde{\alpha},\tilde{\beta}}$ for every $(\tilde{\alpha}, \tilde{\beta}) \in B$.

Now, let $\tilde{x}_1$ be such that

$$f(\tilde{x}_1) = \min_{x \in \Delta_{\alpha 1,\beta 1}} f(x)$$

and let $\tilde{x}_2$ be such that

$$f(\tilde{x}_2) = \min_{x \in \Delta_{\tilde{\alpha}2,\beta 2}} f(x)\,.$$

Also, suppose that

$$\lambda = (\lambda_1, \lambda_2) \in \Lambda_2\,, \quad \tilde{\alpha} = \lambda_1\,\tilde{\alpha}_1 + \lambda_2\,\tilde{\alpha}_2\,, \quad \tilde{\beta} = \lambda_1\,\tilde{\beta}_1 + \lambda_2\,\tilde{\beta}_2\,.$$

Then,

$$\varphi(\lambda_1\,\tilde{x}_1 + \lambda_2\,\tilde{x}_2) = \lambda_1\,\tilde{\alpha}_1 + \lambda_2\,\tilde{\alpha}_2 = \tilde{\alpha}$$

$$\psi(\lambda_1\,\tilde{x}_1 + \lambda_2\,\tilde{x}_2) = \lambda_1\,\psi(\tilde{x}_1) + \lambda_2\,\psi(\tilde{x}_2) \geqslant \lambda_1\,\tilde{\beta}_1 + \lambda_2\,\tilde{\beta}_2 = \tilde{\beta}\,.$$

Consequently,

$$\lambda_1\,\tilde{x}_1 + \lambda_2\,\tilde{x}_2 \in \Delta_{\tilde{\alpha},\tilde{\beta}}\,.$$

Also,

$$f(\lambda_1 \tilde{x}_1 + \lambda_2 \tilde{x}_2) = \lambda_1 f(\tilde{x}_1) + \lambda_2 f(\tilde{x}_2) = \lambda_1\, \theta(\tilde{\alpha}_1, \tilde{\beta}_1) + \lambda_2\, \theta(\tilde{\alpha}_2, \tilde{\beta}_2)$$

and, consequently,

$$\theta(\tilde{\alpha}, \tilde{\beta}) \leqslant \lambda_1\, \theta(\tilde{\alpha}_1, \tilde{\beta}_1) + \lambda_2\, \theta(\tilde{\alpha}_2, \tilde{\beta}_2)\,.$$

Thus, θ is a convex function. This completes the proof.

Theorem 3. *Let $\tilde{\alpha}_0$ and $\tilde{\beta}_0$ denote values of $\tilde{\alpha}$ and $\tilde{\beta}$ for which $\tilde{x}_0$ is an optimal point of $\Delta_{\tilde{\alpha}_0,\tilde{\beta}_0}$. Define $K_0 = K(\tilde{x}_0, \tilde{\beta}_0)$. Suppose also that the forms φ^j (for $j = 1, \ldots, q$) and ψ^k (for $k \in K_0$) are independent and that the number of indices k belonging to K_0 is equal to $m - q$. Then, the point $\tilde{x}$ satisfying the relations*

$$\begin{cases} \varphi^j(\tilde{x}) = \tilde{\alpha}^j & (j = 1, \ldots, q)\,, \\ \psi^k(\tilde{x}) = \tilde{\beta}^k & (k \in K_0)\,, \end{cases}$$

is optimal on $\Delta_{\tilde{\alpha},\tilde{\beta}}$ if it is admissible, which will be the case if $(\tilde{\alpha}, \tilde{\beta})$ belongs to a closed polyhedral convex cone A_{K_0}. If $\mu \in (\mathbf{R}^q)'$ and $\nu \in (\mathbf{R}^r)'$ are such that

$$\nu \geqslant 0, \quad \nu_k = 0 \quad \textit{for} \quad k \notin K_0, \quad f = \sum_{j=1}^{q} \mu_j\, \varphi^j + \sum_{k \in K_0} \nu_k\, \psi^k = \mu\varphi + \nu\psi\,,$$

then

$$\theta(\tilde{\alpha}, \tilde{\beta}) = \mu\tilde{\alpha} + \nu\tilde{\beta} \quad \textit{for} \quad (\tilde{\alpha}, \tilde{\beta}) \in A_{K_0}\,.$$

Proof. The condition for optimality at $\tilde{x}_0$ may be stated as follows. There exists μ and ν such that

$$\nu \geqslant 0\,, \quad \nu_k = 0 \quad \text{for} \quad k \notin K_0\,, \quad f = \mu\varphi + \nu\psi\,.$$

This condition for optimality is again satisfied at the point $\tilde{x}$, which is therefore an optimal point if it is admissible. Now, $\tilde{x}$ depends linearly on $\tilde{\alpha}$ and $\tilde{\beta}$. Consequently, the set A_{K_0}, which is defined by the relations $\psi^k(\tilde{x}) \geqslant \tilde{\beta}^k$ (for $k \notin K_0$), is a closed polyhedral convex cone (each of the preceding relations defining a closed half-space). Finally,

$$\theta(\tilde{\alpha}, \tilde{\beta}) = f(\tilde{x}) = \mu\varphi(\tilde{x}) + \nu\psi(\tilde{x}) = \mu\tilde{\alpha} + \nu\tilde{\beta}\,.$$

We shall make precise with an example of the function θ.

Theorem 4. *Consider the linear program*

$$\begin{cases} ax \leqslant b \\ x \geqslant 0 \\ \text{MAX } ux\,. \end{cases}$$

Let B denote the set of b such that the set Δ_b defined by the inequalities $ax \leqslant b$, $x \geqslant 0$ is nonempty. Suppose that u is bounded above on Δ_b for $b \in B$. Define

$$\theta(b) = \max_{\begin{cases} ax \leqslant b \\ x \geqslant 0 \end{cases}} ux$$

for $b \in B$. Let $v^1, \ldots, v^l, \ldots, v^L$ denote the vertices of the set defined by the inequalities

$$\begin{cases} va \geqslant u \\ v \geqslant 0 \end{cases}$$

(this set is the permitted set of the dual linear program). Then,

$$\theta(b) = \min_l v^l b .$$

Proof. For $b \in B$, the dual program has optimal solutions (in general, only one), and among these solutions there is at least one vertex. Thus,

$$\min_{\begin{cases} vA \geqslant u \\ v \geqslant 0 \end{cases}} vb = \min_l v^l b .$$

From the duality theorem (Theorem 1), we have

$$\theta(b) = \max_{\begin{cases} ax \leqslant b \\ x \geqslant 0 \end{cases}} ux = \min_{\begin{cases} va \geqslant u \\ v > 0 \end{cases}} vb ,$$

from which the desired relation follows.

15

Dynamic Programming

1. EXPOSITION OF THE METHOD

We shall expound the method of dynamic programming in the case in which time is a discrete variable, for example, one assuming integral values.

Consider a system governed by a recursion equation of the form

$$X_{n+1} = f(X_n, U_n, n) \tag{1}$$

where X_n is the state of the system at the instant n and U_n is the value of the control (French "commande") at the instant n.

We denote by H the space of the states X_n and we denote by $K_n(X_n)$ the set of permitted values of U_n at the instant n if the system is in the state X_n (thus, $U_n \in K_n(X_n)$). If $K_n(X_n)$ is fixed, we shall denote it by K.

A finite or infinite sequence $U_0, U_1, \ldots, U_n, \ldots$ (where $U_n \in K_n(X_n)$) will be called a *control*. If we know a control and the initial state of the system, the subsequent states are determined by Eq. (1). The sequence $X_0, X_1, \ldots, X_n, \ldots$ will be called the *trajectory* of the system associated with the control $U_0, U_1, \ldots, U_n, \ldots$. Conversely, every sequence $X_0, X_1, \ldots, X_n, \ldots$ will be called a *permitted trajectory* if it can be realized by one or more controls.

The number N of steps involved will be called the *horizon*. We assume this number to be finite.

Thus, the problem is to find the controls that maximize a criterion of the form

$$v_0(X_0, U_0) + v_1(X_1, U_1) + \cdots + v_{N-1}(X_{N-1}, U_{N-1}) + V_N(X_N)$$

for a given initial state. In the case in which only the last term exists, the criterion will be called *final*.

To introduce the method, let us first consider the case of a system that can assume only finitely many states at each instant. We can draw a graph whose vertices are the possible states of the system, the arcs representing the permitted transitions (see Fig. 1). To each transition and to each possible final state is assigned a gain. It is then a matter of finding the path or paths maximizing the total gain over all the paths leading from the initial state to one of the final states.

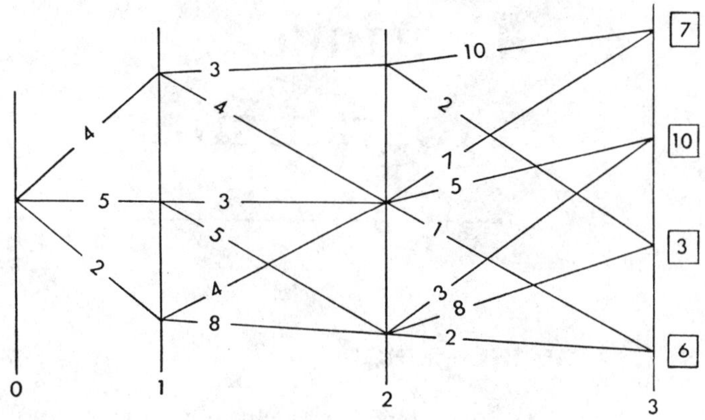

FIG. 1.

Now, consider the different possible states at the instant 2 and assign to each a value equal to the maximum gain that can be extracted from it in the course of future evolution. For this, we need only consider, for each state, the total gains contributed by the different possible transitions (the gain assigned to the transition plus the gain assigned to the state attained as a result of that transition) and not the maximum value of it. At the same time, we can indicate with an arrow the transition or transitions for which the maximum is attained. The arrows indicate the optimal transitions in each state at the instant 2 (see Fig. 2).

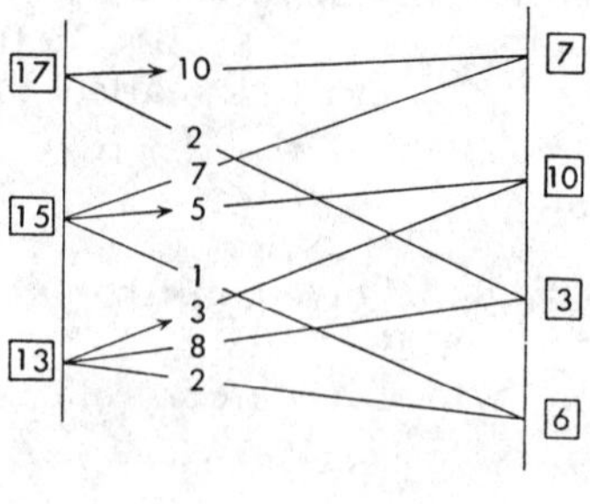

FIG. 2.

Let us now consider the different possible states at the instant 1. It is possible to assign to each of these states a value equal to the maximum of the gain that can be extracted from it in the course of future evolution. For this, it will be sufficient to consider, for each state, the total gains that the different possible transitions (the gain assigned to the transition plus the gain that can be extracted from the state attained as a result of that transition) can contribute and to note the maximum value of it. We mark with an arrow the optimum transition or transitions. In other words, we repeat for the instant 1 the operations performed at the instant 2 by replacing the final gains with the values assigned to the different possible states at the instant 2.

We can continue the operation up to the instant 0. This process yields the following:

(a) at each instant and in each state, the optimal transition or transitions, which enable us to extract the maximum gain from the future evolution of the system;

(b) for each instant and each state, the maximum gain that can be extracted from the future evolution of the system;

(c) the optimum trajectory or trajectories from the initial instant on.

This last (which is unique in the case in question) is indicated by a heavy line (see Fig. 3). Statement (*a*) can be expressed by saying that we have found the optimum strategy.

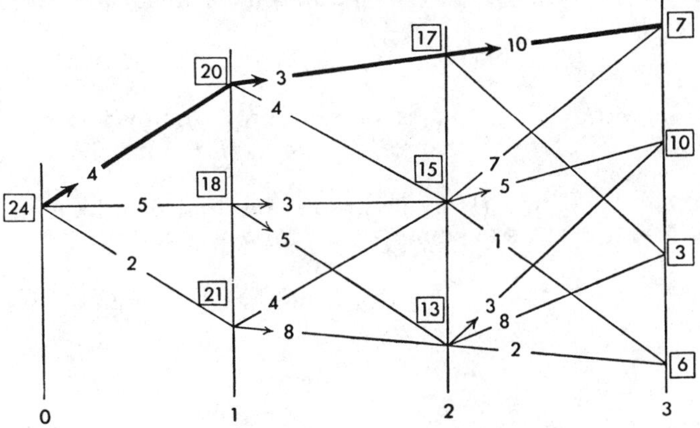

FIG. 3.

We shall now return to the general formulation and state a general principle, known as the *principle of optimality*.*

The optimality principle. If a control $\tilde{U}_0, \tilde{U}_1, ..., \tilde{U}_{N-1}$ is optimal, so is the control $\tilde{U}_p, ..., \tilde{U}_{N-1}$ for the initial state $\tilde{X}_p$ and the criterion

$$v_p(\tilde{X}_p, U_p) + v_{p+1}(X_{p+1}, U_{p+1}) + \cdots + v_{N-1}(X_{N-1}, U_{N-1}) + V_N(X_N).$$

* This term is due to R. Bellman.

We set

$$V_{N-1}(X_{N-1}) = \max_{U_{N-1}} v_{N-1}(X_{N-1}, U_{N-1}) + V_N(X_N)$$

and, in general,

$$V_p(X_p) = \max_{U_p} \left(v_p(X_p, U_p) + V_{p+1}(X_{p+1})\right), \quad \text{for } p = 0, \ldots, N-1 .$$

We assume that the conditions for these maxima to exist are satisfied. The above equations are called the ***realization equations.****

Theorem 1. (1)

$$V_p(X_p) = \max_{U_p, \ldots, U_{N-1}} v_p(X_p, U_p) + \cdots + v_{N-1}(X_{N-1}, U_{N-1}) + V_N(X_N) .$$

(2) *A necessary and sufficient condition for a control* $\tilde{U}_0, \ldots, \tilde{U}_{N-1}$ *to be maximal is that*

$$V_p(\tilde{X}_p) = v_p(\tilde{X}_p, \tilde{U}_p) + V_{p+1}(\tilde{X}_{p+1}) \qquad (\forall p = 0, \ldots, N-1) .$$

(3) *If a control* $\tilde{U}_0, \ldots, \tilde{U}_{N-1}$ *is maximal, then the control* $\tilde{U}_0, \ldots, \tilde{U}_{p-1}$ *is maximal (the initial state remaining unchanged) for the criterion*

$$v_0(X_0, U_0) + \cdots + v_{p-1}(X_{p-1}, U_{p-1}) + V_p(X_p) .$$

Conversely, if $\tilde{U}_0, \ldots, \tilde{U}_{p-1}$ *is maximal for this criterion, it can be extended as an optimal control for the initial criterion.*

Proof. We prove (1) by inverse induction. The formula is true for $p = N - 1$. Let us assume it true for $p + 1$:

$$V_{p+1}(X_{p+1}) = \max_{U_{p+1}, \ldots, U_{N-1}} \{ v_{p+1}(X_{p+1}, U_{p+1}) + \cdots + v_{N-1}(X_{N-1}, U_{N-1}) + V_N(X_N) \} .$$

Then,

$$\begin{aligned} v_p(X_p, U_p) + v_{p+1}(X_{p+1}, U_{p+1}) + \cdots + v_{N-1}(X_{N-1}, U_{N-1}) + V_N(X_N) \\ \leqslant v_p(X_p, U_p) + V_{p+1}(X_{p+1}) \\ \leqslant \max_{U_p} \{ v_p(X_p, U_p) + V_{p+1}(X_{p+1}) \} = V_p(X_p) . \end{aligned}$$

Equality will hold between the first and last members of this condition inequality if and only if

$$v_{p+1}(X_{p+1}, U_{p+1}) + \cdots + v_{N-1}(X_{N-1}, U_{N-1}) + V_N(X_N) = V_{p+1}(X_{p+1})$$

---

*These equations were introduced in an economic text by P. Massé, Les reserves et la regulation de l'avenir, Hermann, 1946.

and

$$v_p(X_p, U_p) + V_{p+1}(X_{p+1}) = V_p(X_p)\,.$$

The second of these equations determines U_p and hence X_{p+1}. The first equation means that the control $U_{p+1}, \ldots, U_{N-1}$ is maximal with respect to the criterion

$$v_{p+1}(X_{p+1}, U_{p+1}) + \cdots + v_{N-1}(X_{N-1}, U_{N-1}) + V_N(X_N)\,.$$

Thus, the two equations are compatible and we have

$$V_p(X_p) = \max_{U_p, \ldots, U_{N-1}} v_p(X_p, U_p) + \cdots + v_{N-1}(X_{N-1}, U_{N-1}) + V_N(X_N).$$

At the same time, we have shown that, for every maximal control,

$$v_p(\tilde{X}_p, \tilde{U}_p) + V_{p+1}(\tilde{X}_{p+1}) = V_p(\tilde{X}_p)$$

for every $p = 0, \ldots, N-1$.

Conversely, if this relation holds for $p = 0, \ldots, N-1$, we obtain, by adding these equations,

$$v_0(X_0, \tilde{U}_0) + \cdots + v_{N-1}(\tilde{X}_{N-1}, \tilde{U}_{N-1}) + V_N(\tilde{X}_N) = V_0(X_0)\,,$$

which shows that the control is maximal.

The first part of (3) is obvious from the characterization of maximal controls that was given in (2). Conversely, if $\tilde{U}_0, \ldots, \tilde{U}_{p-1}$ is maximal for the criterion

$$v_0(X_0, U_0) + \cdots + v_{p-1}(X_{p-1}, U_{p-1}) + V_p(X_p)\,,$$

then there exist $\tilde{U}_p, \ldots, \tilde{U}_{N-1}$ such that

$$V_q(\tilde{X}_q) = v_q(\tilde{X}_q, \tilde{U}_p) + V_{q+1}(\tilde{X}_{q+1})$$

for $q = p, \ldots, N-1$. Consequently, the control $\tilde{U}_0, \ldots, \tilde{U}_{N-1}$ is maximal. This completes the proof.

Remark. In the case in which the criterion needs to be minimized, we replace "max" everywhere by "min."

2. AN EXAMPLE

Consider a system governed by the following recursion equation:

$$X_{n+1} = AX_n + U_n h\,,$$

where

$$\begin{cases} X_n \in \mathbf{R}^m \\ A \in L(\mathbf{R}^m, \mathbf{R}^m) \text{ with } \det(A) \neq 0 . \\ h \in \mathbf{R}^m \\ U_n \in \mathbf{R} . \end{cases}$$

We are given the initial state X_0 and a horizon N that is determined. We seek the controls $\{ U_n \}$ that minimize the criterion

$$\sum_{n=0}^{N} \| X_n \|^2 .$$

First, we shall study the optimization problem that arises if, at any instant, the system is in the state X and we wish to minimize a quadratic form $\bar{Y}MY$ of the next state

$$Y = AX + uh .$$

It is this problem that we shall need to solve at the instant $N-1$ (with $M = 1$) but also at the intermediary instants, as we shall see in the course of the realization process. Thus, we seek the value u that minimizes the quantity $\bar{Y}MY$. We have

$$\begin{aligned} \bar{Y}MY &= (\bar{X}\bar{A} + u\bar{h})\, M(AX + uh) \\ &= \bar{X}\bar{A}MAX + 2\, u\bar{h}MAX + u^2\, \bar{h}Mh . \end{aligned}$$

The minimum is attained at

$$\tilde{u} = -\frac{\bar{h}MAX}{\bar{h}Mh}$$

and it is equal to

$$\begin{aligned} \tilde{m}(X) &= \frac{\bar{h}Mh\, .\, \bar{X}\bar{A}MAX - (\bar{h}MAX)^2}{\bar{h}Mh}, \\ &= \bar{X}NX , \end{aligned}$$

where

$$N = \frac{\bar{h}Mh\, .\, \bar{A}MA - \bar{A}Mh\, .\, \bar{h}MA}{\bar{h}Mh} .$$

Note that N is a positive-semidefinite symmetric matrix.

Let us set

$$\begin{aligned} V_N(X_N) &= \| X_N \|^2 = \bar{X}_N X_N = \bar{X}_N M_N X_N \quad (M_N = 1) \\ V_{N-1}(X_{N-1}) &= \min_{U_{N-1}} \left[V_N(X_N) + \| X_{N-1} \|^2 \right] \\ V_p(X_p) &= \min_{U_p} \left[V_{p+1}(X_{p+1}) + \| X_p \|^2 \right] . \end{aligned}$$

We obtain

$$V_p(X_p) = \overline{X}_p M_p X_p ,$$

with

$$M_p = \frac{\overline{h}M_{p+1}\, h \, . \, \overline{A}M_{p+1}\, A - \overline{A}M_{p+1}\, h \, . \, \overline{h}M_{p+1}\, A}{\overline{h}M_{p+1}\, h} + 1$$

$$= \overline{A}M_{p+1}\, A + 1 - \frac{\overline{A}M_{p+1}\, h \, . \, \overline{h}M_{p+1}\, A}{\overline{h}M_{p+1}\, h} .$$

The matrix M_p is positive-definite and symmetric.

At each instant p, we need to choose U_p so as to minimize $V_{p+1}(X_{p+1})$. Thus, there exists a single optimum control

$$\tilde{U}_p(X_p) = -\frac{\overline{h}M_{p+1}\, AX_p}{\overline{h}M_{p+1}\, h} .$$

Note that the optimum control at the instant p is a linear form of the state X_p at that instant. Also,

$$V_0(X_0) = \overline{X}_0\, M_0\, X_0 = \min_{U_0,\dots,U_{N-1}} \sum_{n=0}^{N} \| X_n \|^2$$

and, more generally,

$$V_p(X_p) = \overline{X}_p\, M_p\, X_p = \min_{U_p,\dots,U_{N-1}} \sum_{n=p}^{N} \| X_n \|^2 .$$

We shall now pass to the case of an infinite horizon. First of all, let us give a definition.

Definition. A system governed by the recursion relation

$$X_{n+1} = AX_n + U_n\, h$$

where

$$\begin{cases} X_n \in \mathbf{R}^m \\ A \in L(\mathbf{R}^m, \mathbf{R}^m) \qquad (\det(A) \neq 0) \\ U_n \in \mathbf{R} \\ h \in \mathbf{R}^m \end{cases}$$

is said to be controllable if for every initial state X_0 there exist an integer N and an N-tuple $(U_0, \dots, U_{N-1})$ such that $X_N = 0$. The system is then said to be adjusted at the origin.

The general formula giving X_N as a function of X_0 and $U_0, \dots, U_{N-1}$ is

$$X_N = A^N X_0 + U_0 A^{N-1} h + U_1 A^{N-2} h + \cdots + U_{N-1} h .$$

Thus, the relation $X_N = 0$ may be written

$$X_0 = - U_0 A^{-1} h - U_1 A^{-2} h \ldots - U_{N-1} A^{-N} h .$$

Theorem 2. *A necessary and sufficient condition for the system to be controllable is that the smallest stable vector subspace under A and containing h be the entire space* $\mathbf{R}^m$.

Proof. For the system to be controllable, it is necessary and sufficient that every vector X be a linear combination of the vectors $A^{-1} h, A^{-2} h, \ldots$, in other words, that the vector subspace generated by the vectors $A^{-1} h, A^{-2} h, \ldots$, be $\mathbf{R}^m$. Now, this subspace is the smallest stable vector subspace under A that contains h.

Corollary. *A necessary and sufficient condition for the system to be controllable is that the vectors* $A^{-1} h, \ldots, A^{-m} h$ *be independent.*

Proof. If $A^{-1} h, \ldots, A^{-m} h$ are independent, they generate the space $\mathbf{R}^m$. Thus, the condition is sufficient. Conversely, if $A^{-1} h, \ldots, A^{-m} h$ are not independent, one of them, $A^{-i} h$, is a linear combination of the preceding ones:

$$A^{-i} h = \lambda_1 A^{-1} h + \cdots + \lambda_{i-1} A^{-i+1} h .$$

From this we conclude that, for every $k \geqslant 0$,

$$A^{-i-k} h = \lambda_1 A^{-(1+k)} h + \cdots + \lambda_{i-1} A^{-(i-1+k)} h .$$

Consequently, by setting successively $k = 1, 2, \ldots$, we see that all the vectors $A^{-(i+k)} h$ (for $k = 0, 1, \ldots$) are linear combinations of the $A^{-1} h, \ldots, A^{-i+1} h$. Therefore, the vector subspace generated by the vectors $A^{-1} h$, $A^{-2} h$, ..., is identical to the vector space generated by the vectors $A^{-1} h$, $A^{-2} h$, ..., $A^{-i+1} h$. Thus, it is different from $\mathbf{R}^m$.

Theorem 3. *If the system is controllable, there exists exactly one m-tuple* $\hat{U}_0, \ldots, \hat{U}_{m-1}$ *such that* $X_m = 0$.

Proof. For X_m to be equal to 0, it is necessary and sufficient that

$$X_0 = - \hat{U}_0 A^{-1} h - \hat{U}_1 A^{-2} h - \cdots - \hat{U}_{m-1} A^{-m} h .$$

Now, the vectors $A^{-1} h$, $A^{-2} h$, ..., $A^{-m} h$ constitute a basis for $\mathbf{R}^m$, from which the condition of the theorem follows.

Note that $\hat{U}_0, \ldots, \hat{U}_{m-1}$ are linear forms with respect to X_0. From this we conclude immediately that, for this particular control, the successive states $\hat{X}_1, \ldots, \hat{X}_{m-1}$ are linear with respect to X_0. Therefore, we set

$$\hat{V}(X_0) = \| X_0 \|^2 + \| \hat{X}_1 \|^2 + \cdots + \| \hat{X}_{m-1} \|^2 .$$

The function $\hat{V}$ is a positive-definite quadratic form.

Now we shall seek sequences $U_0, U_1, \ldots,$ such that $\sum_{n=0}^{\infty} \| X_n \|^2$ is minimized. The system is assumed controllable.

Let us set

$$\begin{aligned} V^{(N)}(\xi) &= \min_{\left\{ \begin{smallmatrix} U_0,\ldots,U_{N-1} \\ X_0=\xi \end{smallmatrix} \right.} [\| X_0 \|^2 + \cdots + \| X_N \|^2] \\ &= \min_{\left\{ \begin{smallmatrix} U_k,\ldots,U_{k+N-1} \\ X_k=\xi \end{smallmatrix} \right.} [\| X_k \|^2 + \cdots + \| X_{N+k} \|^2] . \end{aligned}$$

We know that $V^{(N)}$ is a quadratic form. Furthermore, we have, for $N \geqslant m$,

$$V^{(N)}(\xi) \leqslant \hat{V}(\xi) .$$

Therefore, since $\{V^{(N)}(\xi)\}$ is a bounded increasing sequence for every ξ, it has a limit $V(\xi)$ and this limit is a quadratic form in ξ. Also, by virtue of the realization equations,

$$V^{(k)}(X_n) = \| X_n \|^2 + \min_{U_n} V^{(k-1)}(X_{n+1}) .$$

(The induction is direct instead of inverse since we have substituted a system of indexing with respect to the horizon by a system of indexing with respect to time.)

We have $V^{(k)}(\xi) = \bar{\xi} M^{(k)} \xi$, where $M^{(k)}$ is a positive-definite symmetric matrix satisfying the recursion equation

$$\begin{cases} M^{(k)} = \bar{A} M^{(k-1)} A + \mathbf{1} - \dfrac{\bar{A} M^{(k-1)} h \bar{h} M^{(k-1)} A}{\bar{h} M^{(k-1)} h} . \\ M^{(0)} = \mathbf{1} . \end{cases}$$

Theorem 4. *Let $\tilde{X}_0 = X_0, \tilde{X}_1, \ldots, \tilde{X}_n, \ldots,$ denote the trajectory such that*

$$V(\tilde{X}_{n+1}) = \min_{U_n} V(X_{n+1}) .$$

Then, this trajectory is minimal with respect to the criterion $\sum_{n=0}^{+\infty} \| X_n \|^2$ and conversely. Also,

$$V(X_0) = \sum_{n=0}^{+\infty} \| \tilde{X}_n \|^2 .$$

Proof. We need to show that

$$\sum_{n=0}^{+\infty} \| \tilde{X}_n \|^2 \leqslant \sum_{n=0}^{+\infty} \| \dot{X}_n \|^2$$

for an arbitrary trajectory $\dot{X}_0 = X_0, \dot{X}_1, ..., \dot{X}_n, ...$. We can confine ourselves to the case in which

$$\lim_{n\to\infty} \dot{X}_n = 0 .$$

Now, for every k,

$$V^{(k)}(\dot{X}_n) = \| \dot{X}_n \|^2 + \min_{U_n} V^{(k-1)}(X_{n+1})$$
$$\leqslant \| \dot{X}_n \|^2 + V^{(k-1)}(\dot{X}_{n+1}) .$$

By letting k approach $+\infty$, we get

$$V(\dot{X}_n) \leqslant \| \dot{X}_n \|^2 + V(\dot{X}_{n+1})$$

and, consequently (we set $n = 0, 1, ..., p$ in the preceding inequality and sum with respect to n),

$$V(X_0) \leqslant \sum_{n=0}^{p} \| \dot{X}_n \|^2 + V(\dot{X}_{p+1}) .$$

If we let p approach $+\infty$, we get

$$(\alpha) \qquad V(X_0) \leqslant \sum_{n=0}^{+\infty} \| \dot{X}_n \|^2 .$$

Also,

$$V^{(k)}(\tilde{X}_n) = \| \tilde{X}_n \|^2 + \min_{U_n} V^{(k-1)}(X_{n+1}) .$$

Here, if we let k approach $+\infty$, we get

$$V(\tilde{X}_n) = \| \tilde{X}_n \|^2 + \min_{U_n} V(X_{n+1}) ,$$

that is,

$$V(\tilde{X}_n) = \| \tilde{X}_n \|^2 + V(\tilde{X}_{n+1})$$

and, consequently,

$$V(\tilde{X}_0) = \sum_{n=0}^{p} \| \tilde{X}_n \|^2 + V(\tilde{X}_{p+1}) .$$

If we let p approach $+\infty$, we obtain

$$(\beta) \qquad V(\tilde{X}_0) = \sum_{n=0}^{+\infty} \| \tilde{X}_n \|^2 .$$

From (α) and (β), we get

$$\sum_{n=0}^{+\infty} \| \tilde{X}_n \|^2 \leqslant \sum_{n=0}^{+\infty} \| \dot{X}_n \|^2 .$$

Conversely, the relation

$$V(X_{n+1}) = \min_{U_{n+1}\ldots} \sum_{p=n+1}^{\infty} \| X_p \|^2$$

shows that every minimal trajectory satisfies the relation

$$V(\widetilde{X}_{n+1}) = \min_{U_n} V(X_{n+1}) .$$

Consequences. If we set $V(\xi) = \bar{\xi} M \xi$, we get $\widetilde{U}_p = \tilde{\sigma}\, \widetilde{X}_p$, where

$$\tilde{\sigma} = -\frac{\bar{h}MA}{\bar{h}Mh} .$$

The mapping $\xi \to \tilde{\sigma}\xi$ yields, as a function of the state of the system, the control to be applied. This mapping is called a *strategy*. This strategy is optimum in the sense that if it is applied at every instant, the trajectory obtained is the optimum one. Note that, subject to this strategy, the system obeys a linear recursion law:

$$X_{n+1} = AX_n - h\,\frac{\bar{h}MA\,X_n}{\bar{h}Mh}$$

$$= \left[A - \frac{h\bar{h}\,MA}{\bar{h}Mh}\right] X_n .$$

A numerical application. If

$$\text{with} \quad A = \begin{bmatrix} 3 & 0 & 0 & 0 \\ 0 & 2 & 0 & 0 \\ 0 & 0 & 1 & 0 \\ 0 & 0 & 1 & 1 \end{bmatrix} \quad \text{and } h = \begin{bmatrix} 1 \\ 1 \\ 1 \\ 0 \end{bmatrix},$$

we get

$$M^{(1)} = \begin{bmatrix} 7 & -2 & -1 & 0 \\ -2 & 3.666 & -0.666 & 0 \\ -1 & -0.666 & 2.666 & 1 \\ 0 & 0 & 1 & 2 \end{bmatrix}$$

$$M^{(2)} = \begin{bmatrix} 40 & -16 & -7 & -2 \\ -16 & 15 & -2 & -0.333 \\ -7 & -2 & 7 & 2.666 \\ -2 & -0.333 & 2.666 & 2.833 \end{bmatrix}$$

$$M^{(3)} = \begin{bmatrix} 144.2 & -70.5 & -19.91 & -7.416 \\ -70.5 & 58 & -5.5 & -0.5 \\ -19.91 & -5.5 & 15.93 & 5.546 \\ -7.416 & -0.5 & 5.546 & 3.824 \end{bmatrix}$$

$$M^{(10)} = \begin{bmatrix} 2\,446 & -2\,365 & 437.0 & 69.77 \\ -2\,365 & 2\,381 & -486.1 & -84.38 \\ 437.0 & -\ 486.1 & 124.1 & 25.68 \\ 69.77 & -\ \ 84.38 & 25.68 & 7.752 \end{bmatrix}$$

$$M = \begin{bmatrix} 2\,454 & -2\,373 & 438.7 & 70.07 \\ -2\,373 & 2\,389 & -487.9 & -84.67 \\ 438.7 & -\ 487.9 & 124.4 & 25.74 \\ 70.07 & -\ \ 84.67 & 25.74 & 7.762 \end{bmatrix}$$

and

$$\tilde{\sigma} = \begin{bmatrix} -52.60 & 39.36 & -3.608 & 0.464\,8 \end{bmatrix}.$$

16

Markov Control Systems

All the systems that we shall consider in this chapter are systems with discrete time and having only a finite number of states. The notations are suggested by those of Chapter 6. In order not to burden the exposition, we shall give the problems a rather "naive" initial formulation. Once this initial formulation is accepted, we shall arrive at mathematically well-posed problems.

1. MARKOV SYSTEMS WITH GAIN

Consider a system including n possible states subjected to a nonstationary Markov chain the transition matrix of which at the instant t is denoted by $P(t)$. To every transition $i \to j$ that takes place between the instant t and the instant $t + 1$ is assigned a gain $r_i^j(t)$.

Consider the evolution of the system between the instant 0 and the instant T. Let us assume that a gain $v_i(T)$ is assigned to the final state i. The total gain in the course of the evolution of the system thus consists of the transition gains and the final gain $v_i(T)$ if the system is in a state i at the instant T.

We denote by $v_i(t)$ the mean gain from the instant t if the system is in the state i at the instant t. We denote by $q_i(t)$ the mean gain in the transition that comes about between the instants t and $t + 1$ if the system is in the state i at the instant t. We have

$$q_i(t) = \sum_j r_i^j(t)\, P_i^j(t)\,.$$

Furthermore (we still assume that the system is in the state i at the instant t), the mean gain after the instant $t + 1$ is equal to

$$\sum_j v_j(t+1)\, P_i^j(t)\,.$$

Thus, we have the recursion relation

$$v_i(t) = \sum_j v_j(t+1)\, P_i^j(t) + q_i(t)\,.$$

If we set

$$v(t) = [v_1(t), \ldots, v_n(t)]\,,$$

$$q(t) = [q_1(t), \ldots, q_n(t)]\,,$$

this recursion equation can be put in the following matrix form:

$$\boxed{v(t) = v(t+1)\, P(t) + q(t)} \qquad (1.1)$$

We recall that $v(T)$ is given. Thus, we have an inverse recursion relation.

Let us now suppose that the system is stationary and let us study the asymptotic behavior of the mean gain as the number of transitions become infinite. Let us denote by P the transition matrix. We suppose that the gain assigned to a transition $i \to j$ is independent of the time. We denote it by r_i^j. Let us suppose that, at an arbitrary instant, the system is in the state i. The mean gain during the following transition is independent of the time and it is equal to

$$q_i = \sum_j r_i^j\, P_i^j\,.$$

We set $q = [q_1, \ldots, q_n]$.

We denote by $v_i(m)$ the gain in the course of m successive transitions when the system begins at the state i. We have the recursion relation

$$v_i(m) = \sum_j v_j(m-1)\, P_i^j + q_i\,.$$

If we set

$$v(m) = [v_1(m), \ldots, v_n(m)]\,,$$

we have

$$\boxed{v(m) = v(m-1)\, P + q} \qquad (1.2)$$

with $v(0) = 0$.

This time, we obtain a direct recursion relation. Let us study the behavior of $v(m)$ as $m \to +\infty$. We have

$$\begin{aligned} v(m) &= v(m-1)\,P + q \\ v(m-1) &= v(m-2)\,P + q \\ &\vdots \\ v(1) &= v(0)\,P + q \qquad \text{with } v(0) = 0 \end{aligned}$$

from which we get

$$v(m) = q(1 + P + \cdots + \overset{m-1}{P})$$

and

$$\frac{v(m)}{m} = q\,\frac{1 + P + \cdots + \overset{m-1}{P}}{m}.$$

Let π_1 denote the spectral projection associated with the eigenvalue 1 of the matrix P. We have

$$\lim_{m \to +\infty} \frac{1 + P + \cdots + \overset{m-1}{P}}{m} = \pi_1 .$$

We set

$$g = [g_1, \ldots, g_n] = q\pi_1 .$$

We have

$$\lim_{m \to +\infty} \frac{v(m)}{m} = g$$

that is,

$$\boxed{v(m) \sim mg} \qquad (1.3)$$

or

$$v_i(m) \sim mg_i .$$

The case in which the matrix *P* is ergodic. If P is ergodic, 1 is a simple eigenvalue and the rank of π_1 is 1. We have

$$\pi_1 = [\omega \mid \cdots \mid \omega]$$

where ω is the limiting probability distribution. Thus, $g_i = q\omega$ for every $i = 1, \ldots, n$. We denote by γ the common value of g_i. We then have

$$\boxed{v_i(m) \sim m\gamma} \,. \tag{1.4}$$

This γ is called the *asymptotic mean gain in a period.*

The case in which the matrix P is primitive. If P is primitive, its only eigenvalue of unit absolute value is 1. We may write $P = \pi_1 + Q$, where Q is a matrix all of the eigenvalues of which are less than 1 in absolute value and that has the property that $\pi_1 Q = Q\pi_1 = 0$. We then have

$$\overset{h}{P} = \pi_1 + \overset{h}{Q}.$$

Therefore,

$$\mathbf{1} + P + \cdots + \overset{m-1}{P} = \mathbf{1} + Q + \cdots + \overset{m-1}{Q} + (m-1)\pi_1 .$$

Since $\mathbf{1} - Q$ is nonsingular,

$$\mathbf{1} + Q + \cdots + \overset{m-1}{Q} + \cdots = (\mathbf{1} - Q)^{-1} ,$$

and, consequently,*

$$\mathbf{1} + P + \cdots + \overset{m-1}{P} \approx (\mathbf{1} - Q)^{-1} - \pi_1 + m\pi_1 ,$$

so that

$$v(m) \approx q\,[(\mathbf{1} - Q)^{-1} - \pi_1] + mq\pi_1 .$$

We have set $q\pi_1 = g$. Let us also set

$$W = [W_1, ..., W_n] = q\,[(\mathbf{1} - Q)^{-1} - \pi_1] .$$

We finally obtain

$$\boxed{v(m) \approx mg + W} \,, \tag{1.5}$$

or

$$v_i(m) \approx mg_i + W_i .$$

The case in which the matrix P is regular. If P is regular, it is both ergodic and primitive. We then have

$$v_i(m) \approx m\gamma + W_i .$$

If we set $\eta = [1, ..., 1]$, this relation can be written

---

*The symbol $\approx$ means that the difference between the two sides approaches 0 as $m \to +\infty$.

$$\boxed{v(m) \approx m\gamma\eta + W} \ . \tag{1.6}$$

We shall now establish an interesting relationship between γ and W by comparing (1.6) with the recursion equation.

$$v(m) = v(m - 1)\,P + q\,.$$

From (1.6),

$$v(m - 1) \approx (m - 1)\,\gamma\eta + W\,.$$

From this we derive

$$v(m) \approx \left[(m - 1)\,\gamma\eta + W\right]P + q = (m - 1)\,\gamma\eta + WP + q\,.$$

Comparing this result with (1.6), we obtain

$$WP + q - \gamma\eta = W$$

that is,

$$\boxed{W(1 - P) + \gamma\eta = q} \ . \tag{1.7}$$

Study of the equation $X(1 - P) + \lambda\eta = q$, where P is an ergodic matrix. Here, P and q are assumed given and X and λ are the unknowns. We denote by ω the limiting probability distribution.

We mention first of all that the equation $X(1 - P) = c$ has solutions if and only if $c\omega = 0$. We derive this by carrying over to the present situation the classical property that a necessary and sufficient condition for the equation $ax = b$ to have solutions is that $ub = 0$ for every u satisfying the equation $ua = 0$.

Consequently, for the present equation to have solutions, it is necessary and sufficient that $(q - \lambda\eta)\,\omega = 0$, that is, that $q\omega - \lambda = 0$ or, finally, that $\lambda = \gamma$. The solutions in X are then defined modulo the solutions of the equation $X(1 - P) = 0$, that is, modulo η. If, in addition, the matrix P is regular, the solutions in X obtained for $\lambda = \gamma$ include the form W.

2. MARKOV SYSTEMS WITH CONTROL AND GAIN

Let us consider a system that assumes n states numbered from 1 to n. The probability $P_i^j(t, k)$ that the system in the state i at the instant t will be in the state j at the instant $t + 1$ depends on the time t and a parameter k, known as the control, which assumes a finite number of values.* We have

---

* The possible values of k can *a priori* depend on t and i. In order not to burden the exposition, we disregard this possibility in our notation.

$$P_i^j(t, k) \geqslant 0\,, \qquad \sum_j P_i^j(t, k) = 1\,.$$

A gain $r_i^j(t, k)$ is assigned to the transition $i \to j$ that occurs between the instants t and $t + 1$ if the control k was applied at the instant t.

A strategy is defined as any mapping $t, i \to \sigma(t, i)$ that fixes the control to be chosen as a function of the instant t and the state i of the system. We make the following hypothesis: If we choose a strategy σ, the system obeys a Markov chain, in general, nonstationary, the transition matrix $P[\sigma]\,(t)$ of which has coefficients

$$P_i^j[\sigma]\,(t) = P_i^j(t, \sigma(t, \mathrm{i}))\,.$$

We shall consider first the case of a finite horizon T. We can assign a gain $v_i(T)$ to the final instant i. Let $v_i(t)$ denote the maximum in the mean gain between the instants t and T (the maximum is taken over all possible strategies between t and T).

Let us set

$$q_i(t, k) = \sum_j P_i^j(t, k)\, r_i^j(t, k)\,.$$

This quantity represents the mean gain in the transition that occurs between the instants t and $t + 1$ if the system is in the state i at the instant t and we apply to it the control k. We have the following realization relation:

$$\boxed{v_i(t) = \max_k \Big[\sum_j v_j(t + 1)\, P_i^j(t, k) + q_i(t, k)\Big]}\,. \tag{2.1}$$

This relation enables us to calculate $v_i(t)$ when we know $v_i(T)$. Also, if $\sigma(t, i)$ is a value of k at which the bracketed expression attains its maximum, then σ is an optimum strategy in the following sense: If it is applied at every instant t to a system that is in the state i, it assures a mean gain equal to the maximum mean gain. Of course, there may exist several optimum strategies.

Henceforth, we shall assume the system to be stationary in the following sense: The probabilities $P_i^j(t, k)$ are independent of t and will be denoted by $P_i^j(k)$. The gains $r_i^j(t, k)$ are independent of t and will be denoted by $r_i^j(k)$. We are interested only in the "stationary" strategies σ that assign to every state i a control $k = \sigma(i)$ that is to be applied to the system if it is in the state i.

If we apply the strategy σ to the system, it obeys a stationary Markov chain whose transition matrix $P[\sigma]$ has coefficients

$$P_i^j[\sigma] = P_i^j(\sigma(i))\,.$$

We shall also make the following hypothesis: For every strategy σ, the matrix $P[\sigma]$ is ergodic.* Then, associated with every strategy σ will be

* For cases in which this hypothesis is not satisfied, see [41].

the limiting probability distribution $\omega[\sigma]$,
the transition gains $r_i^j[\sigma] = r_i^j(\sigma(i))$,
the mean gains in a transition from the state i

$$q_i[\sigma] = \sum_j r_i^j[\sigma] \, P_i^j[\sigma]$$

the asymptotic mean gain in a period $\gamma[\sigma] = q[\sigma]\,\omega[\sigma]$, where

$$q[\sigma] = [q_1[\sigma], \ldots, q_n[\sigma]].$$

Let us now consider a solution $X[\sigma]$ of the equation

$$X[\sigma]\,(\mathbf{1} - P[\sigma]) + \eta\gamma[\sigma] = q[\sigma]\,.$$

If $P[\sigma]$ is regular, a particular solution in $X[\sigma]$ will be the form $W[\sigma]$ such that the mean gain $v[\sigma]\,(m)$ in m stages satisfies the relation

$$v[\sigma]\,(m) \approx m\gamma[\sigma]\,\eta + W[\sigma]\,.$$

Proposition 1. (1) *Let σ and $\tilde{\sigma}$ denote two strategies such that, for every i,*

$$q_i[\sigma] + \sum_j X_j[\tilde{\sigma}]\,P_i^j[\sigma] \geqslant q_i[\tilde{\sigma}] + \sum_j X_j[\tilde{\sigma}]\,P_i^j[\tilde{\sigma}]\,.$$

Then, $\gamma[\sigma] \geqslant \gamma[\tilde{\sigma}]$.

(2) *If, for every i,*

$$q_i[\sigma] + \sum_j X_j[\tilde{\sigma}]\,P_i^j[\sigma] \leqslant q_i[\tilde{\sigma}] + \sum_j X_j[\tilde{\sigma}]\,P_i^j[\tilde{\sigma}]\,,$$

then $\gamma[\sigma] \leqslant \gamma[\tilde{\sigma}]$.

More precisely, if we set

$$u = q[\sigma] - q[\tilde{\sigma}] + X[\tilde{\sigma}]\,P(\sigma) - X[\tilde{\sigma}]\,P[\tilde{\sigma}]\,,$$

that is,

$$u_i = q_i[\sigma] + \sum_j X_j[\tilde{\sigma}]\,P_i^j[\sigma] - q_i[\tilde{\sigma}] - \sum_j X_j[\tilde{\sigma}]\,P_i^j[\tilde{\sigma}]\,,$$

then,

$$\gamma[\sigma] - \gamma[\tilde{\sigma}] = u\omega[\sigma]\,.$$

Proof. We have

$$X[\sigma] + \eta\gamma[\sigma] = q[\sigma] + X[\sigma]\,P[\sigma]$$

and

$$X[\tilde{\sigma}] + \eta\gamma[\tilde{\sigma}] = q[\tilde{\sigma}] + X[\tilde{\sigma}]\,P[\tilde{\sigma}]\,.$$

From this we derive

$$u = [X[\sigma] - X[\tilde{\sigma}]]\,(\mathbf{1} - P[\sigma]) + \eta[\gamma[\sigma] - \gamma[\tilde{\sigma}]]\,.$$

If we multiply both sides on the right by the limiting probability distribution $\omega[\sigma]$ relative to the matrix $P[\sigma]$, we obtain the desired relation

$$u\omega[\sigma] = \gamma[\sigma] - \gamma[\tilde{\sigma}]\,.$$

We can also formulate the preceding results in the form of the following rule, where we designate a strategy σ that maximizes $\gamma(\sigma)$ as an *optimum strategy.*

Rule. *If*

$$q_i[\tilde{\sigma}] + \sum_j X_j[\tilde{\sigma}]\,P_i^j[\tilde{\sigma}] = \max_k q_i(k) + \sum_j X_j[\tilde{\sigma}]\,P_i^j(k)$$

the strategy $\tilde{\sigma}$ is optimum.

Let us suppose the opposite, namely, that $I \neq \varnothing$, where I is the set of indices i such that the preceding relation is false. If $\sigma(i)$ is a value such that

$$\max_k q_i(k) + \sum_j X_j[\tilde{\sigma}]\,P_i^j(k) = q_i(\sigma(i)) + \sum_j X_j[\tilde{\sigma}]\,P_i^j(\sigma(i))\,,$$

then the strategy σ satisfies the inequality $\gamma[\sigma] \geqslant \gamma[\tilde{\sigma}]$. Furthermore, if there exists an $i \in I$ such that $\omega^i[\sigma] \neq 0$, then $\gamma[\sigma] > \gamma[\tilde{\sigma}]$.

This will be the case, in particular, if $P[\sigma]$ is irreducible since all the components of $\omega[\sigma]$ are then positive.

Let us assume that $P[\sigma]$ is irreducible for every strategy σ. Then, we derive from the preceding rule a procedure for finding the strategies σ that maximize $\gamma[\sigma]$.

Let $\tilde{\sigma}$ denote a strategy. We calculate simultaneously $\gamma[\tilde{\sigma}]$ and a particular solution $X[\tilde{\sigma}]$ of the equation

$$X[\tilde{\sigma}]\,(\mathbf{1} - P[\tilde{\sigma}]) + \eta\gamma[\tilde{\sigma}] = q[\tilde{\sigma}]\,.$$

Then, for every i, we calculate the quantity

$$\max_k q_i(k) + \sum_j X_j[\tilde{\sigma}]\,P_i^j(k)\,.$$

(1) If this maximum is attained for $k = \tilde{\sigma}(i)$, the strategy $\tilde{\sigma}$ is optimum.

(2) Otherwise, we denote by $\sigma(i)$ a value at which the maximum is attained and we make the calculation again, replacing $\tilde{\sigma}$ with σ.

Example. In the following system, k may assume the two values 1 and 2. There are two states. The tables for $P_i^j(k)$ and $r_i^j(k)$ are as follows.

$j \backslash i$	1		2		$j \backslash i$	1		2	
1	0.5	0.3	0.2	0.6	1	10	3	2	12
2	0.5	0.7	0.8	0.4	2	5	6	8	1
	$k = 1$	$k = 2$	$k = 1$	$k = 2$		$k = 1$	$k = 2$	$k = 1$	$k = 2$

Let us calculate the $q_i(k)$. We find

i	1		2	
	7.5	5.1	6.8	7.6
	$k = 1$	$k = 2$	$k = 1$	$k = 2$

Let us study, for example, the strategy σ_1 such that $\sigma_1(1) = 1$ and $\sigma_1(2) = 1$. We have

$$P[\sigma_1] = \begin{bmatrix} 0.5 & 0.2 \\ 0.5 & 0.8 \end{bmatrix}$$

$$q[\sigma_1] = [7.5 \quad 6.8] .$$

Let us solve the equation

$$X[\sigma_1]\,(\mathbf{1} - P[\sigma_1]) + \eta\gamma[\sigma_1] = q[\sigma_1]) .$$

By setting $X[\sigma_1] = [X_1, X_2]$ and $\gamma[\sigma_1] = \gamma$, we find

$$\begin{aligned} 0.5\,X_1 - 0.5\,X_2 + \gamma &= 7.5 \\ -\,0.2\,X_1 + 0.2\,X_2 + \gamma &= 6.8 , \end{aligned}$$

which yields

$$\gamma = 7$$

and, for example,

$$X_1 = 1, \quad X_2 = 0 .$$

For each i, we need to find the control k that maximizes

$$X[\sigma_1]\,P_i(k) + q(k).$$

For $i = 1$,

$$\max_k \{\, X[\sigma_1]\,P_1(k) + q_1(k) \,\} = \max \{\, 0.5 + 7.5 \,;\, 0.3 + 5.1 \,\} = 8$$

which is attained with $\sigma_2(1) = 1$.

For $i = 2$,

$$\max_k \{\, X[\sigma_1]\,P_2(k) + q_2(k) \,\} = \max \{\, 0.2 + 6.8 \,;\, 0.6 + 7.6 \,\} = 8.2$$

which is attained with $\sigma_2(2) = 2$.

For the strategy σ_2, we have

$$P[\sigma_2] = \begin{bmatrix} 0.5 & 0.6 \\ 0.5 & 0.4 \end{bmatrix}$$

$$q[\sigma_2] = [7.5 \quad 7.6] .$$

Let us now solve the equation

$$X[\sigma_2]\,(\mathbf{1} - P[\sigma_2]) + \eta\gamma[\sigma_2] = q[\sigma_2] .$$

If we set

$$X[\sigma_2] = [X_1, X_2] \quad \text{and} \quad \gamma[\sigma_2] = \gamma ,$$

we find

$$\begin{aligned} 0{,}5\,X_1 - 0.5\,X_2 + \gamma &= 7.5 \\ -\,0{,}6\,X_1 + 0.6\,X_2 + \gamma &= 7.6 , \end{aligned}$$

which yields

$$\gamma = 7.\,545\,5$$

and, for example,

$$X_1 = 0, \qquad X_2 = 1/11 .$$

For each i, we need to find the control k that maximizes

$$X[\sigma_2]\, P_i(k) + q(k).$$

For $i = 1$,

$$\max_k \{ X[\sigma_2]\, P_1(k) + q_1(k) \} = \max \left\{ \frac{0.5}{11} + 7.5 ; \frac{0.7}{11} + 5.1 \right\} = 0.5/11 + 7.5$$

which is attained with $\sigma_3(1) = 1$.

For $i = 2$,

$$\max_k \{ X[\sigma_2]\, P_2(k) + q_2(k) \} = \max \left\{ \frac{0.8}{11} + 6.8 ; \frac{0.4}{11} + 7.6 \right\} = 0.4/11 + 7.6$$

which is attained with $\sigma_3(2) = 2$.

We find $\sigma_3 = \sigma_2$. Consequently, the strategy σ_2 is optimum. In sum, the optimum strategy is $\tilde{\sigma}$ defined by $\tilde{\sigma}(1) = 1$. $\tilde{\sigma}(2) = 2$. We have

$$\gamma[\sigma_2] = 7.545\,5 .$$

3. MARKOV SYSTEMS WITH TARGETS AND TRANSITION COSTS

Let us consider a system that can assume N states and that obeys a stationary Markov chain with transition matrix P. Consider a subset Σ known as a target of the set of states. The complement of Σ is denoted by Δ. Let us suppose that $\Delta = \{1, ..., n\}$ and $\Sigma = \{n+1, ..., N\}$. We are interested only in the evolution of the system beginning with a state belonging to Δ up to the instant at which its state belongs to Σ. Therefore, without changing the nature of the study, we can assume that Σ is closed and that the transition submatrix induced on Σ is the identity matrix, in other words, that P is partitioned as follows:

$$P = \left[\begin{array}{c|c} Q & 0 \\ \hline R & \mathbf{1} \end{array}\right] \begin{array}{l} \}\,\Delta \\ \}\,\Sigma \end{array}$$

Proposition 2. *The following properties are equivalent:*

(1) *From every state $i \in \Delta$ there leads a path ending in Σ.*
(2) *Δ is the union of transitory classes.*
(3) $\|\overset{n}{Q}\| < 1$.

Proof. Note first of all that each of the states $n+1, ..., N$ defines a final class reduced to that state. Therefore, Δ is a union of classes.

Let us assume that (1) is satisfied. Then Δ contains no final class. Therefore, Δ is the union of transitory classes. Thus, (1) implies (2). Also, (2) implies (3) (cf. Chapter 6, proof of Proposition 8).

Suppose now that (3) is satisfied. Let α denote the set of states from which one cannot reach Σ. The set α is closed by definition. Therefore, if α is nonempty, we have

$$\| Q_\alpha^\alpha \| = 1\,, \qquad \| (\overset{n}{Q})_\alpha^\alpha \| = 1$$

which is contradictory.

Proposition 3. *If properties equivalent to Proposition 2 obtain, the probability of reaching Σ from a state i in at most m transitions approaches 1 as m approaches $+\infty$.*

Proof. The probability that the system that is initially in the state i will be in Δ at the end of m stages is equal to $\sum_j (\overset{m}{Q})_i^j$. Since $\overset{m}{Q}$ tends to 0 as $m \to +\infty$, this probability approaches 0.

Corollary. *The probability that the system leaving a state $i \in \Delta$ will attain Σ is equal to 1.*

Let us suppose now that to each transition from the state i to the state j is assigned a cost c_i^j for $i = 1, ..., n$ and $j = 1, ..., N$. We may set $c_i^j = 0$ for $i \in \Sigma$. Let $\chi_i^{(p)}$ denote the mean cost in the course of p transitions if the system starts from the state $i \in \Delta$. Define $q = [q_1, ..., q_n]$, where

$$q_i = \sum_{j=1}^{N} P_i^j c_i^j .$$

We have the recursion relation

$$\chi_i^{(p)} = \sum_{j=1}^{n} \chi_j^{(p-1)} Q_i^j + q_i ,$$

that is,

$$\boxed{\chi^{(p)} = \chi^{(p-1)} Q + q}$$

where $\chi^{(0)} = 0$.

Lemma 1. *Suppose that $\| \overset{n}{Q} \| < 1$. Let T denote the mapping defined for $X \in (\mathbf{R}^n)'$ by*

$$T(X) = XQ + q .$$

The mapping $\overset{n}{T}$ is a contraction.

Proof. We have

$$\overset{n}{T}(X) = X\overset{n}{Q} + q\overset{n-1}{Q} + q\overset{n-2}{Q} + \cdots + q .$$

Therefore,

$$\overset{n}{T}(X) - \overset{n}{T}(Y) = (X - Y)\overset{n}{Q} ,$$

and, consequently,

$$\| \overset{n}{T}(X) - \overset{n}{T}(Y) \| \leqslant \| \overset{n}{Q} \| \, \| X - Y \| .$$

Corollary. *The mapping T has exactly one fixed point. In other words, the equation $TX = X$ has exactly one solution. For every $X^{(0)} \in (\mathbf{R}^n)$, the sequence $\{X^{(p)}\}$ defined by $X^{(p)} = TX^{(p-1)}$ converges to the unique solution of the equation $TX = X$.*

In particular, we have the following result:

Proposition 4. *Suppose that $\| \overset{n}{Q} \| < 1$. Then, the sequence $\{\chi^{(p)}\}$ converges to a limit χ as $p \to +\infty$, where χ is the unique solution of the equation*

$$\chi = \chi Q + q .$$

Lemma 2. *Suppose that for some element* $r \gg 0$ *of* $(\mathbf{R}^n)'$, *the equation* $X = XQ + r$ *can be solved for* X. *Then* $\| \overset{n}{Q} \| < 1$.

Proof. If this were not the case, Δ would contain a final class δ. There would exist an $\omega \in \mathbf{R}^n$ (with support δ) such that

$$\begin{cases} \omega \geqslant 0 \\ \sum_{i=1}^{n} \omega^i = 1 \\ Q\omega = \omega . \end{cases}$$

If we multiply both sides of the equation $X = XQ + r$ on the right by ω, we obtain $r\omega = 0$, which is incompatible with the hypothesis that $r \gg 0$.

4. MARKOV SYSTEMS WITH CONTROLS, TARGETS, AND TRANSITION COSTS

Let us now suppose that the probability that the system in the state i (for $i = 1, \ldots, n$) will be in the state j (for $j = 1, \ldots, n, \ldots, N$) the following instant depends on a parameter k, called the *control* parameter, that assumes a finite number of values. We denote this probability by $P_i^j(k)$. For every k,

$$\begin{cases} P_i^j(k) \geqslant 0 , \\ \sum_{j=1}^{N} P_i^j(k) = 1 . \end{cases}$$

We still assume that, if the system is in a state belonging to Σ, it remains there indefinitely. Thus, we have

$$P_i^j(k) = \begin{cases} 0 & \text{if} \quad i \in \Sigma \quad \text{and} \quad j \neq i \\ 1 & \text{if} \quad i = j \in \Sigma. \end{cases}$$

We term a *strategy* any mapping $i \to \sigma(i)$ defined for $i \in \Delta$ that fixes the control to be chosen as a function of the state of the system.

We make the following hypothesis: If we choose a strategy σ, the system obeys a stationary Markov chain whose transition matrix is $P[\sigma]$ with

$$P_i^j[\sigma] = P_i^j(\sigma(i)) \qquad \text{for } i \in \Delta$$

$$P_i^j[\sigma] = \begin{Bmatrix} 1 \text{ si } j = i \\ 0 \text{ si } j \neq i \end{Bmatrix} \qquad \text{for } i \in \Sigma .$$

We can partition $P[\sigma]$ as follows:

$$P[\sigma] = \left[\begin{array}{c|c} Q[\sigma] & \ulcorner\!\urcorner \\ \hline R[\sigma] & \mathbf{1} \end{array}\right] \begin{array}{l} \}\,\Delta \\ \}\,\Sigma \end{array} .$$

Definition. A strategy will be called *proper* if Δ is the union of transitory classes.

A necessary and sufficient condition for σ to be a proper strategy is that $\| \overset{n}{Q}[\sigma] \| < 1$.

To each transition is assigned a cost $c_i^j(k)$. It depends on the control k applied to the state i.

We set

$$q_i(k) = \sum_{j=i}^{N} P_i^j(k) c_i^j(k),$$

$$q_i[\sigma] = q_i(\sigma(i)),$$

and

$$q[\sigma] = [q_1[\sigma], \ldots, q_n[\sigma]].$$

We make the hypothesis that $q_i(k) > 0$ for every k and every $i \in \Delta$.

For every proper strategy σ, we denote by $\chi[\sigma]$ the unique solution of the equation

$$\chi[\sigma] = \chi[\sigma]\, Q[\sigma] + q[\sigma].$$

For every strategy σ, let $T[\sigma]$ denote the mapping defined by

$$T[\sigma]\, X = X\, Q[\sigma] + q[\sigma] \quad \text{for} \quad X \in (\mathbf{R}^n)'.$$

If $T[\sigma]$ has a fixed point, the σ is a proper strategy (by Lemma 2). We shall say that a proper strategy σ is *better* than a proper strategy σ' is $\chi[\sigma] \leqslant \chi[\sigma']$, that is, if

$$\chi_i[\sigma] \leqslant \chi_i[\sigma'] \quad \text{for every} \quad i \in \Delta.$$

We shall call a strategy $\tilde{\sigma}$ *optimum* if it is a proper strategy and if $\chi(\tilde{\sigma}) \leqslant \chi(\sigma)$ for every proper strategy σ.

Lemma 3. *For every strategy* σ,

$$X \leqslant Y \Rightarrow T[\sigma](X) \leqslant T[\sigma](Y).$$

In other words, $T[\sigma]$ *is an increasing function.*

Proof. This follows from the fact that $Q[\sigma] \geqslant 0$.

Proposition 5. *For two proper strategies* σ' *and* σ'', *there exists a proper strategy* $\tilde{\sigma}$ *that is better than either* σ' *or* σ''; *that is,* $\tilde{\sigma}$ *satisfies the inequality*

$$\chi[\tilde{\sigma}] \leqslant \inf(\chi[\sigma'], \chi[\sigma'']).$$

Proof. Let us assume that Δ is indexed in such a way that

$$\chi_i[\sigma'] \leqslant \chi_i[\sigma''], \quad \text{for } 0 \leqslant i \leqslant h,$$
$$\chi_i[\sigma'] \geqslant \chi_i[\sigma''], \quad \text{for } h < i \leqslant n.$$

Let $\tilde{\sigma}$ denote the strategy defined by

$$\tilde{\sigma}(i) = \begin{cases} \sigma'(i) & \text{for } 0 \leqslant i \leqslant h, \\ \sigma''(i) & \text{for } h < i \leqslant n. \end{cases}$$

Define

$$Y^{(0)} = \inf(\chi[\sigma'], \chi[\sigma'']),$$

that is,

$$Y_i^{(0)} = \begin{cases} \chi_i[\sigma'], & \text{for } 0 \leqslant i \leqslant h, \\ \chi_i[\sigma''], & \text{for } h < i \leqslant n. \end{cases}$$

Let us set

$$Y^{(1)} = T[\tilde{\sigma}]\, Y^{(0)} = Y^{(0)}\, Q[\tilde{\sigma}] + q[\tilde{\sigma}].$$

We have

$$Y_i^{(1)} = Y^{(0)}\, Q_i[\tilde{\sigma}] + q_i[\tilde{\sigma}].$$

If $i \leqslant h$,

$$Y_i^{(1)} = Y^{(0)}\, Q_i[\sigma'] + q_i[\sigma'] \leqslant \chi[\sigma']\, Q_i[\sigma'] + q_i[\sigma'] = \chi_i[\sigma'].$$

If $i > h$,

$$Y_i^{(1)} = Y^{(0)}\, Q_i[\sigma''] + q_i[\sigma''] \leqslant \chi[\sigma'']\, Q_i[\sigma''] + q_i[\sigma''] = \chi_i[\sigma''].$$

Thus,

$$Y^{(1)} \leqslant \inf(\chi[\sigma'], \chi[\sigma'']) = Y^{(0)}.$$

Let us define inductively

$$Y^{(k)} = T[\tilde{\sigma}]\; Y^{(k-1)}.$$

Since $T(\tilde{\sigma})$ is an increasing function and since $Y^{(1)} \leqslant Y^{(0)}$, we have

$$Y^{(0)} \geqslant Y^{(1)} \geqslant \cdots \geqslant Y^{(k)} \geqslant \cdots.$$

Since the sequence $k \to Y^{(k)}$ is minorized by 0, it is convergent. Define

$$Y = \lim_{k \to +\infty} Y^{(k)}.$$

Thus, we have

$$Y = T[\tilde{\sigma}]\, Y = YQ[\tilde{\sigma}] + q[\tilde{\sigma}].$$

Therefore, the strategy $\tilde{\sigma}$ is a proper strategy and

$$Y = \chi[\tilde{\sigma}].$$

Furthermore,

$$\chi[\tilde{\sigma}] = Y \leqslant Y^{(0)} = \inf\left(\chi[\sigma'], \chi[\sigma'']\right).$$

Therefore, $\tilde{\sigma}$ is better than σ' or σ''.

Corollary. *If there exists proper strategies, there exist optimum strategies.*

Proposition 6. *Let $\tilde{\sigma}$ denote a proper strategy. A necessary and sufficient condition for $\tilde{\sigma}$ to be optimum is that*

$$\chi_i[\tilde{\sigma}] = \min_k \left[\chi[\tilde{\sigma}]\, Q_i(k) + q_i(k)\right], \qquad \forall i = 1, \ldots, n.$$

If this condition is not satisfied, then every strategy σ such that

$$\min_k \left[\chi[\tilde{\sigma}]\, Q_i(k) + q_i(k)\right] = \chi\,[\tilde{\sigma}]\, Q_i[\sigma] + q_i[\sigma]$$

is better than $\tilde{\sigma}$.

Proof. Note first that for every proper strategy,

$$\chi_i[\tilde{\sigma}] \geqslant \min_k \left[\chi[\tilde{\sigma}]\, Q_i(k) + q_i(k)\right].$$

(1) To prove the necessity of the condition, let us set

$$X_i = \min_k \left[\chi[\tilde{\sigma}]\, Q_i(k) + q_i(k)\right].$$

Let $\sigma(i)$ be such that

$$X_i = \min_k \left[\chi[\tilde{\sigma}]\, Q_i(k) + q_i(k)\right] = \chi[\tilde{\sigma}]\, Q_i(\sigma(i)) + q_i(\sigma(i)).$$

The mapping $\sigma : i \to \sigma(i)$ defines a strategy. We may write

$$X_i = \chi[\tilde{\sigma}]\, Q_i[\sigma] + q_i[\sigma],$$

that is,

$$X = \chi[\tilde{\sigma}]\, Q[\sigma] + q[\sigma].$$

We have $X \leqslant \chi[\tilde{\sigma}]$. Let us set $u = \chi[\tilde{\sigma}] - X$. Then,

$$\chi[\tilde{\sigma}] = \chi[\tilde{\sigma}]\, Q[\sigma] + q[\sigma] + u, \quad \text{with} \quad u \geqslant 0,$$

which implies that σ is a proper strategy (by Lemma 2). We then have

$$\begin{aligned}\chi[\sigma] &= \chi[\sigma]\, Q[\sigma] + q[\sigma] \\ &= \chi[\sigma]\, Q[\sigma] + X - \chi[\tilde{\sigma}]\, Q[\sigma] \\ &\leqslant \chi[\sigma]\, Q[\sigma] + \chi[\tilde{\sigma}] - \chi[\tilde{\sigma}]\, Q[\sigma],\end{aligned}$$

that is,

$$(\chi[\tilde{\sigma}] - \chi[\sigma])\,(\mathbf{1} - Q[\sigma]) \geqslant 0\,.$$

If we multiply this inequality by $(\mathbf{1} - Q[\sigma])^{-1}$, noting that

$$(\mathbf{1} - Q[\sigma])^{-1} = \mathbf{1} + Q[\sigma] + \cdots + \overset{p}{Q}[\sigma] + \cdots \geqslant 0\,,$$

we get

$$\chi[\tilde{\sigma}] - \chi[\sigma] \geqslant 0\,.$$

Consequently,

$$\chi[\sigma] = X + (\chi[\sigma] - \chi[\tilde{\sigma}])\,Q[\sigma] \leqslant X\,.$$

If we assume that $X < \chi[\tilde{\sigma}]$, then $\tilde{\sigma}$ is not optimum. Therefore, the condition is necessary and we have proven the second part of the theorem.

(2) To prove that the condition is sufficient, let $\tilde{\sigma}$ denote an optimum strategy and let σ denote a nonoptimum strategy. There exists an i such that

$$\chi_i[\sigma] > \chi[\sigma]\,Q_i[\tilde{\sigma}] + q_i[\tilde{\sigma}]\,,$$

because otherwise we have

$$\chi[\sigma] \leqslant \chi[\sigma]\,Q[\tilde{\sigma}] + q[\tilde{\sigma}]\,,$$

from which we get

$$\chi[\sigma]\,(\mathbf{1} - Q[\tilde{\sigma}]) \leqslant q[\tilde{\sigma}]\,.$$

Multiplying by $(\mathbf{1} - Q[\tilde{\sigma}])^{-1}$, we get

$$\chi[\sigma] \leqslant q[\tilde{\sigma}]\,(\mathbf{1} - Q[\tilde{\sigma}])^{-1} = \chi[\tilde{\sigma}]\,,$$

which contradicts the hypothesis that σ is nonoptimum. Therefore,

$$\chi_i[\sigma] > \min_k\,[\chi[\sigma]\,Q_i(k) + q_i(k)]\,.$$

Thus, the strategy σ does not satisfy the condition.

Corollary I. *If $X \in (\mathbf{R}^n)'$ is a solution of the equation*

$$X_i = \min_k\,[XQ_i(k) + q_i(k)]\,,$$

then every strategy σ such that

$$\min_k\,[XQ_i(k) + q_i(k)] = XQ_i(\sigma(i)) + q_i(\sigma(i))$$

is optimum.

Proof. We have $X = XQ[\sigma] + q[\sigma]$. Therefore, σ is proper and $X = \chi[\sigma]$. According to the preceding theorem, σ is optimum.

Corollary II. *The equation*

$$X_i = \min_k \left[XQ_i(k) + q_i(k)\right]$$

has a unique solution.

Consider now the mapping U defined on $(\mathbf{R}^n)'$ by

$$U(X) = \min_k \left[XQ_i(k) + q_i(k)\right].$$

Lemma 4. *The mapping U is increasing.*

Proof. Suppose that $X \leqslant X'$. For every k,

$$XQ_i(k) + q_i(k) \leqslant X'\,Q_i(k) + q_i(k),$$

so that

$$\min_k \left(XQ_i(k) + q_i(k)\right) \leqslant \min_k \left(X'\,Q_i(k) + q_i(k)\right),$$

that is,

$$U(X) \leqslant U(X').$$

An application. Let $\tilde{\sigma}$ denote an optimum strategy. Suppose that $X^{(0)} \in (\mathbf{R}^n)'$ and $X^{(0)} \leqslant \chi[\tilde{\sigma}]$. Define a sequence $\{X^{(p)}\}$ by $X^{(p)} = U(X^{(p-1)})$. Suppose also that $X^{(0)} \leqslant X^{(1)}$. Then,

$$\lim_{p\to+\infty} X^{(p)} = \chi[\tilde{\sigma}].$$

From the inequality $X^{(0)} \leqslant X^{(1)}$, we derive the inequality $U(X^{(0)}) \leqslant U(X^{(1)})$, that is, $X^{(1)} \leqslant X^{(2)}$ and, by induction, $X^{(p)} \leqslant X^{(p+1)}$. Also, the inequality $X^{(0)} \leqslant \chi[\tilde{\sigma}]$ implies $U(X^{(0)}) \leqslant U\chi[\tilde{\sigma}])$, that is, $X^{(1)} \leqslant \chi[\tilde{\sigma}]$ and, by induction, $X^{(p)} \leqslant \chi[\tilde{\sigma}]$. Therefore, the sequence $\{X^{(p)}\}$ is a majorized increasing sequence (that is, for every $i \in \Delta$, the sequence $\{X_i(p)\}$ is a bounded increasing numerical sequence). Therefore, it is convergent. If we set

$$\tilde{X} = \lim_{p\to+\infty} X^{(p)},$$

we have $U(\tilde{X}) = \tilde{X}$ and, consequently, $\tilde{X} = \chi[\tilde{\sigma}]$.
Similarly, if $X^{(0)} \geqslant \chi[\tilde{\sigma}]$ and $X^{(1)} \leqslant X^{(0)}$, then

$$\lim_{p\to+\infty} X^{(p)} = \chi[\tilde{\sigma}].$$

An example. In the following system, k can assume the two values 1 and 2. Therefore, there are four states, the fourth of which constitutes the target ($n = 3$, $N = 4$). The tables of the $Q_i^j(k)$ and $c_i^j(k)$ are as follows.

j \ i	1		2		3	
1	0.2	0.1	0.1	0.4	0.5	0.1
2	0.4	0.3	0.2	0.2	0.4	0.3
3	0.2	0.5	0.3	0.4	0.1	0
4	0.2	0.1	0.4	0	0	0.6
	$k = 1$	$k = 2$	$k = 1$	$k = 2$	$k = 1$	$k = 2$

j \ i	1		2		3	
1	1	5	2	3	1	1
2	3	3	3	6	4	5
3	3	5	5	1	2	
4	5	2	2			5

Calculation of the $q_i(k)$:

i	1		2		3	
	3	4.1	3.1	2.8	2.3	4.6
	$k = 1$	$k = 2$	$k = 1$	$k = 2$	$k = 1$	$k = 2$

By induction, we calculate

$$X^{(p+1)} = U(X^{(p)}) = \min_k \left[X^{(p)}\, Q_i(k) + q_i(k)\right]$$

beginning, for example, with

$$X^{(0)} = [1, 1, 1] .$$

If we set $X^{(p)} = [x^{(p)}, y^{(p)}, z^{(p)}]$, this equation may be written

$$x^{(p+1)} = \min (0.2\, x^{(p)} + 0.4\, y^{(p)} + 0.2\, z^{(p)} + 3\,;\quad 0.1\, x^{(p)} + 0.3\, y^{(p)} + 0.5\, z^{(p)} + 4.1)$$

$$y^{(p+1)} = \min (0.1\, x^{(p)} + 0.2\, y^{(p)} + 0.3\, z^{(p)} + 3.1\,;\quad 0.4\, x^{(p)} + 0.2\, y^{(p)} + 0.4\, z^{(p)} + 2.8)$$

$$z^{(p+1)} = \min (0{,}5\, x^{(p)} + 0{,}4\, y^{(p)} + 0.1\, z^{(p)} + 2.3\,;\quad 0.1\, x^{(p)} + 0.3\, y^{(p)} + 4.6) .$$

We denote by $k = \sigma_p(i)$ the value at which the minimum is attained.

Calculation of $X^{(1)}$:

$$x^{(1)} = \min(3.8\,;\,5.0) = 3.8$$
$$y^{(1)} = \min(3.7\,;\,3.8) = 3.7$$
$$z^{(1)} = \min(3.3\,;\,5.0) = 3.3$$

$$\sigma_0(1) = 1,\ \sigma_0(2) = 1,\ \sigma_0(3) = 1\,.$$

Calculation of $X^{(2)}$:

$$x^{(2)} = \min(5.9\ \ ;\,7.24) = 5.9$$
$$y^{(2)} = \min(5.21\,;\,6.38) = 5.21$$
$$z^{(2)} = \min(6.01\,;\,6.09) = 6.01$$

$$\sigma_1(1) = 1,\ \sigma_1(2) = 1,\ \sigma_1(3) = 1\,.$$

Calculation of $X^{(3)}$:

$$x^{(3)} = \min(8.466\,;\,9.258) = 7.466$$
$$y^{(3)} = \min(6.535\,;\,8.606) = 6.535$$
$$z^{(3)} = \min(7.935\,;\,6.753) = 6.753$$

$$\sigma_2(1) = 1,\ \sigma_2(2) = 1,\ \sigma_2(3) = 2\,.$$

Calculation of $X^{(4)}$:

$$x^{(4)} = \min(8.457\,8\,;\,10.183\,6) = 8.458$$
$$y^{(4)} = \min(7.179\,5\,;\ \ 9.794\,6) = 7.180$$
$$z^{(4)} = \min(9.322\,3\,;\ \ 7.307\,1) = 7.307$$

$$\sigma_3(1) = 1,\ \sigma_3(2) = 1,\ \sigma_3(3) = 2\,.$$

Calculation of $X^{(5)}$:

$$x^{(5)} = \min\ (9.025\ \ \ ;\,10.753\,3) = 9.025$$
$$y^{(5)} = \min\ (7.575\,9\,;\,10.542)\ \ = 7.574$$
$$z^{(5)} = \min(10.131\,7\,;\ \ 7.599\,8) = 7.6$$

$$\sigma_4(1) = 1,\ \sigma_4(2) = 1,\ \sigma_4(3) = 2\,.$$

We have found $\sigma_2 = \sigma_3 = \sigma_4$.

We can stop the iteration and see if this strategy is optimum. Thus, let us set

$$\tilde{\sigma}(1) = 1,\ \tilde{\sigma}(2) = 1,\ \tilde{\sigma}(3) = 2\,.$$

We then have

$$P[\tilde{\sigma}] = \left[\begin{array}{ccc|c} 0.2 & 0.1 & 0.1 & 0 \\ 0.4 & 0.2 & 0.3 & 0 \\ 0.2 & 0.3 & 0 & 0 \\ \hline 0.2 & 0.4 & 0.6 & 1 \end{array}\right] = \left[\begin{array}{c|c} Q[\tilde{\sigma}] & 0 \\ \hline R[\tilde{\sigma}] & 1 \end{array}\right]$$

$$q[\tilde{\sigma}] = [3,\ \ 3.1,\ \ 4.6]\,.$$

Therefore, let us solve the equation

$$\chi[\tilde{\sigma}] = \chi[\tilde{\sigma}]\, Q[\tilde{\sigma}] + q[\tilde{\sigma}]\,.$$

We find

$$\chi[\tilde{\sigma}] = [9.805 \mid 8.105 \mid 8.012]\,.$$

We have

$$\begin{aligned} U(\chi[\tilde{\sigma}]) &= [\min(9.805\,,\, 11.518) \mid \min(8.105\,,\, 11.548) \mid \min(11.246\,;\, 8.012)] \\ &= [9.805 \mid 8.105 \mid 8.012] = \chi\,[\tilde{\sigma}]\,. \end{aligned}$$

Consequently, $\tilde{\sigma}$ is optimum.

17

Minimum-Time Control of Linear Systems

1. DISCRETE SYSTEMS

Let us again consider the system governed by the difference equation

$$X_{n+1} = AX_n + U_n h$$

which we studied in Chapter 15, section 2. We still impose the hypotheses

$$\begin{cases} X_n \in \mathbf{R}^m \\ A \in L(\mathbf{R}^m, \mathbf{R}^m), \text{ with } \det(A) \neq 0. \\ h \in \mathbf{R}^m \\ U_n \in \mathbf{R}. \end{cases}$$

Throughout this section, we shall assume that the system is controllable. We also suppose that we have the constraint

$$|U_n| \leqslant K, \qquad (K > 0).$$

By replacing h with a proportional vector, we can assume that $K = 1$. The constraint then becomes

$$|U_n| \leqslant 1.$$

Suppose that X_0 is given. We shall say that a control $\tilde{U}$ is *optimum* if there exists an integer N such that $X_N = 0$ and if there is no control U such that $X_n = 0$, where $n < N$. We then say that $\tilde{U}$ adjusts the system (to the margin) in minimum time.

We recall that

$$X_N = A^N X_0 + U_0 A^{N-1} h + U_1 A^{N-2} h + \cdots + U_{N-1} h,$$

and hence that the condition $X_N = 0$ may be written

$$X_0 = -U_0 A^{-1} h - U_1 A^{-2} h - \cdots - U_{N-1} A^{-N} h,$$

which, if we set $-A^{-i} h = e_i$, becomes

$$X_0 = U_0 e_1 + \cdots + U_{N-1} e_N.$$

Let us denote by M_N the set of states that are adjustable to the origin in the time N.

Theorem 1. *The set M_N is a symmetric compact convex set. For $N \geqslant m$, the set M_N has 0 as an interior point.*

Proof. The set $U = \{U_0, \ldots, U_{N-1}\}$ of elements U_n such that $|U_n| \leqslant 1$ for every $n = 0, \ldots, N-1$ is a convex subset of $\mathbf{R}^N$. The set M_N is the image of this set under the mapping

$$U_0, \ldots, U_{N-1} \to U_0 e_1 + \cdots + U_{N-1} e_N.$$

Thus, it is a compact convex set. For $N \geqslant m$, the vectors $e_1, \ldots, e_N$ generate $\mathbf{R}^m$, so that $\dim(M_N) = m$. Since M_N is symmetric, 0 is an interior point of it.

Remarks. I. The set M_N is the convex hull of the $2N$ vectors $\pm e_1, \ldots, \pm e_N$ (see Fig. 1).

II. The sets M_N are such that $M_N \subset M_{N+1}$.

Suppose that X_0 is a member of M_N but not a member of M_{N-1}. Therefore, state X_0 is adjustable in the time N without being adjustable in the time $N-1$. Every control $U = \{U_0, \ldots, U_{N-1}\}$ such that

$$X_0 = U_0 e_1 + \cdots + U_{N-1} e_N,$$
$$|U_n| \leqslant 1, \quad (n = 0, \ldots, N-1)$$

is an optimum control. In general, since $e_1, \ldots, e_N$ are not independent, the optimum control will not be unique.

Theorem 2. *If the absolute value of every eigenvalue of A is less than 1, then*

$$\bigcup_N M_N = \mathbf{R}^m.$$

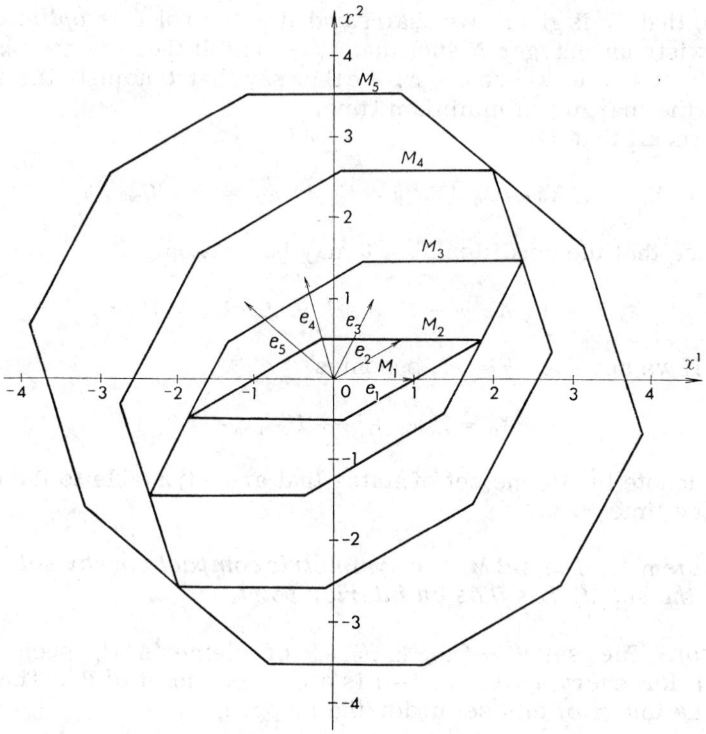

FIG. 1.

Proof. Let X_0 denote a member of $\mathbf{R}^m$. Let us apply the control

$$U_n = 0 \ (n = 0, 1, 2, \dots).$$

We have $\lim_{n \to \infty} X_n = 0$. Since M_m is a neighborhood of 0, there exists an N such that $X_N \in M_m$, from which we conclude that $X_0 \in M_{N+m}$.

For each $X_0 \in \bigcup_N M_N$, we shall choose a particular optimum control $\varphi(X_0)$:

$$\varphi(X_0) = \{ \tilde{U}_0, \tilde{U}_1, \dots, \tilde{U}_{N-1}, 0, 0, \dots \} \quad \text{for} \quad X_0 \in M_N - M_{N-1}.$$

Let us set $\varphi_i(X_0) = \tilde{U}_i$.

The function φ is defined for $N \leqslant m$ since the vectors $e_1, \dots, e_N$ are then independent. Let us assume that φ is defined on M_{N-1}. Let us define φ on $M_N - M_{N-1}$. The line with direction e_N that passes through X_0 intersects M_{N-1} along a line segment (see Fig. 2). On that line there exists a point $\xi^\star \in M_{N-1}$ such that

$$]\xi^\star, X_0] \cap M_{N-1} = \varnothing;$$

in other words, there exists a $\xi^\star$ and a $u \in [-1, +1]$ such that

$$\begin{cases} X_0 = \xi^\star + u e_N\,, \\ \xi^\star \in M_{N-1}\,, \\ \forall\theta \in]0,1], \quad \xi^\star + \theta u\, e_N \in M_N - M_{N-1}\,. \end{cases}$$

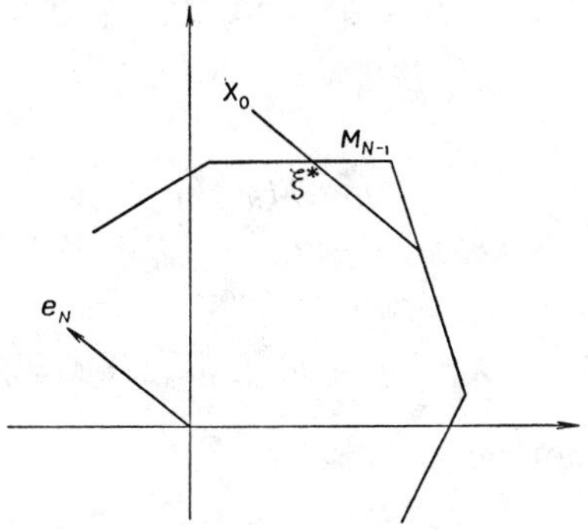

FIG. 2.

We now set

$$\varphi(X_0) = \{\, \varphi_0(\xi^\star), \varphi_1(\xi^\star), \dots, \varphi_{N-2}(\xi^\star), u, 0, 0, \dots \}\,.$$

Let $\widetilde{X}_0 = X_0, \widetilde{X}_1, \dots, \widetilde{X}_n, \dots,$ be the trajectory of the system under the influence of the control $\varphi(X_0)$. Let us show by induction that

$$\widetilde{X}_n = \widetilde{U}_n\, e_1 + \widetilde{U}_{n+1}\, e_2 + \cdots + \widetilde{U}_{N-2}\, e_{N-n-1} + \widetilde{U}_{N-1}\, e_{N-n}\,.$$

We assume this formula true at the time n. Then (using the equations $Ae_1 = -h$ and $Ae_i = e_{i-1}$ for $i \geqslant 2$) we obtain

$$\begin{aligned} \widetilde{X}_{n+1} &= A\widetilde{X}_n + \widetilde{U}_n\, h \\ &= \widetilde{U}_n\, Ae_1 + \widetilde{U}_{n+1}\, Ae_2 + \cdots + \widetilde{U}_{N-2}\, Ae_{N-n-1} + \widetilde{U}_{N-1}\, Ae_{N-n} + \widetilde{U}_n\, h \\ &= -\widetilde{U}_n\, h + \widetilde{U}_{n+1}\, e_1 + \cdots + \widetilde{U}_{N-2}\, e_{N-n-2} + \widetilde{U}_{N-1}\, e_{N-n-1} + \widetilde{U}_n\, h \\ &= \widetilde{U}_{n+1}\, e_1 + \cdots + \widetilde{U}_{N-2}\, e_{N-n-2} + \widetilde{U}_{N-1}\, e_{N-n-1}\,. \end{aligned}$$

Let $\widetilde{X}'_0, \widetilde{X}'_1, \dots, \widetilde{X}'_n, \dots,$ denote the trajectory of the system for the initial position $\widetilde{X}'_0 = \xi^\star + \theta u\, e_N$, where $\theta \in]0,1]$ and under the influence of the control

$$\begin{aligned} \varphi(\widetilde{X}'_0) &= \{\, \varphi_0(\xi^\star), \varphi_1(\xi^\star), \dots, \varphi_{N-2}(\xi^\star), \theta u, 0, 0 \dots \} \\ &= \{\, \widetilde{U}_0, \widetilde{U}_1, \dots, \widetilde{U}_{N-2}, \theta u, 0, 0 \dots \}\,. \end{aligned}$$

We have

$$\widetilde{X}'_n = \widetilde{U}_n\, e_1 + \widetilde{U}_{n+1}\, e_2 + \cdots + \widetilde{U}_{N-2}\, e_{N-n-1} + \theta u\, e_{N-n}\,.$$

Also, $\widetilde{X}_n \in M_{N-n}$. We set

$$\eta^{*} = \widetilde{U}_n e_1 + \widetilde{U}_{n+1} e_2 + \cdots + \widetilde{U}_{N-2} e_{N-n-1} .$$

Therefore, we may write

$$\widetilde{X}_n = \eta^{*} + \widetilde{U}_{N-1} e_{N-n}$$

where

$$\eta^{*} \in M_{N-n-1}$$

and

$$\forall \theta \in]0, 1] , \qquad \widetilde{X}'_n = \eta^{*} + \theta \widetilde{U}_{N-1} e_{N-n} \in M_{N-n} - M_{N-n-1} ,$$

from which we conclude

$$\varphi(\widetilde{X}_n) = \{ \widetilde{U}_n, \widetilde{U}_{n+1}, \ldots, \widetilde{U}_{N-2}, \widetilde{U}_{N-1}, 0, 0 \ldots \} ,$$

that is,

$$\varphi_0(\widetilde{X}_n) = \varphi_n(\widetilde{X}_0), \ \ \varphi_1(\widetilde{X}_n) = \varphi_{n+1}(\widetilde{X}_0), \ldots .$$

Thus, we have

Theorem 3. *The function φ defined above assigns to each initial state $X_0 \in \bigcup_N M_N$ an optimum control*

$$\varphi(X_0) = \{ \widetilde{U}_0, \widetilde{U}_1, \ldots, \widetilde{U}_n, \ldots \} ,$$

that vanishes from some n on. The value $\widetilde{U}_n$ of the control to be applied at the instant n is independent of the state of the system at that instant:

$$\widetilde{U}_n = \varphi_0(\widetilde{X}_n) ,$$

where $\varphi_0(\xi)$ is the control parameter to be applied at the instant 0 if the system is in the initial state ξ.

In general, we term a *strategy* any mapping σ of $\mathbf{R}^m$ (the state space) into $\mathbf{R}$ (the control parameter space). A strategy will be called *admissible* if $\sigma(\xi) \in [-1, +1]$ for every $\xi \in \mathbf{R}^m$. If we apply to the system the strategy σ, its evolution is governed by the recursion equation

$$X_{n+1} = AX_n + \sigma(X_n)\, h .$$

We see that if to each initial state X_0 is assigned the control $\varphi(X_0)$, the evolution of the system is the one that will be obtained if we apply the strategy φ_0.

2. CONTINUOUS SYSTEMS

Let us consider a dynamic system governed by the differential equation

$$\frac{dX(t)}{dt} = AX(t) + U(t)\,h\,, \tag{2.1}$$

where

$$X(t) \in H,\ \dim(H) = m,\ A \in L(H, H),\ h \in H,\ U(t) \in \mathbf{R}.$$

The function U is called the control. To begin with, let us assume that U is piecewise-continuous, in other words, that U has only finitely many discontinuities in any bounded interval and that U has a left-hand and a right-hand limit at every point of discontinuity. The response $t \to X(t)$ to the control U will be called the *trajectory* of the system. Also, $X(t)$ is called the *state* of the system at the instant t. The space H is the space of states. We can represent the value of $X(t)$ as a function of $X(0)$ and U:

$$X(t) = e^{tA} X(0) + \int_0^t e^{(t-\tau)A}\, hU(\tau)\, d\tau\,. \tag{2.2}$$

Definition. A state $\xi \in H$ is said to be *adjustable to the origin* (or, for short, *adjustable*) in the time T if there exists a control U and an instant $T > 0$ such that $X(T) = 0$ for $X(0) = \xi$. We note that taking $U(t) = 0$ for $t > T$ gives us $X(t) = 0$ for $t > T$. Therefore, every adjustable state in time T is adjustable in the time $t > T$. Let us express in writing that the state ξ has been adjusted in the time T by the control U:

$$e^{TA}\xi + \int_0^T e^{(T-\tau)A}\, hU(\tau)\, d\tau = 0\,;$$

that is,

$$\xi = -\int_0^T e^{-\tau A}\, hU(\tau)\, d\tau\,. \tag{2.3}$$

Theorem 4. *The set of states that are adjustable in the time $T > 0$ is independent of T. This is the smallest subspace of H that contains h and is stable under A.*

Proof. Let $\mathfrak{M}_T$ denote the set of states that are adjustable in the time T. The formula (2.3) shows that $\mathfrak{M}_T$ is a vector subspace of H.

Let H' denote the smallest subspace of H that contains h and is stable under A. Let ξ denote a state that is adjustable in the time T by the control U. We have $e^{-\tau A} h \in H'$ for every $\tau \in [0, T]$ and hence $\xi \in H'$. Thus, $\mathfrak{M}_T \subset H'$.

Also, every state of the form

$$Y(t) = \int_0^t e^{-\tau A} hU(\tau)\,d\tau$$

is, for $t \leqslant T$, a state that is adjustable to the origin in the time t and *a fortiori* belongs to $\mathfrak{M}_T$. Consequently, $dY(t)/dt \in \mathfrak{M}_T$; that is,

$$e^{-tA} hU(t) \in \mathfrak{M}_T$$

and, hence, since $U(t)$ is arbitrary,

$$e^{-tA} h \in \mathfrak{M}_T \quad \text{for every} \quad t \in [0, T]\,.$$

If we take the kth derivative at $t = 0$, we obtain

$$A^k h \in \mathfrak{M}_T \quad \text{for every} \quad k = 0, 1, 2, \ldots$$

which implies that $H' \subset \mathfrak{M}_T$. Thus, $H' = \mathfrak{M}_T$.

Definition. We shall call the system governed by Eq. (2.1) *controllable* if every state is adjustable to the origin.

Theorem 4 shows that a necessary and sufficient condition for the system governed by Eq. (2.1) to be controllable is that the smallest stable subspace containing h and stable under A be the entire space H.

For the remainder of this section, we shall assume that the system in question is controllable.

Theorem 4 states, in particular, that the problem of adjustment in minimum time cannot be posed in the continuous case if all controls that are piecewise-continuous are admissible because then, every adjustable state could be adjusted in an arbitrarily small interval of time. To make the problem of adjustment in minimum time meaningful, we shall modify the set of admissible controls.

In what follows, we shall consider the sequence of controls satisfying a constraint of the form $|U(t)| \leqslant K$, where $K > 0$. By replacing h with a proportional vector, we may assume that $K = 1$. The constraint is then written

$$|U(t)| \leqslant 1\,.$$

Let us consider the space $\mathbf{L}^\infty$ of real functions on $\mathbf{R}_+$ that are Lebesgue-measurable and essentially bounded. Formula (2.2) remains meaningful for $U \in \mathbf{L}^\infty$. Therefore, we may replace the requirement that U be piecewise-continuous with the requirement that U belong to $\mathbf{L}^\infty$. It should be understood that two controls U_1 and U_2 are considered equal if $U_1(t) = U_2(t)$ almost everywhere.

In view of the constraint $|U(t)| \leqslant 1$, the function U must belong to the unit ball of $\mathbf{L}^\infty$: $\|U\| \leqslant 1$.

We shall also need to consider the space $\mathbf{L}^\infty[0, t]$ of measurable functions that are essentially bounded on $[0, t]$. We still denote by $\| U \|$ the norm of the element U of that space.

We denote by M_t the set of states $\xi \in H$ that are adjustable to the origin in the time t (for $t \geqslant 0$) by a control $U \in \mathbf{L}^\infty$ and satisfying the inequality $\| U \| \leqslant 1$. According to Eq. (2.3), the set M_t consists of elements of the form

$$\xi = -\int_0^t e^{-\tau A} hU(\tau)\, d\tau, \quad \text{with} \quad \| U \| \leqslant 1 .$$

We have $M_0 = \{ 0 \}$.

Theorem 5. *The set M_t is a symmetric compact convex set.*

Proof. Let t denote a positive number. The set of all $U \in \mathbf{L}^\infty_{[0,t]}$ such that $\| U \| \leqslant 1$ is the unit ball of $\mathbf{L}^\infty_{[0,T]}$. Now, $\mathbf{L}^\infty_{[0,t]}$ is the dual of the space $\mathbf{L}^1_{[0,t]}$ of measurable functions that are integrable over $[0, t]$. Consequently, the unit ball of $\mathbf{L}^\infty_{[0,t]}$ is weakly compact (that is, compact for the topology of simple convergence on $\mathbf{L}^1_{[0,t]}$). Now, the mapping

$$U \to \xi = -\int_0^t e^{-\tau A} hU(\tau)\, d\tau$$

is a continuous mapping of $\mathbf{L}^\infty_{[0,t]}$, equipped with the weak topology, in H. This is true because if $H = \mathbf{R}^m$, we have

$$\xi^i = -\int_0^t (e^{-\tau A} h)^i\, U(\tau)\, d\tau , \qquad (i = 1, ..., m) .$$

Since the function $\tau \to (e^{-\tau A} h)^i$ is continuous, it belongs to $\mathbf{L}^1_{[0,t]}$. Hence, the mapping $U \to \xi^i$ is a continuous mapping of $\mathbf{L}^\infty_{[0,t]}$, equipped with the weak topology, in H. Thus, the set M_t is the continuous (linear) image of a compact convex set. Therefore, it is convex and compact.

Theorem 6. *The dimension of the set M_t, where $t > 0$, is equal to the dimension of H.*

Proof. The set M_t contains all the elements of the form

$$Y(\theta) = -\int_0^t e^{-\tau A} hU(\tau)\, d\tau ,$$

if U is piecewise-continuous and satisfies the inequality $| U(t) | \leqslant 1$.

The vector subspace generated by M_t is simply the set of homothetic transformations of M_t (since M_t is symmetric). Therefore, this subspace contains all elements of the preceding form when U is piecewise-continuous. Therefore, it is identical to H since the system studied is assumed controllable.

A consequence. For every positive t, the set M_t is identical to the closure of its interior $\mathring{M}_t$.

We note that, if $t' \leqslant t$, then $M_{t'} \subset M_t$.

Theorem 7. *If the real part of every eigenvalue of A is negative, then*

$$\bigcup_t M_t = H\,.$$

Proof. Suppose that $X(0) \in H$. Let us apply the control $U(t) = 0$. We have $\lim_{t \to +\infty} X(t) = 0$. Since M_θ is a neighborhood of 0 for arbitrary positive θ, there exists a T such that $X(T) \in M_\theta$, from which we conclude that $X(0) \in M_{T+\theta}$.

Let us equip H with a norm and let us equip the operators Z on H with the associated norm

$$\| Z \| = \sup_{\|x\| \leqslant 1} \| Zx \|\,.$$

The following lemma provides a continuity property of M_t with respect to t.

Lemma. *For every positive T and ε, there exists a positive η such that for every t, t' and ξ satisfying the relations*

$$\begin{cases} 0 \leqslant t' \leqslant t \leqslant T, \\ t - t' \leqslant \eta\,, \\ \xi \in M_t\,, \end{cases}$$

there exists a $\xi' \in M_{t'}$ such that $\| \xi - \xi' \| \leqslant \varepsilon$.

Proof. Suppose that $0 < t' \leqslant t \leqslant T$ and that $\xi \in M_t$. Let U denote a member of $\mathbf{L}^\infty_{[0,t]}$ such that $\| U \| \leqslant 1$ and

$$\xi = -\int_0^t e^{-\tau A}\, hU(\tau)\, d\tau\,.$$

Let us set

$$\xi' = -\int_0^{t'} e^{-\tau A}\, hU(\tau)\, d\tau\,.$$

We have $\xi' \in M_{t'}$ and

$$\xi - \xi' = -\int_{t'}^t e^{-\tau A}\, hU(\tau)\, d\tau\,,$$

from which we derive

$$\| \xi - \xi' \| \leqslant K(t - t') \| h \|, \quad \text{with} \quad K = \sup_{\tau \in [0,T]} \| e^{-\tau A} \| .$$

For given ε, we take

$$\eta = \frac{\varepsilon}{K \| h \|} .$$

For every t' and t such that $0 < t' \leqslant t \leqslant T$ and $t - t' \leqslant \eta$, we have thus assigned to every $\xi \in M_t$ an element $\xi' \in M_{t'}$ such that $\| \xi - \xi' \| \leqslant \varepsilon$.

Corollary.

$$\mathring{M}_t = \bigcup_{0 < t' < t} \mathring{M}_{t'} .$$

Proof. For every positive ε and every $\xi \in M_t$, there exist a $t' < t$ and a $\xi' \in M_{t'}$ such that $\| \xi - \xi' \| \leqslant \varepsilon/2$. Also, there exists a $\xi'' \in \mathring{M}_{t'}$ such that $\| \xi' - \xi'' \| \leqslant \varepsilon/2$. Therefore, $\| \xi - \xi'' \| \leqslant \varepsilon$. Thus, ξ belongs to the closure of the set

$$\bigcup_{0 < t' < t} \mathring{M}_t .$$

Therefore, M_t is the closure of the set. Furthermore, since M_t is the closure of its interior, the corollary is proven.

The set of adjustable states is $\bigcup_t M_t$. It is a convex set. The following theorem assures us that for every $\xi \in \bigcup_t M_t$, there exists a time t_0 such that ξ is adjustable in the time t_0 but not adjustable in any time $t < t_0$.

Theorem 8. *Suppose that $\xi \in \bigcup_t M_t$. Then, the set of values of t such that ξ is adjustable in the time t has a minimum t_0.*

Proof. Let t_0 denote the greatest lower bound of values of t such that $\xi \in M_t$. We need to prove that $\xi \in M_{t_0}$. Let us suppose that $\xi \notin M_{t_0}$ and let δ denote the distance from ξ to M_{t_0}. Since M_{t_0} is closed, we have $\delta > 0$. There exists a $t > t_0$ such that, for every $u \in M_t$, there exists a $u_0 \in M_{t_0}$ that satisfies the inequality $\| u - u_0 \| \leqslant \delta/2$. Then,

$$\| \xi - u \| \geqslant \| \xi - u_0 \| - \| u_0 - u \| \geqslant \delta - \frac{\delta}{2} = \frac{\delta}{2} .$$

From this we conclude that $\xi \notin M_t$, contrary to the definition of t_0.

Theorem 9. *In the notation of Theorem 8, ξ is a boundary point of M_{t_0}.*

Proof. If ξ were a member of $\mathring{M}_{t_0}$, we would (in accordance with the corollary to the lemma) have $\xi \in M_t$ for a value $t < t_0$, contrary to the definition of t_0.

Now, we are interested in the controls that adjust the system in minimum time. For a good interpretation of the following theorem, let us recall that two controls are considered equal if they assume the same values almost everywhere.

Theorem 10. *If ξ is an adjustable state, there exists a unique "optimum" control $\tilde{U}$ adjusting the system in minimum time beginning from the initial state ξ. There exists a linear form W such that almost everywhere*

$$\tilde{U}(t) = \operatorname{sgn}(W\,e^{-tA}\,h)\,.$$

The function $\varphi(t) = W\,e^{-tA}\,h$ has only finitely many zeros in any bounded interval.

Proof. Since ξ is a boundary point of M_{t_0}, there exists a support hyperplane to M_{t_0} at ξ. Hence, there exists a nonzero linear form W such that $W\xi' \geqslant W\xi$ for every $\xi' \in M_{t_0}$. Suppose that

$$\xi = -\int_0^{t_0} e^{-\tau A}\,hU(\tau)\,d\tau\,.$$

To every $U' \in \mathbf{L}^\infty_{[0,t_0]}$ such that $\| U' \| \leqslant 1$ is assigned the element

$$\xi' = -\int_0^{t_0} e^{-\tau A}\,hU'(\tau)\,d\tau$$

which belongs to M_{t_0}. Thus,

$$W\xi' = -\int_0^{t_0} W\,e^{-\tau A}\,hU'(\tau)\,d\tau \geqslant -\int_0^{t_0} W\,e^{-\tau A}\,hU(\tau)\,d\tau = W\xi\,,$$

that is,

$$\int_0^{t_0} W\,e^{-\tau A}\,hU(\tau)\,d\tau \geqslant \int_0^{t_0} W e^{-\tau A}\,hU'(\tau)\,d\tau\,,$$

for every $U' \in \mathbf{L}^\infty_{[0,t_0]}$ that satisfies the inequality $\| U' \| \leqslant 1$. Therefore, we must have

$$U(\tau) = \operatorname{sgn}(W\,e^{-\tau A}\,h)$$

almost everywhere.

Note now that, for every linear form W, the function $\tilde{\varphi}(\tau) = W e^{\tau A}\,h$ is the sum of exponential polynomials. Specifically, for every polynomial P, we have

$$\begin{aligned} P(D)\,(W\,e^{\tau A}\,h) &= WP(D)\,e^{\tau A}\,h \\ &= WP(A)\,e^{\tau A}\,h \end{aligned}$$

where D is the operator denoting differentiation. In particular, if P_0 is the characteristic polynomial of the A, we have, from the Cayley-Hamilton theorem, $P_0(A) = 0$ and, hence,

$$P_0(D)\,(W\,e^{\tau A}\,h) = 0\,.$$

Therefore, for every linear form W, the quantity $\tilde{\varphi}(\tau) = W e^{\tau A} h$ is a solution of the differential equation $P_0(D)\,y = 0$. Thus, this function has only finitely many zeros in any bounded interval unless it is identically equal to 0. The same is true of φ.

Also, since W is assumed nonzero, φ cannot be identically zero. If it were identically zero, we would have $e^{tA}\,h \in \ker(W)$ for every t and this would imply that $M_t \subset \ker(W)$ for every positive t, contrary to the hypothesis that the system is controllable. This completes the proof of the theorem.

We see that the condition

$$U(t) = \operatorname{sgn}(W\,e^{-tA}\,h) \quad \text{(p.p.)}$$

unambiguously defines U since we consider two controls equal if they assume the same value almost everywhere.

Now it is possible and normal to take

$$U(t) = \operatorname{sgn}(W\,e^{-tA}\,h)$$

at every t at which the right-hand member is defined. At points at which $W\,e^{-tA}\,h = 0$, it is expedient to take $U(t)$ equal to its right-hand limit. Thus, we have a unique continuous determination of the optimum command that is piecewise-continuous and that is always equal to ± 1. The instants at which the sign changes are called *switching instants* and the corresponding states are called *switching points*.

Remark. If all the eigenvalues of A are real, the number of zeros of any solution of the equation $P_0(D)\,y = 0$ (where P_0 still denotes the characteristic polynomial of A) is less than the degree of P_0, that is, the dimension of the space. Therefore, the number of switching instants is less than the dimension of the space H.

From every point x of the space H there leaves exactly one optimum trajectory, and the value of $\tilde{U}(0)$ of the optimum command at the time $t = 0$ for an initial state x is a function σ of the point x that is always equal to ± 1 (with the conventions made above still holding). For every optimum trajectory $\tilde{X}$, no matter what its origin, we have $\tilde{U}(\theta) = \sigma(\tilde{X}(\theta))$. Specifically, the control

$$t \to \tilde{U}(t + \theta)$$

is the optimum control for the initial state $\tilde{X}(\theta)$. The function σ is a mapping of H into the space of control parameters. We shall call it a *strategy*. This strategy is optimum in the sense that all the optimum trajectories are solutions of the differential equation

$$\frac{dX(t)}{dt} = AX(t) + \sigma(X(t))\,h\,.$$

Example. Consider the differential equation

$$\frac{d^2X^1(t)}{dt^2} + 2b\frac{dX^1(t)}{dt} + X^1(t) = U(t)$$

where $b > 0$, with constraint $|U(t)| \leqslant 1$.

If we set $dX^1/dt = X^2$, we obtain the system of differential equations

$$\begin{cases} \dfrac{dX^1(t)}{dt} = X^2(t) \\ \dfrac{dX^2(t)}{dt} = -X^1(t) - 2bX^2(t) + U(t) \end{cases}$$

that is,

$$\frac{dX(t)}{dt} = AX(t) + U(t)h$$

where

$$X(t) = \begin{bmatrix} X^1(t) \\ X^2(t) \end{bmatrix}, \quad A = \begin{bmatrix} 0 & 1 \\ -1 & -2b \end{bmatrix}, \quad h = \begin{bmatrix} 0 \\ 1 \end{bmatrix}.$$

The eigenvalues of the matrix A are given by the characteristic equation

$$s^2 + 2bs + 1 = 0$$

the discriminant of which is $4(b^2 - 1)$. We shall study the two principal cases $b > 1$ and $b < 1$.

Case I: $b > 1$. The eigenvalues are then

$$s_1 = -b + \sqrt{b^2 - 1}, \quad s_2 = -b - \sqrt{b^2 - 1}.$$

We can take for associated eigenvectors

$$K_1 = \begin{bmatrix} b + \sqrt{b^2 - 1} \\ -1 \end{bmatrix} \quad \text{and} \quad K_2 = \begin{bmatrix} b - \sqrt{b^2 - 1} \\ -1 \end{bmatrix}.$$

The associated matrix is then

$$K = [K_1 \mid K_2] = \begin{bmatrix} b + \sqrt{b^2 - 1} & b - \sqrt{b^2 - 1} \\ -1 & -1 \end{bmatrix}.$$

Then, by setting $X(t) = KY(t)$,

$$\begin{cases} \dfrac{dY^1}{dt} = s_1 Y^1 + \underline{h}^1 U(t), \\ \dfrac{dY^2}{dt} = s_2 Y^2 + \underline{h}^2 U(t), \end{cases}$$

where

$$\underline{h} = \begin{bmatrix} \underline{h}^1 \\ \underline{h}^2 \end{bmatrix} = \frac{1}{2\sqrt{b^2 - 1}} \begin{bmatrix} b - \sqrt{b^2 - 1} \\ - b - \quad b^2 - 1 \end{bmatrix}.$$

For $U(t) = 0$, the trajectories tend asymptotically to zero (see Fig. 3).

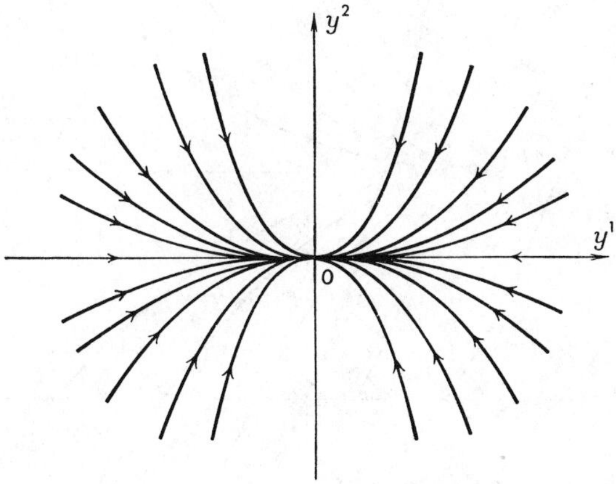

FIG. 3.

For $U(t) = +1$ (resp. -1), the trajectories (derived by translation from the preceding ones) tend asymptotically to the point

$$\omega_1 = \frac{1}{2\sqrt{b^2 - 1}} \begin{bmatrix} 1 \\ -1 \end{bmatrix} \quad \left(\text{resp. } \omega_{-1} = -\omega_1 = \frac{1}{2\sqrt{b^2 - 1}} \begin{bmatrix} -1 \\ 1 \end{bmatrix}\right).$$

Let us denote by $\mathcal{C}_1$ (resp. $\mathcal{C}_{-1}$) the family of these trajectories. Figure 4 is drawn for $b = 1.1$ ($s_1 = -0.642$ and $s_2 = -1.558$).

Through O passes a trajectory of $\mathcal{C}_1$. We shall denote by γ_1 that portion of the trajectory prior to passage through O. For the trajectory $\mathcal{C}_{-1}$ passing through O, we denote by γ_{-1} that portion prior to passage through O.

Every optimum trajectory terminates either in an arc of γ_1 or in an arc of γ_{-1}. Since there is at most one switching point, the trajectories other than γ_1 and γ_{-1} are made up as follows:

an arc of trajectory $\mathcal{C}_{-1}$ with termination on γ_1
followed by an arc of γ_1

or

an arc of trajectory of $\mathcal{C}_1$ with termination on γ_{-1}
followed by an arc of γ_{-1}.

Thus, the locus of the switching points is $\gamma_1 \cup \gamma_{-1}$.

The optimum strategy γ assumes the values $\sigma(x) = +1$ or $\sigma(x) = -1$ according to the position of x with respect to $\gamma_1 \cup \gamma_{-1}$.

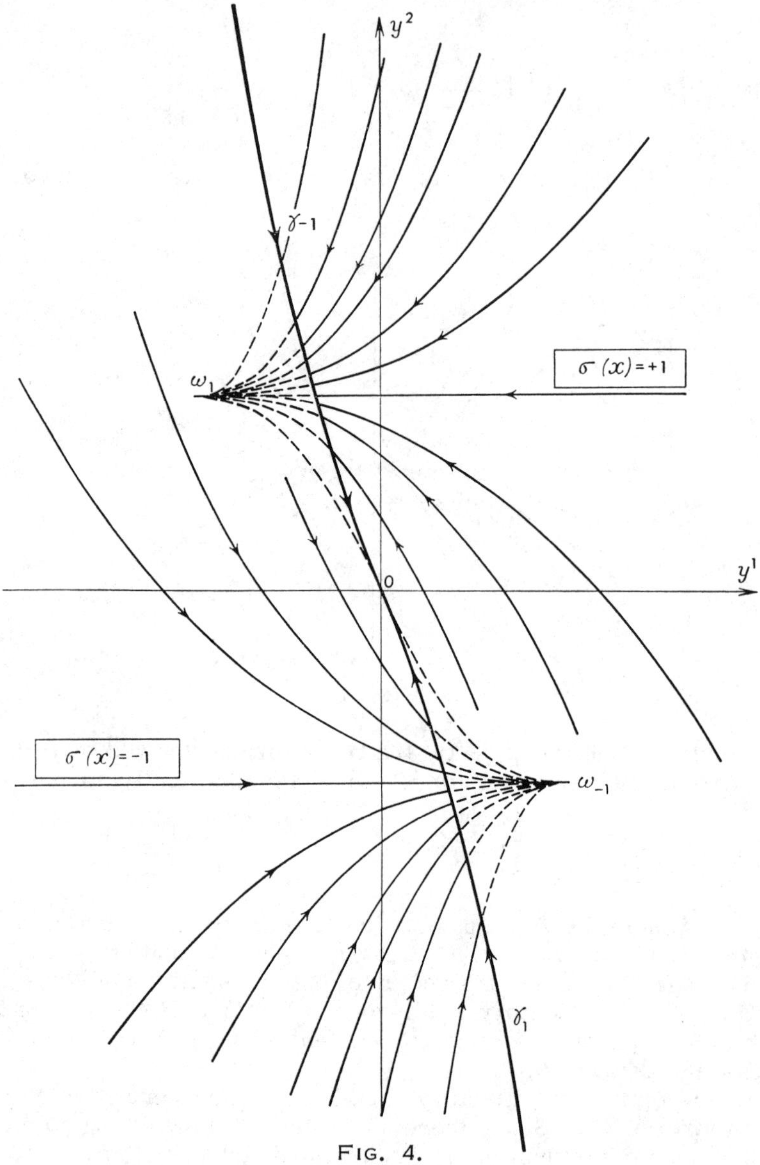

FIG. 4.

Case II. $b < 1$. Let us make the change of basis defined by

$$y^1 = x^1 + bx^2 ,$$
$$y^2 = \alpha x^2 ,$$

where $\alpha = \sqrt{1 - b^2}$. The system of differential equations becomes

$$\begin{cases} \dfrac{\mathrm{d}Y^1(t)}{\mathrm{d}t} = -\, bY^1(t) + \alpha Y^2(t) + bU(t) \\[2ex] \dfrac{\mathrm{d}Y^2(t)}{\mathrm{d}t} = -\, \alpha Y^1(t) - bY^2(t) + aU(t) . \end{cases}$$

For $U(t) = 0$, the trajectories are logarithmic spirals (see Fig. 5). The trajectory Γ_0 passing through the point $\begin{bmatrix} 1 \\ 0 \end{bmatrix}$ for $t = 0$ is given by the parametric equations in polar coordinates

$$\begin{cases} \theta = -t\sqrt{1-b^2} \\ \rho = e^{-bt} . \end{cases}$$

The other trajectories are derived from this one by an arbitrary rotation about O. We note that the polar angle varies by an amount $-\Delta t\sqrt{1-b^2}$ during the interval Δt.

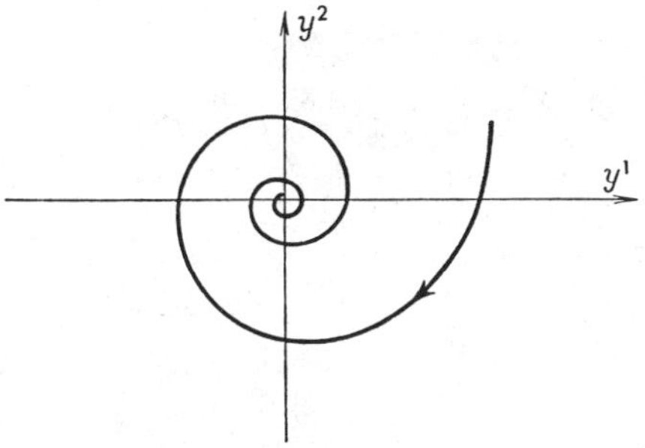

FIG. 5.

In what follows, we shall designate by "arc of a spiral" any arc of a spiral derived from Γ_0 by a displacement (or, what is equivalent, by any similarity transformation) and corresponding to a variation of π in the polar angle with respect to the asymptotic point.

For $U(t) = +1$ (resp. $U(t) = -1$), the trajectories are logarithmic spirals derived from Γ_0 by displacement and having the points

$$\omega_1 = \begin{bmatrix} 1 \\ 0 \end{bmatrix} \quad \left(\text{resp. } \omega_{-1} = -\omega_1 = \begin{bmatrix} -1 \\ 0 \end{bmatrix}\right)$$

as asymptotic points. We denote by $\mathcal{C}_1$ (resp. $\mathcal{C}_{-1}$) the family of these spirals.

Consider an optimum trajectory $\tilde{X}$ defined by the optimum control $\tilde{U}$. In accordance with the general results, we have $\tilde{U}(t) = \operatorname{sgn}(\varphi(t))$, where $\varphi(t)$ is a solution of the differential equation

$$Z'' - 2bZ' + Z = 0 .$$

Thus,

$$\varphi(t) = e^{+bt}(c\cos\alpha t + d\sin\alpha t) ,$$

where $\alpha = \sqrt{1-b^2}$, and, consequently,

$$\tilde{U}(t) = \operatorname{sgn}(c\cos\alpha t + d\sin\alpha t) .$$

Let us set $\Theta = 2\pi/\alpha$. From this, we conclude that the switching instants constitute an arithmetic progression with difference

$$\frac{\Theta}{2} = \frac{\pi}{\alpha} = \frac{\pi}{\sqrt{1 - b^2}}.$$

Between any two consecutive switching instants, the point describes in the negative sense an arc of a spiral (belonging to the family $\mathcal{C}_1$ if $U(t) = +1$ and belonging to the family $\mathcal{C}_{-1}$ if $U(t) = -1$).

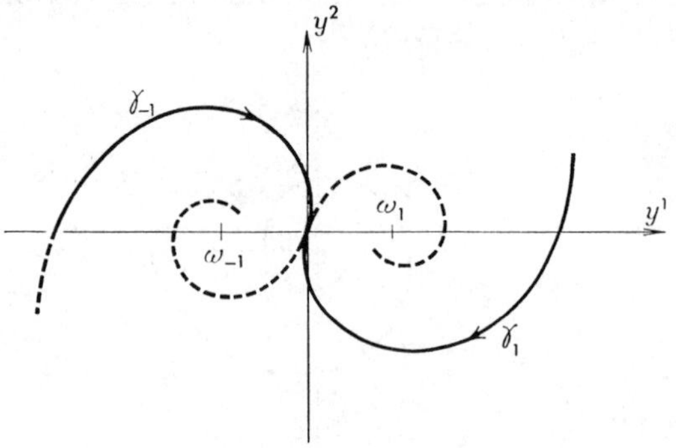

FIG. 6.

Through O there passes a trajectory of the family $\mathcal{C}_1$ of which we denote by γ_1 the arc preceding passage through O (see Fig. 6). Similarly, through O also passes a trajectory of the family $\mathcal{C}_{-1}$ of which we denote the arc preceding passage through O by γ_{-1}. Every optimum trajectory that does not reduce to an arc of γ_1 or γ_{-1} necessarily has its last switching point m_1 either on γ_1 or γ_{-1}. If $m_1 \in \gamma_1$, the trajectory terminates in an arc of γ_1 with $U(t) = +1$. If $m_1 \in \gamma_{-1}$, the trajectory terminates in an arc of γ_{-1} with $U(t) = -1$. Suppose,

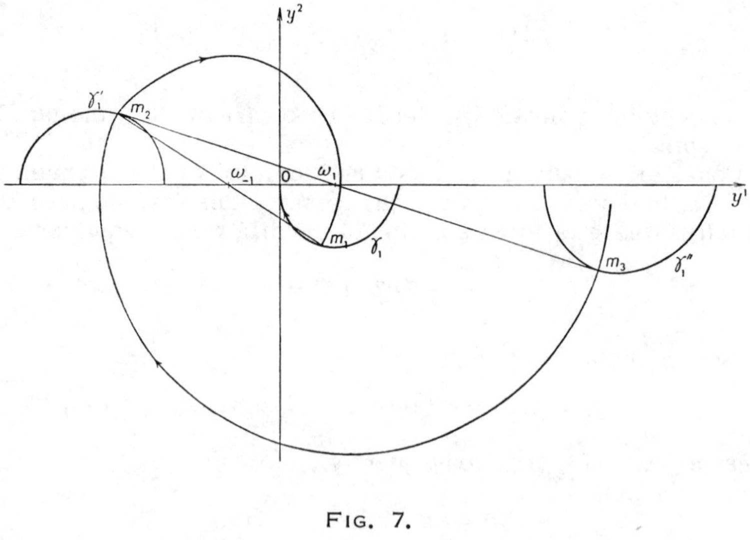

FIG. 7.

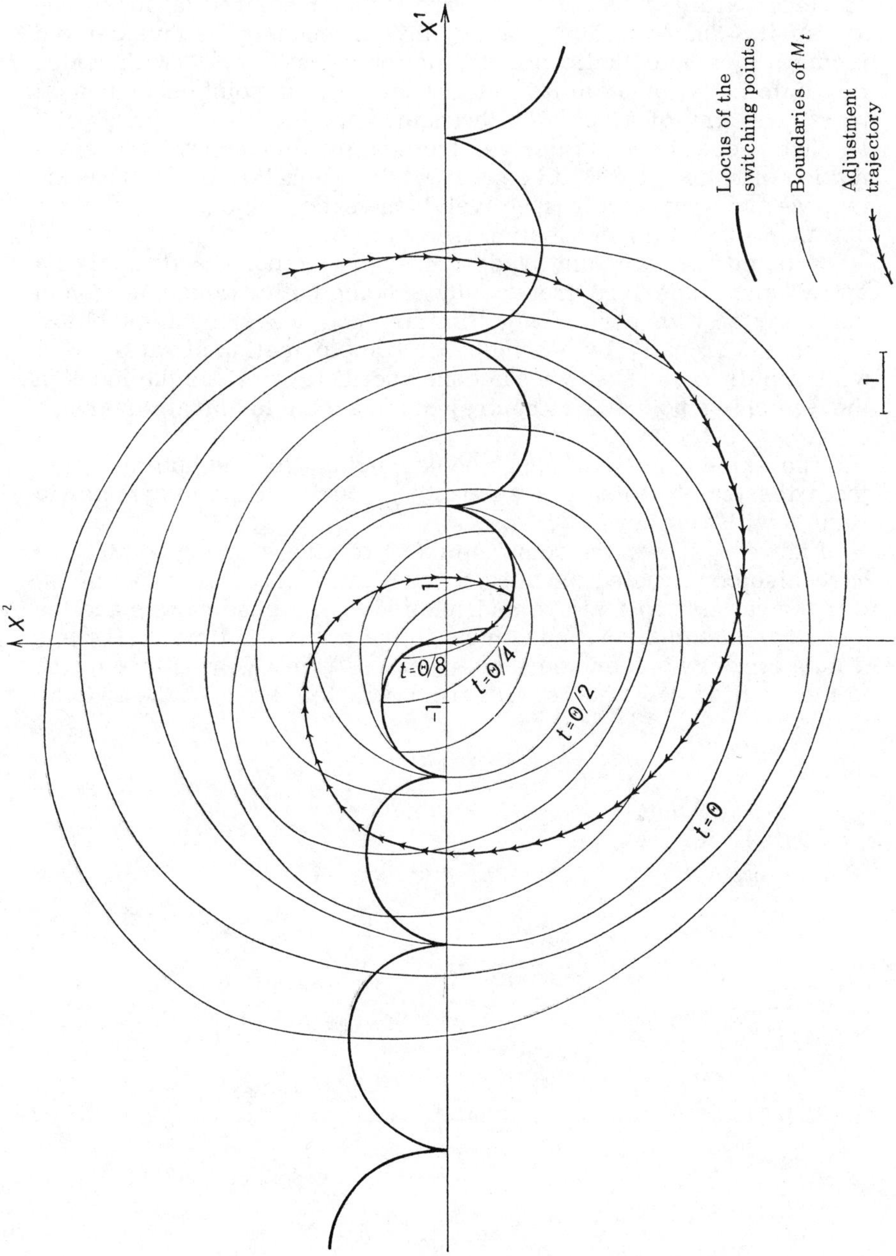

x^1
x^2
Locus of the switching points
Boundaries of M_t
Adjustment trajectory
1
1
-1
$t=\Theta/8$
$t=\Theta/4$
$t=\Theta/2$
$t=\Theta$

FIG. 8.

for example, that $m_1 \in \gamma_1$. Prior to its passage at m_1, the point is displaced along a spiral of $\mathcal{C}_{-1}$ and it has described (at most) one arc of it. The switching point m_2 that preceeds m_1 is thus derived from m_1 by a homothetic transformation of ratio $-e^{b\Theta/2}$ with center ω_{-1}. Similarly, prior to its passage into m_2, the point describes (at most) one arc of a spiral of the family $\mathcal{C}_1$, and the switching point m_3 that preceeds m_2 is derived from m_2 by a homothetic transformation of ratio $-e^{b\Theta/2}$ with center ω_1. Pursuing this line of reasoning, we construct "retrospectively" the successive portions of the optimum trajectory in question (see Fig. 7).

Note that as the point m_1 describes γ_1, the point m_2 describes a spiral arc γ_1' derived from γ_1 by a homothetic transformation of ratio $-e^{b\Theta/2}$ with center ω_{-1}. Similarly, m_3 describes a spiral arc γ_1'' derived from γ_1' by a homothetic transformation of ratio $-e^{b\Theta/2}$ with center ω_1. Thus, we can construct little by little the locus of the switching points for the trajectories that terminate in an arc of γ_1.

The same construction can be carried out for the optimum trajectories that terminate in an arc of γ_{-1} and it leads to symmetric results with respect to O.

Figure 8 shows the complete locus of the switching points. We have also constructed the curves constituting the boundaries of M_t (for the values of t that are multiples of $\Theta/8$). These curves are the loci of points that are adjustable in the minimum time t. The optimum strategy σ assumes values $\sigma(x) = +1$ or $\sigma(x) = -1$ according to the position of x with respect to the complete locus of the switching points.

18

Pontryagin's Principle

1. REVIEW OF DIFFERENTIAL EQUATIONS

Consider a linear differential equation of the form

$$\frac{\mathrm{d}X(t)}{\mathrm{d}t} = \Phi(t)\,X(t)\,,$$

where $X(t)$ is an element of a Banach space H and $\Phi(t)$ is a continuous linear operator on H that depends continuously on t [for the uniform topology of $L(H, H)$].

For every t_0 and every $\xi \in H$, there exists exactly one solution $X(t)$ such that $X(t_0) = \xi$. The value of $X(t)$ at the instant t depends linearly and continuously on $X(t_0)$. The operator $A(t, t_0)$ such that

$$X(t) = A(t, t_0)\,X(t_0)$$

is called the *resolvent* of the equation. We have

$$A(t_0, t_0) = \mathbf{1}\,,$$

$$A(t_2, t_1)\,A(t_1, t_0) = A(t_2, t_0)\,,$$

$$A(t_1, t_0) = \overset{-1}{A}(t_0, t_1)\,.$$

Also,

$$\frac{\mathrm{d}}{\mathrm{d}t}A(t, t_0) = \Phi(t)\,.\,A(t, t_0)\,,$$

and this differential equation with values in $L(H, H)$ determines $A(t, t_0)$ with the aid of the initial condition $A(t_0, t_0) = \mathbf{1}$.

Consider now a differential equation of the form

$$\frac{dX(t)}{dt} = F(X(t), t)$$

where $X(t)$ belongs to the Banach space H and the function $F : x, t \to F(x, t)$ is continuous and has a continuous partial derivative F'_x with respect to the first variable.*

Let $\tilde{X}$ denote a solution of this equation that is defined in a closed interval $[t_0, t_1]$. For ξ in a neighborhood of $\tilde{X}(t_0)$, there exists a solution $X(t)$ defined on $[t_0, t_1]$ and satisfying the equation $X(t_0) = \xi$. The mapping $X(t_0) \to X(t)$ is differentiable, and its derivative at the point $\tilde{X}(t_0)$ is the resolvent $A(t, t_0)$ of the homogeneous linear equation (called the *linearized equation*)

$$\frac{d\delta X(t)}{dt} = F'_x(\tilde{X}(t), t)\,\delta X(t)\,.$$

This property can be generalized to the case in which F is piecewise-continuous with respect to t in the following precise sense: For every interval $[t_0, t_1]$, there exists finitely many points $\theta_0 = t_0$, $\theta_1, \ldots, \theta_p = t_1$ such that $\theta_{i-1} < \theta_i$ and functions $F_1, \ldots, F_p$ such that, for each i, the function F_i is defined in $H \times [\theta_{i-1}, \theta_i]$, is continuous, has partial derivative $(F_i)'_x$ with respect to the first variable, and satisfies the equation

$$F(x, t) = F_i(x, t) \quad \text{for} \quad t \in]\theta_{i-1}, \theta_i[\,.$$

Here, a solution of the equation

$$\frac{dX(t)}{dt} = F(X(t), t)$$

defined on the interval $[t_0, t_1]$ means any continuous mapping $t \to X(t)$ such that

$$\frac{dX}{dt} = F_i(X(t), t) \quad \text{for} \quad t \in]\theta_{i-1}, \theta_i[\,.$$

In what follows, we shall need the following proposition:

Proposition 1. *Of the two linear differential equations*

$$\frac{dX(t)}{dt} = \Phi(t)\,X(t)\,,$$

---

* When we say that a function of several variables has a continuous derivative with respect to one of the variables, we assume this derivative exists and is continuous with respect to the set of variables.

$$\frac{dP(t)}{dt} = -\Phi^\star(t)\,P(t)\,,$$

suppose that the first is applicable to a Banach space H and the second to the dual H' of H. Let $\Phi^\star(t)$ denote the transpose of the operator $\Phi(t)$. Suppose that the mapping $t \to \Phi(t)$ is piecewise-continuous. Let $A(t_0, t_1)$ and $B(t_0, t_1)$ denote the resolvents of these two equations. Then,

$$B(t_0, t_1) = A^\star(t_1, t_0)\,.$$

Proof.

$$\frac{d}{dt}\langle X(t), P(t)\rangle = \left\langle \frac{dX(t)}{dt}, P(t)\right\rangle + \left\langle X(t), \frac{dP(t)}{dt}\right\rangle$$

$$= \langle \Phi(t)\,X(t), P(t)\rangle - \langle X(t), \Phi^\star(t)\,P(t)\rangle = 0\,,$$

so that

$$\langle X(t_0), P(t_0)\rangle = \langle X(t_1), P(t_1)\rangle\,,$$

that is,

$$\langle X(t_0), B(t_0, t_1)\,P(t_1)\rangle = \langle A(t_1, t_0)\,X(t_0), P(t_1)\rangle$$

for every $X(t_0) \in H$ and $P(t_1) \in H'$. Consequently,

$$B(t_0, t_1) = A^\star(t_1, t_0)\,.$$

2. THE PROBLEMS OF OPTIMUM CONTROL

Let H denote a Banach space, K a topological space, and I an interval (finite or infinite) of $\mathbf{R}$. Consider a system governed by the equation

$$\frac{dX(t)}{dt} = f\left(X(t), U(t), t\right).$$

The state $X(t)$ of the system at the instant t is an element of H. The mapping $f : x, u, t \to f(x, u, t)$ is a mapping of $H \times K \times I$ into H. The function $U : t \to U(t)$, called the *control*, is a mapping of an interval $[t_0, t_1] \subset I$ into K. Once a control is chosen, the system is governed by an ordinary differential equation.

Throughout this chapter, we shall make the following hypotheses:

(a) f is a continuous mapping with a derivative f'_x with respect to the first variable; this derivative is continuous on $H \times K \times I$.

(b) The controls in question, which we shall call *admissible*, are piecewise-continuous, In other words, they have only finitely many discontinuities on any finite interval and they have left- and right-hand limits at every point of discontinuity.

Once an admissible control U is chosen, the mapping $x, t \rightarrow f(x, U(t), t)$ is piecewise-continuous with respect to t (in the sense of section 1). For every given initial condition $X(t_0) = x_0$, Eq. (1) then has a unique solution $X : t \rightarrow X(t)$ defined on an interval $[t_0, t_0 + \theta]$. We shall call this solution X the *trajectory associated with the control* U.

We shall refer to any problem of finding controls optimizing a certain criterion involving the trajectory as a *control problem*.

We distinguish between the following:

fixed-horizon problems for which we are interested only in controls defined in a fixed interval, let us say, $[0, T]$, and for which the trajectory is (for a given initial condition) defined on that same interval,

indeterminate-horizon problems, such as the minimum-time problems studied in the preceding chapter.

On the basis of the form of the criterion, we distinguish between the following:

final-criterion problems for which the criterion is expressed as a function of the final state of the system,

integral-criterion problems, in which the criterion is of the form

$$\int_{t_0}^{t_0+\theta} g(X(t), U(t), t)\, dt .$$

Finally, we shall need to distinguish between two cases according to whether the final state of the system is free or subjected to relations. Actually, as we shall see, some of these distinctions can be removed by mathematical transformations that change the apparent form of the problem. On the other hand, other distinctions persist and compel us in certain cases to make supplementary hypotheses.

3. THE CASE OF A FIXED HORIZON AND A FREE FINAL STATE

This is the simplest case to treat. Although we may consider it as a particular subcase of the case in which the final state is subjected to relations, we shall treat it separately. No supplementary hypothesis is necessary here and certain generalizations can be made for it. Also, its treatment is a good introduction to the case treated in the following section.

Consider a system governed by the equation

$$\frac{dX}{dt} = f(X(t), U(t), t), \tag{1}$$

with initial condition $X(0) = x_0$.

Assume that the controls U in question are such that the corresponding trajectory is defined in a fixed interval of time $[0, T]$.

Let g denote a differentiable function on H known as the *final criterion.* A control $\tilde{U}$ will be called ***maximal*** (for given x_0 and g) if $g(X(T))$ (which is a function of U that is defined on the set of admissible controls) has a global maximum for $U = \tilde{U}$. We seek a necessary condition for $\tilde{U}$ to be maximal.

In what follows, we shall assume that $g'(\tilde{X}(T)) \neq 0$. Note that $g'(\tilde{X}(T))$ is an element of the topological dual H' of H. We shall call the element $g'(\tilde{X}(T))$ the *linearized final criterion.* Except in the case in which g is linear, $g'(\tilde{X}(T))$ depends on the final state $\tilde{X}(T)$.

The homogeneous linearized equation corresponding to (1) in a neighborhood of the trajectory $\tilde{X}$ may be written

$$\frac{\mathrm{d}\delta X(t)}{\mathrm{d}t} = f'_x(\tilde{X}(t), \tilde{U}(t), t)\, \delta X(t)\,. \tag{1'}$$

In connection with it, consider the equation

$$\frac{\mathrm{d}\tilde{P}(t)}{\mathrm{d}t} = -\, f'^{\star}_x(\tilde{X}(t), \tilde{U}(t), t)\, \tilde{P}(t)\,, \tag{2}$$

called the *adjoint equation,* where $\tilde{P}(t) \in H'$, with the final condition $\tilde{P}(T) = g'(\tilde{X}(T))$.

Equation (2) is a homogeneous linear equation with range in H'. The function $t \to \tilde{P}(t)$ is a function defined on $[0, T]$ into H' that is piecewise-differentiable (that is, $\mathrm{d}\tilde{P}(t)/\mathrm{d}t$ cannot have points of discontinuity different from those of $\tilde{U}$).

If the final criterion is a linear form w, then $\tilde{P}(t)$ is defined by

$$\begin{cases} \dfrac{\mathrm{d}\tilde{P}}{\mathrm{d}t} = -\, f'^{\star}_x(\tilde{X}(t), \tilde{U}(t), t)\, \tilde{P}(t) \\ \tilde{P}(T) = w\,. \end{cases}$$

Theorem 1 (Pontryagin's principle, first formulation). *For $\tilde{U}$ to be maximal, it is necessary that*

$$\langle f(\tilde{X}(t), \tilde{U}(t), t), \tilde{P}(t) \rangle = \max_{u \in K} \langle f(\tilde{X}(t), u, t), \tilde{P}(t) \rangle \tag{3}$$

at every instant t at which $\tilde{U}$ is continuous.

Proof. (a) Let t_0 denote a member of the half-open interval $]0, T]$ at which $\tilde{U}$ is continuous. We shall replace the control $\tilde{U}$ with a modified control U_τ depending on a nonnegative parameter τ and we shall study the corresponding modification in the final state of the system.

Let u denote a member of K. Let U_τ denote the control defined by

$$U_\tau(t) = \begin{cases} u & \text{for } t \in [t_0 - \tau, t_0[\\ \tilde{U}(t) & \text{for } t \in [0, t_0 - \tau[\cup [t_0, T]\,. \end{cases}$$

Suppose that τ is so small that $\tilde{U}$ is continuous in $[t_0 - \tau, t_0]$.

For $\tau = 0$, we have $U_\tau = \tilde{U}$.

Let us show that, for sufficiently small τ, the corresponding trajectory X_τ relative to the same initial condition $X_\tau(0) = x_0$ is defined. Let ε denote a positive number such that the equation

$$\frac{dX(t)}{dt} = f(X(t), u, t)$$

has a solution corresponding to the final condition $X(t_0) = \tilde{X}(t_0)$ defined in the interval $[t_0 - \varepsilon, t_0]$. Since

$$\lim_{\tau \to 0} \tilde{X}(t_0 - \tau) = \tilde{X}(t_0),$$

the equation will, for $\tau \leqslant \varepsilon$ and sufficiently small, have a solution defined on $[t_0 - \varepsilon, t_0]$ for the initial condition $X(t_0 - \tau) = \tilde{X}(t_0 - \tau)$. In other words, for sufficiently small τ, the function X_τ is defined for $t \in [0, t_0]$. Also,

$$\lim_{\tau \to 0} X_\tau(t_0) = \tilde{X}(t_0),$$

which implies that, for sufficiently small τ, the function X_τ is also defined on $[t_0, T]$.

(b) Let us examine $X_\tau(t_0)$. For $\tau = 0$, we have $X_\tau(t_0) = \tilde{X}(t_0)$. Let us calculate the derivative of $X_\tau(t_0)$ with respect to τ at $\tau = 0$ (the calculation will show at the same time that this derivative exists). We have

$$X_\tau(t_0) = \tilde{X}(t_0 - \tau) + \int_{t_0 - \tau}^{t_0} f(X_\tau(\theta), u, \theta)\, d\theta .$$

Now

$$\begin{aligned}\tilde{X}(t_0 - \tau) &= \tilde{X}(t_0) - \tau \left(\frac{d\tilde{X}}{dt}\right)_{t_0} + o_1(\tau) \\ &= \tilde{X}(t_0) - \tau f(\tilde{X}(t_0), \tilde{U}(t_0), t_0) + o_1(\tau),\end{aligned}$$

and

$$\int_{t_0 - \tau}^{t_0} f(X_\tau(\theta), u, \theta)\, d\theta = \tau f(\tilde{X}(t_0)\, u, t_0) + o_2(\tau),$$

where

$$\lim_{\tau \to 0} \frac{\| o_1(\tau) \|}{\tau} = \lim_{\tau \to 0} \frac{\| o_2(\tau) \|}{\tau} = 0 .$$

Consequently,

$$X_\tau(t_0) = \tilde{X}(t_0) + \tau[f(\tilde{X}(t_0), u, t_0) - f(\tilde{X}(t_0), \tilde{U}(t_0), t_0)] + o(\tau)$$

where

$$\lim_{\tau \to 0} \frac{\| o(\tau) \|}{\tau} = 0 .$$

Thus, for the derivatives of $X_\tau(t_0)$ at $\tau = 0$, we obtain the following value:

$$\left[\frac{dX_\tau(t_0)}{d\tau}\right]_{\tau=0} = f(\tilde{X}(t_0), u, t_0) - f(\tilde{X}(t_0), \tilde{U}(t_0), t_0).$$

(c) In the interval $[t_0, T]$, the control applied is $\tilde{U}$. Therefore, to study $X_\tau(T)$, we may use the resolvent of Eq. (1′), which we denote by $A(\bullet, \bullet)$. Thus, we obtain

$$\left[\frac{dX_\tau(T)}{d\tau}\right]_{\tau=0} = A(T, t_0)\left[f(\tilde{X}(t_0), u, t_0) - f(\tilde{X}(t_0), \tilde{U}(t_0), t_0)\right].$$

(d) Let us now express the fact that $\tilde{U}$ is maximal. We have

$$\left[\frac{d}{d\tau} g(X_\tau(T))\right]_{\tau=0} \leqslant 0,$$

that is,

$$\left\langle \left[\frac{dX_\tau(T)}{d\tau}\right]_{\tau=0}, g'(\tilde{X}(T)) \right\rangle \leqslant 0,$$

or

$$\left\langle \left[\frac{dX_\tau(T)}{d\tau}\right]_{\tau=0}, \tilde{P}(T) \right\rangle \leqslant 0.$$

This gives us successively

$$\langle A(T, t_0)\left[f(\tilde{X}(t_0), u, t_0) - f(\tilde{X}(t_0), \tilde{U}(t_0), t_0)\right], \tilde{P}(T) \rangle \leqslant 0$$

$$\langle f(\tilde{X}(t_0), u, t_0) - f(\tilde{X}(t_0), \tilde{U}(t_0), t_0), A^*(T, t_0)\, \tilde{P}(T) \rangle \leqslant 0.$$

In accordance with Proposition 1, we finally obtain

$$\langle f(\tilde{X}(t_0), u, t_0) - f(\tilde{X}(t_0), \tilde{U}(t_0), t_0), \tilde{P}(t_0) \rangle \leqslant 0,$$

that is,

$$\langle f(\tilde{X}(t_0), u, t_0), \tilde{P}(t_0) \rangle \leqslant \langle f(\tilde{X}(t_0), \tilde{U}(t_0), t_0), \tilde{P}(t_0) \rangle \qquad (\forall u \in K)$$

which completes the proof for $t_0 \neq 0$.

To obtain the same conclusion for $t_0 = 0$, we need only let t_0 approach 0 in the above inequality.

We shall now interpret the above proof in terms of the concept of a contingent.

Definition. Let E denote a subset of a vector space H and let x denote a member of E. An arc with origin x traced in E is defined as a continuous mapping $\tau \to \xi_\tau$ of an interval $[0, \tau_0]$ into E that satisfies the equation $\xi_0 = x$ and that is differentiable at $\tau = 0$. Its derivative at $\tau = 0$ is called a *tangent vector.*

The *vector contingent* of E at x is defined as the set of tangent vectors at x to an arc with origin x traced in E.

The vector contingent of E at x is a pointed cone, that is, one containing 0.

Notation. We shall denote by $\Gamma(t)$ (where $t \in [0, T]$) the set of states $X(t)$, where X is any trajectory of the system satisfying the initial condition $X(t_0) = x_0$. We shall call the elements of $\Gamma(t)$ *accessible states* at the time t (from the initial state x_0). For every trajectory X, we obviously have $X(t) \in \Gamma(t)$. The purpose of part (b) of the preceding proof was to show the following property:

Let $\tilde{U}$ denote a control, $\tilde{X}$ the corresponding trajectory, t_0 an instant at which $\tilde{U}$ is continuous. Then, all vectors of the form

$$f(\tilde{X}(t_0), u, t_0) - f(\tilde{X}(t_0), \tilde{U}(t_0), t_0)$$

where $u \in K$, belong to the vector contingent of $\Gamma(t_0)$ at the point $\tilde{X}(t_0)$.

Also, since $\tilde{U}$ is a control and $\tilde{X}$ is the corresponding trajectory and since t_0 is any instant in $[0, T]$, *if the vector h belongs to the vector contingent of $\Gamma(t_0)$ at $\tilde{X}(t_0)$, then the vector $A(T, t_0)\, h$ belongs to the vector contingent of $\Gamma(T)$ at $\tilde{X}(T)$.*

Proof. Suppose that $h = \left[\dfrac{d\xi\tau}{d\tau}\right]_{\tau=0}$ where $\xi_\tau \in \Gamma(t_0)$. Let us apply the control $\tilde{U}$ in the interval $[t_0, T]$. Let X_τ denote the corresponding trajectory for the initial condition $X_\tau(t_0) = \xi_\tau$. This trajectory is defined in $[t_0, T]$ for sufficiently small τ, and we have

$$\begin{cases} X_\tau(T) \in \Gamma(T) \\ \left[\dfrac{dX_\tau(T)}{d\tau}\right]_{\tau=0} = A(T, t_0) \left[\dfrac{dX_\tau(t_0)}{d\tau}\right]_{\tau=0} = A(T, t_0)\, h\,. \end{cases}$$

Step (c) of the above proof of Pontryagin's theorem amounts to applying this remark to the vectors

$$h = f(\tilde{X}(t_0), u, t_0) - f(\tilde{X}(t_0), \tilde{U}(t_0), t_0)\,.$$

Finally, Step (d) consists in expressing the fact that since $\tilde{U}$ is assumed maximal, every vector k of the vector contingent of $\Gamma(T)$ at $\tilde{X}(T)$ satisfies the inequality

$$\langle k, g'(\tilde{X}(T)) \rangle \leqslant 0\,;$$

in other words, the vector contingent of $\Gamma(T)$ at $\tilde{X}(T)$ is contained in one of the closed half-spaces defined by the linear form $g'(\tilde{X}(T))$.

Remarks. I. The interpretation that we have just given shows that in the proof of Theorem 1, we may replace the hypothesis of maximality by a weaker hypothesis. Specifically, we use only the fact that, for every vector k belonging to the vector contingent of $\Gamma(T)$ at $\tilde{X}(T)$, we have

$$\langle k, g'(\tilde{X}(T)) \rangle \leqslant 0 .$$

If we agree to express this property by saying that g is *quasi-maximal* on $\Gamma(T)$ at $\tilde{X}(T)$, we can state Theorem 1 in the following stronger form:

If $\tilde{U}$ is a control such that g is quasi-maximal on $\Gamma(T)$ at $\tilde{X}(T)$, then, at every point of continuity of $\tilde{U}$,

$$\langle f(\tilde{X}(t), \tilde{U}(t), t), \tilde{P}(t) \rangle = \max_{u \in K} \langle f(\tilde{X}(t), u, t), \tilde{P}(t) \rangle ,$$

where $\tilde{P}(t)$ is defined by

$$\begin{cases} \dfrac{d\tilde{P}}{dt} = - f_x'^{*}(\tilde{X}(t), \tilde{U}(t), t)\, \tilde{P}(t) \\ \tilde{P}(T) = g'(\tilde{X}(T)) . \end{cases}$$

II. The proof given in the case in which the space H is a Banach space can be followed without essential modification in the case in which H is an abstract differentiable variety of the class C^2. Since the final criterion g is always a differentiable function on H, the linearized criterion $g'(\tilde{X}(T))$ is an element of the cotangent space at $\tilde{X}(T)$ at H. Similarly, $\tilde{P}(t)$ is an element of the cotangent space at $\tilde{X}(t)$ at H. To interpret Eq. (2), we can consider a finite number of maps of the images of which cover the locus of $\tilde{X}(t)$ for $t \in [0, T]$. In each of these maps, Eq. (2) can be written as a system of ordinary differential equations.

4. THE CASE OF A FIXED-HORIZON FINAL CRITERION WITH RELATIONS ON THE FINAL STATE

Consider the system

$$\frac{dX}{dt} = f(X(t), U(t), t) ,$$

and let $[0, T]$ denote a fixed interval of time.

Proposition 2. *Suppose that $U_0, \ldots, U_{p-1}$ are p admissible controls and $t_1, \ldots, t_{p-1}$ (where $0 \leqslant t_1 \leqslant \cdots \leqslant t_{p-1} \leqslant T$) are $p-1$ instants of continuity of $U_0, \ldots, U_{p-1}$. Set $t_0 = 0, \ldots, t_p = T$. Let U denote the control defined by*

$$U(t) = U_i(t) \quad for \quad t \in]\, t_i, t_{i+1}] .$$

Suppose that for particular values $t_1^0, \ldots, t_{p-1}^0$ *of* $t_1, \ldots, t_{p-1}$ *and for the initial state* $X^0(0)$, *the trajectory* X^0 *is defined on* $[0, T]$. *Then, there exists an* $\varepsilon > 0$ *and a neighborhood* V *of* $X^0(0)$ *such that, for* $|t_i - t_i^0| \leqslant \varepsilon$ *and* $X(0) \in V$, *the trajectory* X *is defined on* $[0, T]$ *and depends continuously on* $t_1, \ldots, t_{p-1}, X(0)$ *for the topology of uniform convergence.*

Proof. Let us show that, for every $k = 1, \ldots, p$ and every positive number ρ_k, there exists an $\varepsilon_k > 0$ and a neighborhood V of $X^0(0)$ such that

$$\left\{\begin{array}{l} |t_i - t_i^0| \leqslant \varepsilon_k \\ (i = 1, \ldots, k) \\ X(0) \in V \end{array}\right\} \Rightarrow \left\{\begin{array}{l} X \text{ is defined on } [0, t_k] \text{ and} \\ \| X(t) - X^0(t) \| \leqslant \rho_k, \forall t \in [0, t_k] \end{array}\right\}.$$

This property is true for $k = 1$. Suppose it true for an arbitrary value k. We shall show that it is true for $k + 1$. Suppose that $\rho_{k+1} > 0$. There exist ε_k' and $\rho_k > 0$ such that

$$\left\{\begin{array}{l} |t_k - t_k^0| \leqslant \varepsilon_k' \\ |t_{k+1} - t_{k+1}^0| \leqslant \varepsilon_k' \\ \| X(t_k) - X^0(t_k) \| \leqslant \rho_k \end{array}\right\} \Rightarrow \left\{\begin{array}{l} X \text{ is defined on } [t_k, t_{k+1}] \\ \text{and } \| X(t) - X^0(t) \| \leqslant \rho_{k+1}, \\ \forall t \in [t_k, t_{k+1}] \end{array}\right\}.$$

Let ε_k and V be associated with ρ_k. Let us set $\varepsilon_{k+1} = \min(\varepsilon_k, \varepsilon_k')$. Then,

$$\left\{\begin{array}{l} |t_i - t_i^0| \leqslant \varepsilon_{k+1} \\ (i = 1, \ldots, k+1) \\ X(0) \in V \end{array}\right\} \Rightarrow \left\{\begin{array}{l} X \text{ is defined on } [0, t_{k+1}] \text{ and} \\ \| X(t) - X^0(t) \| \leqslant \rho_{k+1}, \forall t \in [0, t_{k+1}] \end{array}\right\}.$$

This completes the proof. Note that it remains valid even if the numbers t_i^0 are not all distinct.

Henceforth, we shall denote by $\tilde{U}$ an admissible control defined on $[0, T]$ such that the corresponding trajectory $\tilde{X}$ associated with the initial condition $\tilde{X}(0) = x_0$ is also defined on $[0, T]$. We still denote by $A(\cdot, \cdot)$ the resolvent of the equation

$$\frac{d\delta X}{dt} = f_x'(\tilde{X}(t), \tilde{U}(t), t)\, \delta X(t).$$

Let us study the effect that "small" variations in U from $\tilde{U}$ have on the final state and thus derive more precise information on the vector contingent of $\Gamma(T)$ at $\tilde{X}(T)$ than we obtained in the preceding section.

Theorem 2. *Suppose that*

$$\left\{\begin{array}{l} h \in H \\ \tau \geqslant 0 \\ \varepsilon(\tau) \text{ is a function with range in } H \text{ such that } \lim_{\tau \to 0} \varepsilon(\tau) = 0 \\ \lambda_1, \ldots, \lambda_J \geqslant 0 \\ u_1, \ldots, u_J \in K \\ t_0 \in \,]0, T] \text{ is a point of continuity of } \tilde{U}, \end{array}\right.$$

Set $\lambda'_j = \lambda_1 + \cdots + \lambda_j$, $\lambda'_0 = 0$. *Consider the control* U_τ *defined on* $[0, t_0]$ *by*

$$U_\tau(t) = \begin{cases} u_j & \textit{for} \quad t \in [t_0 - \lambda'_j \tau, t_0 - \lambda'_{j-1} \tau[\\ \tilde{U}(t) & \textit{otherwise} \end{cases}$$

and the trajectory X_τ *associated with this control that satisfies the initial condition*

$$X_\tau(0) = \tilde{X}(0) + \tau h + \tau \varepsilon(\tau) .$$

For sufficiently small τ, *the trajectory* X_τ *is defined on* $[0, t_0]$ *and*

$$\begin{aligned} X_\tau(t_0) = \tilde{X}(t_0) &+ \tau A(t_0, 0)\, h \\ &+ \tau \sum_{j=1}^{J} \lambda_j \left[f\left(\tilde{X}(t_0), u_j, t_0\right) - f\left(\tilde{X}(t_0), \tilde{U}(t_0), t_0\right)\right] \\ &+ \tau \varepsilon_1(\tau) , \end{aligned}$$

where $\lim_{\tau \to 0} \varepsilon_1(\tau) = 0$.

If the numbers λ_j *vary in such a way that* $\sum_{j=1}^{J} \lambda_j \leqslant 1$, *if the vector* h *varies in a bounded set, and if the function* ε *varies but converges uniformly to 0 as* τ *approaches 0, then the function* ε_1 *converges uniformly to 0 as* τ *approaches 0.*

Proof. The existence of X_τ on $[0, t_0]$ for sufficiently small τ is assured by Proposition 2. Let us therefore consider the differential equation

$$\frac{\mathrm{d}X}{\mathrm{d}t} = f\left(X(t), \tilde{U}(t), t\right) .$$

Let φ denote the mapping that assigns to the pair (ξ, θ) the value $X(\theta)$ of the solution X such that $X(0) = \xi$. We know that φ is a continuously differentiable mapping and we have

$$\varphi'_\xi(\tilde{X}(0), t_0) = A(t_0, 0)$$

$$\varphi'_\theta(\tilde{X}(0), t_0) = f\left(\tilde{X}(t_0), \tilde{U}(t_0), t_0\right) .$$

Therefore,

$$\begin{aligned} X_\tau(t_0 - \lambda'_J \tau) &= \varphi(\tilde{X}(0) + \tau h + \tau\varepsilon(\tau), t_0 - \lambda'_J \tau) \\ &= \tilde{X}(t_0) + \tau A(t_0, 0)\, h - \lambda'_J \tau f\left(\tilde{X}(t_0), \tilde{U}(t_0), t_0\right) + \tau \varepsilon_2(\tau) , \end{aligned}$$

where $\varepsilon_2(\tau) \to 0$ as $\tau \to 0$.

Also,

$$X_\tau(t_0) = X_\tau(t_0 - \lambda'_J \tau) + \int_{t_0 - \lambda'_J \tau}^{t_0} f(X_\tau(t), U_\tau(t), t)\, dt$$

$$= X_\tau(t_0 - \lambda'_J \tau) + \sum_{j=1}^{J} \int_{t_0 - \lambda'_j \tau}^{t_0 - \lambda'_{j-1}\tau} f(X_\tau(t), u_j, t)\, dt\,.$$

Now, $X_\tau(t)$ approaches $\widetilde{X}(t_0)$ as $\tau \to 0$ and $t \to t_0$. Therefore,

$$\int_{t_0 - \lambda'_j \tau}^{t_0 - \lambda'_{j-1}\tau} f(X_\tau(t), u_j, t)\, dt = \lambda_j \tau f(\widetilde{X}(t_0), u_j, t_0) + \tau \varepsilon_{3,j}(\tau)$$

where $\varepsilon_{3,j}(\tau) \to 0$ as $\tau \to 0$. From this we get

$$X_\tau(t_0) = X_\tau(t_0 - \lambda'_J \tau) + \tau \sum_{j=1}^{J} \lambda_j f(\widetilde{X}(t_0), u_j, t_0) + \tau \varepsilon_3(\tau)$$

where

$$\varepsilon_3(\tau) = \sum_{j=1}^{J} \varepsilon_{3,j}(\tau)\,, \qquad \varepsilon_3(\tau) \to 0 \quad \text{as} \quad \tau \to 0\,.$$

If we substitute the expression already found for $X_\tau(t_0 - \lambda'_J \tau)$, we obtain the formula in the theorem with $\varepsilon_1 = \varepsilon_2 + \varepsilon_3$.

To prove the second part of the theorem, let us show that the hypotheses imply that ε_2 and ε_{3j} converge uniformly to 0. For ε_2, this follows from application of the following property to the function φ described above: Let φ denote a continuously differentiable mapping. Then

$$\varphi(z_0 + \tau h + \tau\varepsilon(\tau)) = \varphi(z_0) + \tau\varphi'(z_0)(h) + \tau\varepsilon_1(\tau)$$

where ε_1 converges uniformly to 0 as h ranges over a bounded set and ε converges uniformly to 0 as τ approaches 0.

Also,

$$\| \varepsilon_{3,j}(\tau) \| \leqslant \sup_{t \in [t_0 - \lambda_j \tau,\, t_0 - \lambda_{j-1}\tau]} \| f(X_\tau(t), \mu_j, t_0) - f(\widetilde{X}(t_0), \mu_j, t_0) \|.$$

By virtue of Proposition 2, $X_\tau(t)$ converges to $\widetilde{X}(t)$ uniformly with respect to t, h and ε. From this we easily conclude that the right-hand member of the inequality converges to 0 uniformly with respect to ε and h as τ approaches 0. This completes the proof of the theorem.

Theorem 3. *Let $t_1, \ldots, t_i, \ldots, t_I \in]0, T]$ denote an increasing sequence of distinct points of continuity of $\widetilde{U}$. Suppose that, for every i,*

$$u_{i,j} \in K \qquad (j = 1, \ldots, J_i)$$

$$\lambda_{i,j} \geqslant 0$$

where $\sum_{i,j} \lambda_{i,j} = 1$. Set

$$\lambda'_{i,j} = \sum_{k=1}^{j} \lambda_{i,k}, \lambda'_{i,0} = 0 .$$

Let U_τ denote the control defined by

$$U_\tau(t) = \begin{cases} u_{i,j} & \textit{for } t \in [t_i - \lambda'_{i,j}\tau, t_i - \lambda'_{i,j-1}\tau[\\ \tilde{U}(t) & \textit{otherwise.} \end{cases}$$

(1) *There exists a τ_1 such that, for $\tau \leqslant \tau_1$, the corresponding trajectory is defined for the initial condition $X_\tau(0) = \tilde{X}(0)$, and*

$$X_\tau(T) = \tilde{X}(T) + \tau \sum_{i,j} \lambda_{i,j} A(T, t_i) \left[f(\tilde{X}(t_i), u_{i,j}, t_i) - f(\tilde{X}(t_i), \tilde{U}(t_i), t_i) \right] + \tau\varepsilon(\tau)$$

where $\varepsilon(\tau)$ approaches 0 as $\tau \to 0$.

(2) *If, in addition, the numbers $\lambda_{i,j}$ vary (with the instants t_i and the elements $u_{i,j}$ fixed), $\varepsilon(\tau)$ converges uniformly to 0 as $\tau_0 > 0$.*

Proof. We may always assume that $t_I = T$ (because if this is not the case, we can add $t_{I+1} = T$, where $\lambda_{I+1,j} = 0$). The existence of X_τ on $[0, T]$ is assured for $\tau \leqslant \tau_1$ (where τ_1 is independent of the numbers $\lambda_{i,j}$) by Proposition 2. Let us show by induction that

$$(\beta) \quad \begin{cases} X_\tau(t_k) = \tilde{X}(t_k) + \\ \quad + \tau \displaystyle\sum_{\substack{i \leqslant k \\ j=1,\ldots,J_i}} \lambda_{i,j} A(t_k, t_i) \left[f(\tilde{X}(t_i), u_{i,j}, t_i) - f(\tilde{X}(t_i), \tilde{U}(t_i), t_i) \right] \\ \quad + \tau\varepsilon_k(\tau) \end{cases}$$

for every k, where $\varepsilon_k(\tau)$ approaches 0 as $\tau \to 0$. For $k = 1$, this follows from Theorem 2. If the property is valid for one value of k, Theorem 2 shows that X_τ is defined on $[t_k, t_{k+1}]$ for sufficiently small τ, and we have

$$\begin{aligned} X_\tau(t_{k+1}) &= \tilde{X}(t_{k+1}) + \\ &+ \tau A(t_{k+1}, t_k) \sum_{\substack{i \leqslant k \\ j=1,\ldots,Ji}} \lambda_{i,j} A(t_k, t_i) \left[f(\tilde{X}(t_i), u_{i,j}, t_i) - f(\tilde{X}(t_i), \tilde{U}(t_,), t_i) \right] \\ &+ \tau \sum_{j=1,\ldots,J_{k+1}} \lambda_{k+1,j} \left[f(\tilde{X}(t_{k+1}), u_{k+1,j}, t_{k+1}) - f(\tilde{X}(t_{k+1}), \tilde{U}(t_{k+1}), t_{k+1}) \right] \\ &+ \tau\varepsilon_{k+1}(\tau) \end{aligned}$$

where $\varepsilon_{k+1}(\tau)$ approaches 0 as $\tau_0 > 0$. This is the same as relation (β) with k replaced by $k + 1$. This completes the proof of part (1).

To prove part (2), we need only verify by induction based on part 2 of Theorem 2 that ε_k converges uniformly to 0 as τ approaches 0.

We shall continue to denote by $\underset{\sim}{\Gamma(T)}$ the set of final states that are accessible from the initial state $\tilde{X}(0)$. We shall denote by L_T the convex hull of vectors of the form

$$A(T, t_0) \left[f(\tilde{Y}(t_0), u, t_0) - f(\tilde{X}(t_0), \tilde{U}(t_0), t_0) \right] ,$$

where t_0 is an instant of continuity of $\tilde{U}$ and where $u \in K$.

Every $\xi \in L_T$ has a representation of the form

$$(*) \qquad \xi = \sum_{i,j} \lambda_{i,j} A(T, t_i) \left[f(\tilde{X}(t_i), u_{i,j}, t_i) - f(\tilde{X}(t_i), \tilde{U}(t_i), t_i) \right]$$

where

$$\begin{cases} t_i \in]0, T] & (i = 1, \dots, I) \\ u_{i,j} \in K & (j = 1, \dots, J_i) \\ \lambda_{i,j} \geqslant 0 & \sum_{i,j} \lambda_{i,j} = 1 \end{cases}$$

and $\tilde{U}$ is continuous at t_i. Finally, we denote by K_T the cone generated by L_T.

Then, Theorem 3 has the

Corollary. *The cone K_T is contained in the vector contingent $\Gamma(T)$ at $\tilde{X}(T)$.*

Proposition 3. *Suppose that $r \leqslant \dim(K_T)$. Suppose that $x_1, \dots, x_r$, are r independent vectors belonging to K_T. Suppose that*

$$\xi = \sum_{i=1}^{r} \mu_i x_i$$

is a compromise of the vectors x_i (that is, $\mu_i \geqslant 0$ and $\sum_{i=1}^{r} \mu_i = 1$). Then, for every ξ, there exists a family of trajectories $X_{\xi,\tau}$ defined for $\tau \in [0, \tau_1]$ such that

$$\begin{aligned} X_{\xi,\tau} &= \tilde{X} \text{ for } \tau = 0 \\ X_{\xi,\tau}(0) &= \tilde{X}(0) \\ X_{\xi,\tau}(T) &= \tilde{X}(T) + \tau\xi + \tau\varepsilon_\xi(\tau) \end{aligned}$$

where the function $\varepsilon_\xi(\tau)$ tends to 0 as $\tau \to 0$, uniformly with respect to ξ as ξ describes the convex hull Σ of the vectors x_i.

Proof. (a) Suppose that $x_1, \dots, x_r \in L_T$. Each x_k (for $k = 1, \dots, r$) has a representation of the form (*). We can manipulate the notation so that the instants t_i are the same for $x_1, \dots, x_r$ and, for an instant t_i, we can combine the $u_{i,j}$ relative to different x_k into a single family. This leads us to set

$$x_k = \sum_{i,j} \lambda_{i,j}^{(k)} A(T, t_i) \left[f(\tilde{X}(t_i), u_{i,j}, t_i) - f(\tilde{X}(t_i), \tilde{U}(t_i), t_i) \right],$$

where the instants t_i are instants of continuity of $\tilde{U}$ and where

$$\lambda_{i,j}^{(k)} \geqslant 0, \ \sum_{i,j} \lambda_{i,j}^{(k)} = 1 \qquad (\forall k = 1, \dots, r).$$

Suppose then that

$$\xi = \sum_{k=1}^{r} \mu_k x_k .$$

We set

$$\lambda_{i,j}^{[\mu]} = \sum_{k=1}^{r} \lambda_{i,j}^{(k)} \mu_k .$$

In accordance with Theorem 3, with every μ is associated a trajectory defined for $\tau \in [0, \tau_1]$. We denote this trajectory by $X_{\xi,\tau}$. We have

$$\begin{aligned} X_{\xi,\tau}(T) &= \tilde{X}(T) + \tau \sum_{i,j} \lambda_{i,j}^{[\mu]} A(T, t_i) \left[f(\tilde{X}(t_i), u_{i,j}, t_i) - f(\tilde{X}(t_i), \tilde{U}(t_i), t_i) \right] \\ &\quad + \tau \varepsilon_\xi(\tau) \\ &= \tilde{X}(T) + \tau\xi + \tau\varepsilon_\xi(\tau) , \end{aligned}$$

where $\varepsilon_\xi(\tau)$ converges uniformly to 0 as ξ describes Σ.

(b) Suppose that $x_1, \ldots, x_r \in K_T$. There exists a $\rho > 0$ such that $\rho x_1, \ldots, \rho x_r \in L_T$. This implies that $\rho\xi \in L_T$. Therefore, we need only set

$$X_{\xi,\tau} = X_{\rho\xi,\tau/\rho} .$$

Theorem 4. *Suppose that* $r \leqslant \dim(K_T)$. *Let* $x_1, \ldots, x_r$ *denote* r *vectors belonging to* K_T. *Let* Σ *denote the convex hull of the set of the vectors* x_i. *Then, there exists a* $\tau_0 > 0$ *and, for* $\tau \in]0, \tau_0]$, *a continuous mapping* φ_τ *of* Σ *into* H *such that*

(1) φ_τ *converges uniformly to the identity mapping* Σ *as* $\tau \to 0$

and

(2) *the ray with origin* $\tilde{X}(T)$ *and direction* $\varphi_\tau(\xi)$ *intersects* $\Gamma(T)$ *at a point different from* $\tilde{X}(T)$.

Proof. With every

$$\xi = \sum_{i=1}^{r} \mu_i x_i$$

let us associate the trajectory $X_{\xi,\tau}$ defined by Proposition 3. We need only set

$$\varphi_\tau(\xi) = \xi + \varepsilon_\xi(\tau) .$$

In what follows, we shall need the following result from algebraic topology.

Lemma 1. *Let* P *denote an* n*-dimensional affine space. Let* Σ *denote an* n*-dimensional simplex. Let* φ_τ *(for* $0 < \tau \leqslant \tau_0$*) denote a family of continuous mappings of* Σ *into* P *that approaches the iden-*

tity mapping uniformly as $\tau \to 0$. *Then, for every* ξ *in the interior of* Σ, *there exists an* $\eta > 0$ *such that*

$$\tau \leqslant \eta \quad \Rightarrow \quad \xi \in \text{val}(\varphi_\tau)\,.$$

Corollary. *Let* E *denote an* n*-dimensional vector space, let* M *denote a* p*-dimensional affine variety, and let* Σ *denote a* p*-dimensional simplex in* M. *Let* V *denote an affine variety such that there does not exist a hyperplane containing both* V *and* M *and such that* V *contains an internal point of* Σ. *Let* φ_τ *(for* $0 < \tau \leqslant \tau_0$*) denote a family of continuous mappings of* Σ *into* E *that converges uniformly to the identity mappings as* $\tau \to 0$. *Then, there exists an* $\eta > 0$ *such that*

$$\tau \leqslant \eta \Rightarrow V \cap \varphi_\tau(\Sigma) \neq \varnothing\,.$$

Proof. By making a translation if necessary, we may assume that $0 \in V \cap \Sigma$, and that 0 is an internal point of Σ.

(a) Suppose that $\dim(M) + \dim(V) = n$. Then, M and V are complementary vector subspaces. Let P denote the projection onto M and parallel to V. The mapping $P \circ \varphi_\tau$ is a continuous mapping of Σ into M that converges uniformly to the identity mapping as $\tau \to 0$. Thus, there exists an $\eta > 0$ such that, for $\tau \leqslant \eta$, we have $0 \in (P \circ \varphi_\tau)(\Sigma)$, that is, $0 \in P(\varphi_\tau(\Sigma))$. This implies $V \cap \varphi_\tau(\Sigma) \neq \varnothing$.

(b) Suppose that $\dim(M) + \dim(V) > n$. Let V_1 denote a complementary subspace of M that is contained in V. There exists an $\eta > 0$ such that, for $\tau \leqslant \eta$, we have $V_1 \cap \varphi_\tau(\Sigma) \neq \varnothing$. This implies $V \cap \varphi_\tau(\Sigma) \neq \varnothing$.

Proposition 4. *Suppose that* $\dim(H) = \dim(K_T) = n$. *Let* Δ *denote an open ray contained in the interior of* K_T *and with origin* O. *Then, the ray* $\tilde{X}(T) + \Delta$ *intersects* $\Gamma(T)$.

Proof. Suppose that $\xi \in \Delta$ and that ξ is an interior point of K_T. There exist n independent vectors $x_1, \dots, x_n$ belonging to K_T such that ξ is an internal point of the simplex Σ generated by $x_1, \dots, x_n$. Consider the family φ_τ of mappings defined by Theorem 4. According to the corollary of Lemma 1, there exists a $\tau > 0$ such that $\tilde{X}(T) + \Delta$ intersects $\varphi_\tau(\Sigma)$ at a point y. Therefore, there exists a $\rho > 0$ such that $\tilde{X}(T) + \rho y \in \Gamma(T)$. This completes the proof of the proposition.

We shall now apply the preceding results to the problem of finding conditions for optimality when the final state $X(T)$ is constrained to belong to an affine variety H_0 of H. Suppose that H_0 is defined by equations of the form

$$\langle x, v_k \rangle = c_k \qquad (k = 1, \dots, \kappa)\,(v_k \in H')\,.$$

Suppose also that $w \in H'$ and that $v_1, \dots, v_\kappa, w$ are independent. We shall say that a control $\tilde{U}$ is *maximal* if the corresponding trajectory $\tilde{X}$ is defined on $[0, T]$, if $\tilde{X}(T) \in H_0$, and if

$$\langle X(T), w \rangle \leqslant \langle \tilde{X}(T), w \rangle$$

for every admissible control U such that X is defined on $[0, T]$ and satisfies the relation $X(T) \in H_0$. We denote by H_0^+ the intersection of H_0 with the half-space defined by the inequality

$$\langle x, w \rangle > \langle \tilde{X}(T), w \rangle.$$

If $\tilde{U}$ is maximal, then $H_0^+ \cap \Gamma(T) = \emptyset$, and conversely.

Let V_0 denote the vector subspace parallel to H_0 and let V_0^+ denote the intersection of V_0 with the half-space defined by the inequality $\langle x, w \rangle > 0$. Then, $H_0 = V_0 + \tilde{X}(T)$ and $H_0^+ = V_0^+ + \tilde{X}(T)$. The half-space V_0 is defined by the equations $\langle x, v_k \rangle = 0$ (for $k = 1, ..., \kappa$) and the subset V_0^+ is defined by these equations and the inequality $\langle x, w \rangle > 0$.

The space H is assumed to be finite-dimensional.

Theorem 5 (Pontryagin's principle, second formulation). *Let $\tilde{U}$ denote a maximal control with reference to the relation $X(T) \in H_0$. The notation and hypotheses are those given above. There exists a $\nu_0 \geqslant 0$ and $\nu_0, \nu_1, ..., \nu_\kappa \in \mathbf{R}$, not all zero, such that, if $\tilde{P}(t)$ is the solution of the equation*

$$\begin{cases} \dfrac{d\tilde{P}(t)}{dt} = -f_x'^{*}(\tilde{X}(t), \tilde{U}(t), t)\,\tilde{P}(t) \\ \tilde{P}(T) = \nu_0 w + \sum\limits_{k=1}^{\kappa} \nu_k v_k \end{cases}$$

then

$$\max_{u \in K} \langle f(\tilde{X}(t), u, t), \tilde{P}(t) \rangle = \langle f(\tilde{X}(t), \tilde{U}(t), t), \tilde{P}(t) \rangle$$

at every instant t at which $\tilde{U}$ is continuous.

Proof. (a) Let us show that there exists a hyperplane separating V_0^+ and K_T. This follows immediately from Proposition 4 if $\dim(K_T) = \dim(H)$ since, in this case, $V_0^+ \cap \mathring{K}_T = \emptyset$. Let us give a proof that is independent of the above hypothesis. If K_T and V_0 are in the same hyperplane, the property is trivial. Therefore, let us suppose that K_T and V_0 are not in the same hyperplane. We shall show by contradiction that K_T and V_0^+ cannot have a common internal point, that is, that no internal point of K_T can belong to V_0^+. Let $a \in V_0^+$ denote an internal point of K_T. Define $p = \dim(K_T)$. Consider p independent vectors $x_1, ..., x_p$ belonging to K_T such that a is an internal point of the simplex Σ which is the convex hull of $x_1, ..., x_p$. Then Σ and V_0 are not in the same hyperplane (since a hyperplane containing Σ contains K_T). Let us consider the mappings φ_τ defined by Theorem 4. There exists an η such that the inequality $\tau \leqslant \eta$ implies $\varphi_\tau(\Sigma) \cap V_0 \neq \emptyset$ and hence $\varphi_\tau(\Sigma) \cap V_0^+ \neq \emptyset$. Let y denote a nonzero member of $\varphi_\tau(\Sigma) \cap V_0^+$. There exists a $\rho > 0$ such that $\tilde{X}(T) + \rho y \in \Gamma(T)$, which implies that $\tilde{X}(T) + \rho y \in \Gamma(T) \cap H_0^+$, in contradiction with the hypothesis.

Since K_T and V_0^+ have no common internal point, there exists a hyperplane that separates them in the broad sense (see Chapter 12, corollary to Theorem 13).

(b) Let $\tilde{P}(T)$ denote a member of H' such that the hyperplane whose equation is $\langle x, \tilde{P}(T) \rangle = 0$ separates K_T and V_0^+ in the broad sense. We may assume that $x \in K_T \Rightarrow \langle x, \tilde{P}(T) \rangle \leqslant 0$, from which it follows in particular that

$$\langle A(T, t)\left[f(\tilde{X}(t), u, t) - f(\tilde{X}(t), \tilde{U}(t), t)\right], \tilde{P}(T) \rangle \leqslant 0$$

for every $u \in K$ and every t at which $\tilde{U}$ is continuous. Let $\tilde{P}(t)$ denote the solution of the adjoint equation for the final $\tilde{P}(T)$ thus chosen. Since

$$\tilde{P}(t) = A^{\star}(T, t)\, P(T),$$

it follows that

$$\langle f(\tilde{X}(t), u, t) - f(\tilde{X}(t), \tilde{U}(t), t), \tilde{P}(t) \rangle \leqslant 0,$$

which proves the last part of the theorem.

Also,

$$\left.\begin{cases} \langle x, v_k \rangle = 0 & (k = 1, \ldots, \kappa) \\ \langle x, w \rangle \geqslant 0 & \end{cases}\right\} \Rightarrow \langle x, \tilde{P}(T) \rangle \geqslant 0,$$

and hence (cf. Chapter 13, corollary to Theorem 1),

$$P(T) = \nu_0 w + \sum_{k=1}^{\kappa} \nu_k v_k \quad \text{with } \nu_0 \geqslant 0 \text{ and } \nu_k \in \mathbf{R},$$

which completes the proof.

Remark. We have denoted by $\langle x, p \rangle$ the value of the form $p \in H'$ at the point $x \in H$. We also denote it by px. This will be the case in particular if $H = \mathbf{R}^n$, in which case x will be represented as a column with n components

$$x = \begin{bmatrix} x^1 \\ \vdots \\ x^n \end{bmatrix}$$

and p will be represented as a row with n components

$$p = [p_1, \ldots, p_n].$$

The adjoint equation

$$\frac{d\tilde{P}}{dt} = -f_x'^{\star}(\tilde{X}(t), \tilde{U}(t).\, t)\, \tilde{P}(t)$$

will then be written

$$\frac{d\tilde{P}}{dt} = -\tilde{P}(t) f'_x(\tilde{X}(t), \tilde{U}(t), t).$$

These notations are particularly suitable when H is a product of several spaces. For this reason, we shall use them in the following sections.

5. THE PARAMETRIC METHOD. THE CASE OF AN UNDETERMINED HORIZON

Every trajectory $x = \tilde{X}(t)$ can also be described by a parametric representation

$$\left\{ \begin{array}{l} x = \tilde{X}(\tilde{\varphi}(s)) \\ t = \tilde{\varphi}(s) \end{array} \right\}$$

where s is a parameter varying from 0 to 1 and $\tilde{\varphi}$ is an increasing differentiable function such that val $(\tilde{\varphi}) = [0, T]$. We may, for example, suppose that

$$\tilde{\varphi}'(s) > 0, \forall s \in [0, 1].$$

If we set

$$\tilde{X}(\tilde{\varphi}(s)) = \tilde{\mathbf{X}}(s),\ \tilde{U}(\tilde{\varphi}(s)) = \tilde{\mathbf{U}}(s),\ \tilde{\varphi}'(s) = \tilde{\mathbf{U}}_0(s),$$

the equation of evolution of the system can be written by taking as the new state of the system the pair $\begin{bmatrix} x \\ t \end{bmatrix}$:

$$\left\{ \begin{array}{l} \dfrac{d\tilde{\mathbf{X}}(s)}{ds} = f(\tilde{\mathbf{X}}(s), \tilde{\mathbf{U}}(s), \tilde{\varphi}(s)) \, . \, \tilde{\mathbf{U}}_0(s) \\ \dfrac{d\tilde{\varphi}(s)}{ds} = \tilde{\mathbf{U}}_0(s) \, . \end{array} \right.$$

To apply the results of the preceding section, let us suppose that f is continuously differentiable with respect to the pair x, t. Let us linearize the equation of evolution. We obtain

$$\left\{ \begin{array}{l} \dfrac{d\delta\mathbf{X}(s)}{ds} = f'_x(\tilde{\mathbf{X}}(s), \tilde{\mathbf{U}}(s), \tilde{\varphi}(s))\, \tilde{\mathbf{U}}_0(s)\, \delta\mathbf{X}(s) + f'_t(\tilde{\mathbf{X}}(s), \tilde{\mathbf{U}}(s), \tilde{\varphi}(s))\, \tilde{\mathbf{U}}_0(s)\, \delta\varphi(s) \\ \dfrac{d\delta\varphi(s)}{ds} = 0 \, . \end{array} \right.$$

The adjoint equation may be written (with $\tilde{\mathbf{P}}(s) \in H'$ and $\tilde{\mathbf{P}}_0(s) \in \mathbf{R}$)

$$\begin{cases} \dfrac{d\tilde{\mathbf{P}}(s)}{ds} = -\,\tilde{\mathbf{P}}(s)\, f_x'(\tilde{\mathbf{X}}(s), \tilde{\mathbf{U}}(s), \tilde{\varphi}(s))\, \tilde{\mathbf{U}}_0(s) \\[2ex] \dfrac{d\tilde{\mathbf{P}}_0(s)}{ds} = -\,\tilde{\mathbf{P}}(s) f_t'(\tilde{\mathbf{X}}(s), \tilde{\mathbf{U}}(s), \tilde{\varphi}(s))\, \tilde{\mathbf{U}}_0(s)\,. \end{cases}$$

Let us suppose that the trajectory is maximal with respect to a final criterion $[w \mid w_0]$ (where $w \in H'$ and $w_0 \in \mathbf{R}$). Then, the adjoint equation has a nonzero solution $s \to [\tilde{\mathbf{P}}(s) \mid \tilde{\mathbf{P}}_0(s)]$ such that, for every value of s at which $\tilde{\mathbf{U}}(s)$ is continuous,

$$[\tilde{\mathbf{P}}(s) f(\tilde{\mathbf{X}}(s), \tilde{\mathbf{U}}(s), \tilde{\varphi}(s)) + \tilde{\mathbf{P}}_0(s)]\, \tilde{\mathbf{U}}_0(s) =$$

$$= \max_{\substack{u \in K \\ u_0 > 0}} [\tilde{\mathbf{P}}(s) f(\tilde{\mathbf{X}}(s), u, \tilde{\varphi}(s)) + \tilde{\mathbf{P}}_0(s)]\, u_0\,.$$

In the right-hand member, let us set $u = \tilde{\mathbf{U}}(s)$ and let us maximize this member with respect to u_0 and then set $u_0 = \tilde{\mathbf{U}}_0(s)$. We obtain the following two equations:

$$\tilde{\mathbf{P}}(s) f(\tilde{\mathbf{X}}(s), \tilde{\mathbf{U}}(s), \tilde{\varphi}(s)) + \tilde{\mathbf{P}}_0(s) = 0$$

$$\tilde{\mathbf{P}}(s) f(\tilde{\mathbf{X}}(s), \tilde{\mathbf{U}}(s), \tilde{\varphi}(s)) = \max_{u \in K} \tilde{\mathbf{P}}(s) f(\tilde{\mathbf{X}}(s), u, \tilde{\varphi}(s))\,.$$

The first of these equations shows that $\tilde{\mathbf{P}}(s) \neq 0$ since otherwise we would have $\tilde{\mathbf{P}}_0(s) = 0$ and hence $[\tilde{\mathbf{P}}(s) \mid \tilde{\mathbf{P}}_0(s)] = 0$. On the other hand, in certain cases, we can have $\tilde{\mathbf{P}}_0(s) = 0$.

The values of $\tilde{\mathbf{P}}(1)$ and $\tilde{\mathbf{P}}_0(1)$ depend on the relations imposed on the final state. Let us examine a few examples (in all of which the forms v_k are assumed to be linearly independent).

(1) *Undetermined horizon,* $\mathbf{X}(1)$ *subjected to relations of the form* $v_k \mathbf{X}(1) = c_k$ *(for* $k = 1, \ldots, \kappa$*),* w *nonzero and independent of the* v_k, $w_0 = 0$. In this case,

$$\begin{cases} \tilde{\mathbf{P}}(1) = \nu_0\, w + \sum_k \nu_k\, v_k \neq 0 \qquad (\nu_0 > 0, \quad \nu_k \in \mathbf{R}) \\ \mathbf{P}_0(1) = 0\,. \end{cases}$$

(2) *Undetermined horizon,* $\mathbf{X}(1)$ *subjected to relations of the form* $v_k \mathbf{X}(1) = c_k$ *(for* $k = 1, \ldots, \kappa$*),* $w = 0$, $w_0 = -1$. In this case the problem is a minimum-time problem. We have

$$\tilde{\mathbf{P}}(1) = \sum_k \nu_k\, v_k \neq 0$$

$$\tilde{\mathbf{P}}_0(1) = -\,\nu_0 \qquad (\nu_0 \geqslant 0)\,.$$

(3) *Determined horizon,* $\mathbf{X}(1)$ *subjected to relations of the form* $v_k \tilde{\mathbf{X}}(1) = c_k$, $w_0 = 0$, $w \neq 0$ *and independent of the* v_k. In this case

$$\tilde{\mathbf{P}}(1) = \sum_k \nu_k\, v_k + \nu_0\, w \neq 0\,, \qquad (\nu_0 \geqslant 0, \nu_k \in \mathbf{R})\,,$$

where $\tilde{\mathbf{P}}_0(1)$ may assume an arbitrary value.

When the system initially given is stationary, that is, when f is independent of t, we also have

$$\frac{d\tilde{\mathbf{P}}_0(s)}{ds} = 0 .$$

Then, $\tilde{\mathbf{P}}_0(s)$ is equal to a constant, which we denote by $-a$. Thus,

$$\tilde{\mathbf{P}}(s) f(\tilde{\mathbf{X}}(s), \tilde{\mathbf{U}}(s)) = a ,$$

where

$$a = 0 \text{ in case (1)}$$
$$a \geqslant 0 \text{ in case (2)}.$$

We shall now reformulate the results obtained by going back to the system initially given. We need only set $\tilde{P}(t) = \tilde{\mathbf{P}}(\overset{-1}{\varphi}(t))$. The adjoint equation shows that

$$\frac{d\tilde{P}(t)}{dt} = - \tilde{P}(t) f_x'(\tilde{X}(t), \tilde{U}(t), t) .$$

Theorem 6 (Pontryagin's principle, third formulation). *Consider the system*

$$\frac{dX}{dt} = f(X(t), U(t), t) .$$

Let $t \to \tilde{U}(t)$ denote a maximal admissible control (for $t \in [0, T]$) for one of the following criteria:

(1) *a linear form w on the final state $X(T)$, where T is arbitrary;*
(2) *$- T$ (a minimum-time problem);*
(3) *a linear form w on the final state $X(T)$ with fixed T and with the relations*

$$v_k X(T) = c_k \qquad (k = 1, ..., \kappa) .$$

Suppose that $v_1, ..., v_\kappa$ are linearly independent in case (2) and that $v_1, ..., v_\kappa$ are linearly independent in cases (1) and (3). Suppose also that $\kappa = 0$ in case (2). Then, there exists a nonzero solution of the adjoint equation

$$\frac{d\tilde{P}(t)}{dt} = - \tilde{P}(t) f_x'(\tilde{X}(t), \tilde{U}(t), t) ,$$

such that

$$\tilde{P}(t) f(\tilde{X}(t), \tilde{U}(t), t) = \max_{u \in K} \tilde{P}(t) f(\tilde{X}(t), u, t) ,$$

at every instant t at which $\tilde{U}$ is continuous and such that

$$\tilde{P}(T) = \sum_{k=1}^{K} \nu_k v_k \quad (\nu_k \in \mathbf{R}) \qquad \text{[case (2)]}$$

$$\tilde{P}(T) = \sum_{k=1}^{K} \nu_k v_k + v_0 w \quad (\nu_k \in \mathbf{R},\, \nu_0 \geqslant 0) \qquad \text{[cases (1) and (3)]}.$$

If the system is stationary, we have also

$$\tilde{P}(t) f(\tilde{X}(t), \tilde{U}(t)) = a\,,$$

where a *is a constant. In case (1), we have* $a = 0$; *in case (2), we have* $a \geqslant 0$.

Example. Consider the linear system

$$\frac{dX}{dt} = AX(t) + U(t)\,h$$

$$\text{with } \begin{cases} X(t) \in \mathbf{R}^n \\ A \in L(\mathbf{R}^n, \mathbf{R}^n) \\ h \in \mathbf{R}^n \\ U(t) \in [-1, +1]\,. \end{cases}$$

Let $\tilde{U}$ denote a control adjusting the system in minimum time after the initial state $\tilde{X}(0)$ up to 0. Let $\tilde{X}$ denote the optimum trajectory. We use matrix notation. There exists a $\tilde{P}(t)$ such that

$$\frac{d\tilde{P}}{dt} = -\tilde{P}(t)\,A$$

and such that, at every instant at which $\tilde{U}$ is continuous, we have

$$\tilde{P}(t)\left[A\tilde{X}(t) + \tilde{U}(t)\,h\right] = \max_{-1 \leqslant u \leqslant +1} \tilde{P}(t)\left[A\tilde{X}(t) + uh\right]$$

that is,

$$\tilde{U}(t)\,\tilde{P}(t)\,h = \max_{-1 \leqslant u \leqslant +1} u\tilde{P}(t)\,h$$

or

$$\tilde{U}(t) = \operatorname{sgn}(\tilde{P}(t)\,h).$$

We also have

$$\tilde{P}(t) = \tilde{P}(0)\,e^{-tA}\,,$$

so that

$$\tilde{U}(t) = \operatorname{sgn}(\tilde{P}(0)\,e^{-tA}\,h)\,.$$

We again find the characterization of optimum controls that were given in Chapter 17.

6. THE CASE OF AN INTEGRAL CRITERION

In this section, we shall examine the case in which the criterion to be maximized is of the form

$$\int_0^T g(X(t), U(t), t)\,\mathrm{d}t\,.$$

(The horizon T may be either fixed or undetermined.) Here, the mapping $x, u, t \to g(x, u, t)$ is continuous and continuously differentiable with respect to the pair (x, t). We can convert this case to the case of a final criterion by setting

$$Y(t) = \int_0^t g(X(\tau), U(\tau), \tau)\,\mathrm{d}\tau\,.$$

By taking $\begin{bmatrix} X(t) \\ Y(t) \end{bmatrix}$ as the new state at the instant t, we can write the equation of evolution of the system

$$\begin{cases} \dfrac{\mathrm{d}X(t)}{\mathrm{d}t} = f(X(t), U(t), t) \\ \dfrac{\mathrm{d}Y(t)}{\mathrm{d}t} = g(X(t), U(t), t) \end{cases}$$

$$\text{with} \qquad \begin{cases} X(0) = X_0 \\ Y(0) = 0\,. \end{cases}$$

The criterion to maximize then becomes $Y(T)$. Thus, we are led to the case of a final criterion.

We shall apply this procedure to the classical problem of the calculus of variations. Let us see about characterizing the functions $t \to \tilde{X}(t)$ defined on $[0, T]$ into H that maximize the integral

$$\int_0^T g\left(X(t), \frac{\mathrm{d}X(t)}{\mathrm{d}t}, t\right)\mathrm{d}t\,,$$

where g is a twice continuously differentiable function. The initial value $X(0)$ is fixed (specifically, $X(0) = x_0$), and the final value $X(T)$ is subjected to the relations

$$v_k\, X(T) = c_k \qquad (k = 1, \ldots, \kappa)\,.$$

If we set

$$\frac{\mathrm{d}X(t)}{\mathrm{d}t} = U(t), \quad Y(t) = \int_0^t g\left(X(\tau), \frac{\mathrm{d}X(\tau)}{\mathrm{d}\tau}, \tau\right)\mathrm{d}\tau\,,$$

we obtain the system

$$\begin{cases} \dfrac{\mathrm{d}X(t)}{\mathrm{d}t} = U(t) \\ \dfrac{\mathrm{d}Y(t)}{\mathrm{d}t} = g(X(t), U(t), t) \end{cases} \qquad \text{with} \qquad \begin{cases} X(0) = x_0 \\ Y(0) = 0\,. \end{cases}$$

The criterion to maximize is $Y(T)$. The control is the function $t \to U(t)$, where $U(t) \in H$.

If $\tilde{X}, \tilde{Y}$ is an optimum trajectory, there exists a nonzero mapping $t \to [\tilde{P}(t) \mid \tilde{Q}(t)]$ of $[0, T]$ into $H' \times \mathbf{R}$ such that

$$\left\{\begin{array}{l} \dfrac{d\tilde{P}(t)}{dt} = -\tilde{Q}(t)\, g'_x(\tilde{X}(t), \tilde{U}(t), t) \\ \dfrac{d\tilde{Q}(t)}{dt} = 0 \end{array}\right. \quad \text{with} \quad \left\{\begin{array}{l} \tilde{P}(T) = \sum \nu_k\, v_k \\ \tilde{Q}(T) = \nu_0 \geq 0\,, \end{array}\right.$$

and

$$\tilde{P}(t)\,\tilde{U}(t) + \tilde{Q}(t)\, g(\tilde{X}(t), \tilde{U}(t), t) = \max_{u \in H} \left[\tilde{P}(t)\, u + \tilde{Q}(t)\, g(\tilde{X}(t), u, t)\right]. \qquad (\alpha)$$

Necessarily, $\tilde{Q}(t)$ is a constant: $\tilde{Q}(t) = \tilde{Q} \geq 0$. Equation ($\alpha$) implies that the first derivative of the function

$$u \to \tilde{P}(t)\, u + \tilde{Q}\, g(\tilde{X}(t), u, t)$$

vanishes when $u = \tilde{U}(t)$ and that its second derivative is negative-semidefinite; that is:

$$\left\{\begin{array}{ll} \tilde{P}(t) + \tilde{Q} g'_u(\tilde{X}(t), \tilde{U}(t), t) = 0 & (\beta) \\ \tilde{Q} g''_{u^2}(\tilde{X}(t), \tilde{U}(t), t) \quad \text{negative-semidefinite}. & (\gamma) \end{array}\right.$$

The relation (β) shows that $\tilde{Q} \neq 0$ (since, otherwise, we would have $\tilde{P}(t) = 0$ and the mapping $t \to [\tilde{P}(t) \mid \tilde{Q}(t)]$ would be identically zero).

In comparison with the expression for $d\tilde{P}(t)/dt$, Eq. (β) yields

$$g'_x(\tilde{X}(t), \tilde{U}(t), t) - \frac{d}{dt} g'_u(\tilde{X}(t), \tilde{U}(t), t) = 0 \quad \left(\text{with}\, \tilde{U}(t) = \frac{d\tilde{X}(t)}{dt}\right)$$

which is Euler's equation.

Equation (γ) shows that $g''_{u^2}(\tilde{X}(t), \tilde{U}(t), t)$ must be negative-semidefinite. This condition is known as "Legendre's condition."

If g is independent of t, we have, by virtue of Theorem 6,

$$\tilde{P}(t)\, \tilde{U}(t) + \tilde{Q}\, g(\tilde{X}(t), \tilde{U}(t)) = a$$

where a is a constant. In view of (β), this relation becomes

$$g'_u(\tilde{X}(t), \tilde{U}(t))\, \tilde{U}(t) - g(\tilde{X}(t), \tilde{U}(t)) = c$$

$$\left(\text{with } \tilde{U}(t) = \frac{d\tilde{X}(t)}{dt} \text{ and } c = \frac{-a}{\tilde{Q}}\right).$$

Thus, we obtain a first integral of Euler's equation. This first integral is identical to that obtained by application of Noether's theorem.*

---

*See, for example, [27], Chapter 4.

APPENDIX

Proof of Lemma 1. We shall give two proofs of Lemma 1, each of which involves concepts of results not contained in the present book. We note first of all that one can replace the word "simplex" in the statement of the lemma with the expression "closed ball." The result then follows without difficulty from the following fact:

Let B denote the unit ball in a normed finite-dimensional vector space E. Let f denote a continuous mapping of B into E such that $\| f(x) - x \| \leqslant \eta$ for every $x \in B$. Then, every $y \in E$ such that $\| y \| \leqslant 1 - \eta$ belongs to $f(B)$.

Our first proof involves Brouwer's fixed-point theorem.* We need to show that for every y such that $\| y \| \leqslant 1 - \eta$, there exists an $x \in B$ such that $f(x) = y$. This last relation may be written

$$x - f(x) + y = x .$$

Now, since

$$\| x - f(x) + y \| \leqslant \| x - f(x) \| + \| y \| \leqslant \eta + (1 - \eta) = 1 ,$$

the mapping $x \to x - f(x) + y$ maps B continuously into B. According to Brouwer's theorem, this mapping has a fixed point, which proves the property asserted.

Our second proof involves homotopy considerations.** Let h denote the restriction of f to the boundary sphere S of B. The mapping h is homotopic to the identity mapping on the crown C, that is, the set of all $x \in E$ such that $1 - \eta \leqslant \| x \| \leqslant 1 + \eta$. To see this, one need only consider the family of mappings

$$x \to \lambda x + (1 - \lambda) h(x) \quad \text{for} \quad \lambda \in [0, 1] .$$

Also, h is homotopic on $f(B)$ to a constant mapping. To see this, one need only consider the family of mappings $x \to f(\lambda x)$ for $\lambda \in [0, 1]$.

Consequently, the identity mapping of S is homotopic to a constant mapping on $C \cup f(B)$. Now, we know that if z is interior to B, the identity mapping of S is not homotopic to a constant mapping on $\complement \{ z \}$. From this we conclude that $f(B)$ contains the ball of radius $1 - \eta$.

---

*This proof is an adaptation of the proof given to Proposition 9.1 in H. Halkin, "On the necessary condition for optimal control of nonlinear systems," *Journal d'Analyse Mathématique*, Vol. XII, 1964.

**This proof was shown to me by J. Cerf.

Bibliography

Here, we have collected a selection of works that seems to us to constitute a natural continuation of the study of the present book. Reference [22] is an article that inspired sections 3 and 4 of Chapter 16. One should also consult [48] for certain concepts in Chapters 15 and 17.

1. Angot, A., *Compléments de Mathématiques,* Revue d'Optique, Paris, 3rd ed. 1957.
2. Arrow, K. J., L. Hurwicz, and H. Usawa, *Studies in Linear and Nonlinear Programming,* Stanford University Press, California, 1958.
3. Arsac, J., *Transformation de Fourier et théorie des distributions,* Dunod, Paris, 1961.
4. Bartlett, M. S., *An Introduction to Stochastic Processes,* The University Press, Cambridge, 1955.
5. Bass, J., *Eléments de calcul des probabilités,* Masson, Paris, 1962.
6. Beckenbach, E. F., *Modern Mathematics for the Engineer,* 2 vols., McGraw Hill, New York, 1956-1961.
7. Beckenbach, E. F., *Applied Combinatorial Mathematics,* Wiley, New York, 1964.
8. Bellman, R., *Dynamic Programming,* Princeton University Press, 1957.
9. Bellman, R. and S. E. Dreyfus, *Applied Dynamic Programming,* Princeton University Press, 1962.
10. Bellman, R., *Adaptive Control Processes,* Rand Corporation, Princeton University Press, 1961.
11. Berge, C., *Theory of Graphs and its Applications,* Wiley, New York, 1962.
12. Berge, C., *Espaces topologiques et fonctions multivoques,* Dunod, Paris, 2nd ed. 1965. There exists an English translation of the first edition *Topological Spaces,* MacMillan, New York, 1963.
13. Berge, C. and A. Ghouila-Houri, *Programming, Games and Transportation Networks,* Wiley, New York and Methuen, London, 1965.

14. Blanc-Lapierre, A. and R. Fortet, *Théorie des fonctions aléatoires,* Masson, Paris, 1953.
15. Bourbaki, N., *Eléments de mathématiques,* Livre V: Espaces vectoriels topologiques, ASI 1189-1220-1230, Hermann, Paris, 1953-1964.
16. Choquet, G., *Cours d'Analyse.* Tome II: Topologie, Masson, Paris, 1964. An English translation, *Topology,* is in preparation, Academic Press, New York.
17. Decaulne, P., J. C. Gille, and M. Pèlegrin, *Problems d'asservissements,* Dunod, Paris, 1958.
18. Descombes, R., Cours d'analyse pour le certificat de Mathématiques I, Vuibert, Paris, 1962.
19. Doob, J. L., *Stochastic Processes,* Wiley, New York, 1953.
20. Dorfman, R., P. A. Samuelson, and R. M. Solow, *Linear Programming and Economic Analysis,* McGraw Hill, New York, 1958.
21. Dynkin, E. B., *Markov Processes,* 2 vols., Academic Press, New York, 1964.
22. Eaton, J. H. and L. A. Zadeh, *Optimal Pursuit Strategies in Discrete-State Probabilistic Systems,* Jour. Bas. Eng., Vol. 84, series D, March 1962.
23. Feller, W., *An Introduction to Probability Theory and its Applications,* Wiley, New York, 2nd ed. 1959.
24. Ferguson, R. O. and L. F. Sargent, *Linear Programming Fundamentals and Applications,* McGraw Hill, New York, 1958.
25. Garsoux, J., *Espaces vectoriels topologiques et distributions,* Dunod, Paris, 1963. An English translation is in preparation, M. I. T. Press.
26. Gass, S. I., *Linear Programming, Methods and Applications,* McGraw Hill, New York, 1958.
27. Gel'fand, I. M. and S. V. Fomin, *Calculus of Variations,* Prentice Hall, Englewood Cliffs, New Jersey, 1963.
28. Gel'fand, I. M. and G. E. Shilov, *Generalized Functions,* Vol. I, Properties and Operations, Academic Press, New York, 1964.
29. Gel'fand, I. M. and G. E. Shilov, *Les Distributions,* Dunod, Paris, 1964. (This is volume 2 of the series *Obobshchennyye Funktsii [Generalized Functions]* by Gel'fand et al.)
30. Gel'fand, I. M. and N. Ya. Vilenkin, *Generalized Functions,* Vol. 4, *Applications of Harmonic Analysis,* Academic Press, New York, 1964.
31. Gille, J. C., P. Decaulne, and M. Pèlegrin, *Théorie et calcul des asservissements,* Dunod, Paris, 3rd ed. 1963.
32. Gille, J. C., P. Decaulne, and M. Pèlegrin, *Méthodes modernes d'études des systémes asservis,* Dunod, Paris, 1960.
33. Gnedenko, B. V., *The Theory of Probability,* Chelsea, 1962.
34. Gnedenko, B. V. and A. Ya. Khinchin, *Elementary Introduction to the Theory of Probability,* Fifth ed., Dover, 1962 and Freeman, 1961.
35. Gordon, P., *Théorie des chaines de Markov finies et ses applications,* Dunod, Paris, 1965.
36. Greenwald, D. U., *Linear Programming, An Explanation of the Simplex Algorithm,* Ronald Press, New York.

37. Hadley, G., *Linear Programming,* Addison Wesley, Reading, Mass., 1962.
38. Hadley, G., *Nonlinear and Dynamic Programming,* Addison Wesley, Reading, Mass., 1964.
39. Hennequin, P. L. and A. Tortrat, *Théorie des probabilités et quelques applications,* Masson, Paris, 1965.
40. Holbrook, J. G., *Laplace Transforms,* Pergamon Press, Oxford, 1959.
41. Howard, R. A., *Dynamic Programming and Markov Processes,* The Massachussets Institute of Technology, 1960.
42. Ito, K., *Random Processes* (in Japanese). A Russian translation, *Veroyatnost'nyye protsessy,* exists, Iz-vo. In. Lit., Moscow, 1962-1963.
43. Yaglom, A. M. and I. M. Yaglom, *Probability and Information,* Dover, 1962 and Heath, 1965.
44. Yaglom, A. M., *An Introduction to the Theory of Stationary Random Functions,* Prentice Hall, Englewood Cliffs, N. J., 1962.
45. Joksch, H. C., *Lineares Programmieren,* Mohr, Tübingen, 1962.
46. Jury, E. T., *Sampled-Data Control Systems,* Wiley, New York, 1958.
47. Kai Lai Chung, *Markov Chains with Stationary Transition Probabilities,* Springer, Berlin, 1960.
48. Kalman, R. E., *On the General Theory of Control Systems,* 1st congress, IFAC, Moscow, 1960.
49. Karlin, S., *Mathematical Methods and Theory in Games,* Programming and Economics, Vols. I and II, Addison Wesley, Reading, Mass., 1959.
50. Kaufmann, A. and R. Cruon, *La programmation dynamique,* Dunod, Paris, 1965.
51. Kemeny, J. G. and S. L. Snell, *Finite Markov Chains,* Van Nostrand, Princeton, 1960.
52. Kunzi, H. P. and W. Krelle, *Nichtlineare Programmierung,* Springer, Berlin, 1962.
53. Laning, J. H. and R. H. Battin, *Random Processes in Automatic Control,* McGraw Hill, New York.
54. Loeve, M., *Probability Theory,* Van Nostrand, Princeton, 3rd ed., 1963.
55. Neveu, J., *Mathematical Basis of the Calculus of Probability,* Holden-Day, San Francisco, 1965.
56. Papoulis, A., *Probability, Random Variables and Stochastic Processes,* McGraw Hill, New York, 1965.
57. Parzen, E., *Stochastic Processes,* Holden-Day, 1962.
58. Pontryagin, L. S., V. G. Boltyanskiy, R. V. Gamkrelidze and E. F. Mischenko, *The Mathematical Theory of Optimal Processes,* Interscience, New York, 1962.
59. Pugachev, V. S., *Theory of Random Functions and its Application to Control Problems,* Pergamon, Oxford, 1965.
60. Ragazzini, J. R. and G. F. Franklin, *Sampled-Data Control Systems,* McGraw Hill, New York, 1958.

61. Rozanov, Yu. A., *Statsionarnyye sluchaynyye protsessy* (Stationary Random Processes), Fizmatgiz, Moscow, 1963.
62. Saaty, T. and J. Bram, *Nonlinear Mathematics,* McGraw Hill, New York, 1964.
63. Schwartz, L., *Théorie des distributions* (tómes I et II), Hermann, Paris, 1957-1959.
64. Schwartz, L., *Méthodes mathématiques pour les Sciences physiques,* Hermann, Paris, 1961.
65. Simonnard, M., *Programmation linéaire,* Dunod, Paris, 1962.
66. Souriau, J. M., *Calcul linéaire,* Presses Universitaires de France, Paris, 1959.
67. Takacs, L., *Stochastic Processes. Problems and Solutions,* Wiley, New York, 1962.
68. Tortrat, A., *Calcul des probabilités,* Masson, Paris, 1963.
69. Tou, J. T., *Modern Control Theory,* McGraw Hill, New York, 1964.
70. Trèves, F., *Eléments de la théorie des espaces topologiques,* Centre de Documentation Universitaire, Paris, 1959.
71. Vajda, S., *Readings in Mathematical Programming,* Pitman, London, 1958.
72. Widder, D. V., *The Laplace Transforms,* Princeton University Press, 1941.
73. Wiener, N., *Extrapolation, Interpolation and Smoothing of Stationary Time-Series,* The Technology Press of the M. I. T., Harvard, 1957.
74. Zadeh, L. and C. Desoer, *Linear System Theory,* McGraw Hill, New York, 1963.

Index

A CATALOGUE OF SELECTED DOVER BOOKS IN ALL FIELDS OF INTEREST

A CATALOGUE OF SELECTED DOVER BOOKS IN ALL FIELDS OF INTEREST

DRAWINGS OF WILLIAM BLAKE, William Blake. 92 plates from Book of Job, *Divine Comedy, Paradise Lost,* visionary heads, mythological figures, Laocoon, etc. Selection, introduction, commentary by Sir Geoffrey Keynes. 178pp. 8⅛ x 11. **22303-5 Pa. $4.00**

ENGRAVINGS OF HOGARTH, William Hogarth. 101 of Hogarth's greatest works: *Rake's Progress, Harlot's Progress, Illustrations for Hudibras, Before and After, Beer Street and Gin Lane,* many more. Full commentary. 256pp. 11 x 13¾. **22479-1 Pa. $12.95**

DAUMIER: 120 GREAT LITHOGRAPHS, Honore Daumier. Wide-ranging collection of lithographs by the greatest caricaturist of the 19th century. Concentrates on eternally popular series on lawyers, on married life, on liberated women, etc. Selection, introduction, and notes on plates by Charles F. Ramus. Total of 158pp. 9⅜ x 12¼. **23512-2 Pa. $6.00**

DRAWINGS OF MUCHA, Alphonse Maria Mucha. Work reveals draftsman of highest caliber: studies for famous posters and paintings, renderings for book illustrations and ads, etc. 70 works, 9 in color; including 6 items not drawings. Introduction. List of illustrations. 72pp. 9⅜ x 12¼. (Available in U.S. only) **23672-2 Pa. $4.00**

GIOVANNI BATTISTA PIRANESI: DRAWINGS IN THE PIERPONT MORGAN LIBRARY, Giovanni Battista Piranesi. For first time ever all of Morgan Library's collection, world's largest. 167 illustrations of rare Piranesi drawings—archeological, architectural, decorative and visionary. Essay, detailed list of drawings, chronology, captions. Edited by Felice Stampfle. 144pp. 9⅜ x 12¼. **23714-1 Pa. $7.50**

NEW YORK ETCHINGS (1905-1949), John Sloan. All of important American artist's N.Y. life etchings. 67 works include some of his best art; also lively historical record—Greenwich Village, tenement scenes. Edited by Sloan's widow. Introduction and captions. 79pp. 8⅜ x 11¼. **23651-X Pa. $4.00**

CHINESE PAINTING AND CALLIGRAPHY: A PICTORIAL SURVEY, Wan-go Weng. 69 fine examples from John M. Crawford's matchless private collection: landscapes, birds, flowers, human figures, etc., plus calligraphy. Every basic form included: hanging scrolls, handscrolls, album leaves, fans, etc. 109 illustrations. Introduction. Captions. 192pp. 8⅞ x 11¾. **23707-9 Pa. $7.95**

DRAWINGS OF REMBRANDT, edited by Seymour Slive. Updated Lippmann, Hofstede de Groot edition, with definitive scholarly apparatus. All portraits, biblical sketches, landscapes, nudes, Oriental figures, classical studies, together with selection of work by followers. 550 illustrations. Total of 630pp. 9⅛ x 12¼. 21485-0, 21486-9 Pa., Two-vol. set $15.00

THE DISASTERS OF WAR, Francisco Goya. 83 etchings record horrors of Napoleonic wars in Spain and war in general. Reprint of 1st edition, plus 3 additional plates. Introduction by Philip Hofer. 97pp. 9⅜ x 8¼. **21872-4 Pa. $4.00**

CATALOGUE OF DOVER BOOKS

HISTORY OF BACTERIOLOGY, William Bulloch. The only comprehensive history of bacteriology from the beginnings through the 19th century. Special emphasis is given to biography-Leeuwenhoek, etc. Brief accounts of 350 bacteriologists form a separate section. No clearer, fuller study, suitable to scientists and general readers, has yet been written. 52 illustrations. 448pp. 5⅝ x 8¼. 23761-3 Pa. $6.50

THE COMPLETE NONSENSE OF EDWARD LEAR, Edward Lear. All nonsense limericks, zany alphabets, Owl and Pussycat, songs, nonsense botany, etc., illustrated by Lear. Total of 321pp. 5⅜ x 8½. (Available in U.S. only) 20167-8 Pa. $3.95

INGENIOUS MATHEMATICAL PROBLEMS AND METHODS, Louis A. Graham. Sophisticated material from Graham *Dial,* applied and pure; stresses solution methods. Logic, number theory, networks, inversions, etc. 237pp. 5⅜ x 8½. 20545-2 Pa. $4.50

BEST MATHEMATICAL PUZZLES OF SAM LOYD, edited by Martin Gardner. Bizarre, original, whimsical puzzles by America's greatest puzzler. From fabulously rare *Cyclopedia,* including famous 14-15 puzzles, the Horse of a Different Color, 115 more. Elementary math. 150 illustrations. 167pp. 5⅜ x 8½. 20498-7 Pa. $2.75

THE BASIS OF COMBINATION IN CHESS, J. du Mont. Easy-to-follow, instructive book on elements of combination play, with chapters on each piece and every powerful combination team—two knights, bishop and knight, rook and bishop, etc. 250 diagrams. 218pp. 5⅜ x 8½. (Available in U.S. only) 23644-7 Pa. $3.50

MODERN CHESS STRATEGY, Ludek Pachman. The use of the queen, the active king, exchanges, pawn play, the center, weak squares, etc. Section on rook alone worth price of the book. Stress on the moderns. Often considered the most important book on strategy. 314pp. 5⅜ x 8½. 20290-9 Pa. $4.50

LASKER'S MANUAL OF CHESS, Dr. Emanuel Lasker. Great world champion offers very thorough coverage of all aspects of chess. Combinations, position play, openings, end game, aesthetics of chess, philosophy of struggle, much more. Filled with analyzed games. 390pp. 5⅜ x 8½. 20640-8 Pa. $5.00

500 MASTER GAMES OF CHESS, S. Tartakower, J. du Mont. Vast collection of great chess games from 1798-1938, with much material nowhere else readily available. Fully annoted, arranged by opening for easier study. 664pp. 5⅜ x 8½. 23208-5 Pa. $7.50

A GUIDE TO CHESS ENDINGS, Dr. Max Euwe, David Hooper. One of the finest modern works on chess endings. Thorough analysis of the most frequently encountered endings by former world champion. 331 examples, each with diagram. 248pp. 5⅜ x 8½. 23332-4 Pa. $3.75

AMERICAN BIRD ENGRAVINGS, Alexander Wilson et al. All 76 plates. from Wilson's *American Ornithology* (1808-14), most important ornithological work before Audubon, plus 27 plates from the supplement (1825-33) by Charles Bonaparte. Over 250 birds portrayed. 8 plates also reproduced in full color. 111pp. 9⅜ x 12½. **23195-X Pa. $6.00**

CRUICKSHANK'S PHOTOGRAPHS OF BIRDS OF AMERICA, Allan D. Cruickshank. Great ornithologist, photographer presents 177 closeups, groupings, panoramas, flightings, etc., of about 150 different birds. Expanded *Wings in the Wilderness.* Introduction by Helen G. Cruickshank. 191pp. 8¼ x 11. **23497-5 Pa. $6.00**

AMERICAN WILDLIFE AND PLANTS, A. C. Martin, et al. Describes food habits of more than 1000 species of mammals, birds, fish. Special treatment of important food plants. Over 300 illustrations. 500pp. 5⅜ x 8½. **20793-5 Pa. $4.95**

THE PEOPLE CALLED SHAKERS, Edward D. Andrews. Lifetime of research, definitive study of Shakers: origins, beliefs, practices, dances, social organization, furniture and crafts, impact on 19th-century USA, present heritage. Indispensable to student of American history, collector. 33 illustrations. 351pp. 5⅜ x 8½. **21081-2 Pa. $4.50**

OLD NEW YORK IN EARLY PHOTOGRAPHS, Mary Black. New York City as it was in 1853-1901, through 196 wonderful photographs from N.-Y. Historical Society. Great Blizzard, Lincoln's funeral procession, great buildings. 228pp. 9 x 12. **22907-6 Pa. $8.95**

MR. LINCOLN'S CAMERA MAN: MATHEW BRADY, Roy Meredith. Over 300 Brady photos reproduced directly from original negatives, photos. Jackson, Webster, Grant, Lee, Carnegie, Barnum; Lincoln; Battle Smoke, Death of Rebel Sniper, Atlanta Just After Capture. Lively commentary. 368pp. 8⅜ x 11¼. **23021-X Pa. $8.95**

TRAVELS OF WILLIAM BARTRAM, William Bartram. From 1773-8, Bartram explored Northern Florida, Georgia, Carolinas, and reported on wild life, plants, Indians, early settlers. Basic account for period, entertaining reading. Edited by Mark Van Doren. 13 illustrations. 141pp. 5⅜ x 8½. **20013-2 Pa. $5.00**

THE GENTLEMAN AND CABINET MAKER'S DIRECTOR, Thomas Chippendale. Full reprint, 1762 style book, most influential of all time; chairs, tables, sofas, mirrors, cabinets, etc. 200 plates, plus 24 photographs of surviving pieces. 249pp. 9⅞ x 12¾. **21601-2 Pa. $7.95**

AMERICAN CARRIAGES, SLEIGHS, SULKIES AND CARTS, edited by Don H. Berkebile. 168 Victorian illustrations from catalogues, trade journals, fully captioned. Useful for artists. Author is Assoc. Curator, Div. of Transportation of Smithsonian Institution. 168pp. 8½ x 9½. **23328-6 Pa. $5.00**

CATALOGUE OF DOVER BOOKS

THE COMPLETE WOODCUTS OF ALBRECHT DURER, edited by Dr. W. Kurth. 346 in all: "Old Testament," "St. Jerome," "Passion," "Life of Virgin," Apocalypse," many others. Introduction by Campbell Dodgson. 285pp. 8½ x 12¼. 21097-9 Pa. $7.50

DRAWINGS OF ALBRECHT DURER, edited by Heinrich Wolfflin. 81 plates show development from youth to full style. Many favorites; many new. Introduction by Alfred Werner. 96pp. 8⅛ x 11. 22352-3 Pa. $5.00

THE HUMAN FIGURE, Albrecht Dürer. Experiments in various techniques—stereometric, progressive proportional, and others. Also life studies that rank among finest ever done. Complete reprinting of *Dresden Sketchbook*. 170 plates. 355pp. 8⅜ x 11¼. 21042-1 Pa. $7.95

OF THE JUST SHAPING OF LETTERS, Albrecht Dürer. Renaissance artist explains design of Roman majuscules by geometry, also Gothic lower and capitals. Grolier Club edition. 43pp. 7⅞ x 10¾ 21306-4 Pa. $3.00

TEN BOOKS ON ARCHITECTURE, Vitruvius. The most important book ever written on architecture. Early Roman aesthetics, technology, classical orders, site selection, all other aspects. Stands behind everything since. Morgan translation. 331pp. 5⅜ x 8½. 20645-9 Pa. $4.50

THE FOUR BOOKS OF ARCHITECTURE, Andrea Palladio. 16th-century classic responsible for Palladian movement and style. Covers classical architectural remains, Renaissance revivals, classical orders, etc. 1738 Ware English edition. Introduction by A. Placzek. 216 plates. 110pp. of text. 9½ x 12¾. 21308-0 Pa. $10.00

HORIZONS, Norman Bel Geddes. Great industrialist stage designer, "father of streamlining," on application of aesthetics to transportation, amusement, architecture, etc. 1932 prophetic account; function, theory, specific projects. 222 illustrations. 312pp. 7⅞ x 10¾. 23514-9 Pa. $6.95

FRANK LLOYD WRIGHT'S FALLINGWATER, Donald Hoffmann. Full, illustrated story of conception and building of Wright's masterwork at Bear Run, Pa. 100 photographs of site, construction, and details of completed structure. 112pp. 9¼ x 10. 23671-4 Pa. $5.50

THE ELEMENTS OF DRAWING, John Ruskin. Timeless classic by great Viltorian; starts with basic ideas, works through more difficult. Many practical exercises. 48 illustrations. Introduction by Lawrence Campbell. 228pp. 5⅜ x 8½. 22730-8 Pa. $3.75

GIST OF ART, John Sloan. Greatest modern American teacher, Art Students League, offers innumerable hints, instructions, guided comments to help you in painting. Not a formal course. 46 illustrations. Introduction by Helen Sloan. 200pp. 5⅜ x 8½. 23435-5 Pa. $4.00

THE ANATOMY OF THE HORSE, George Stubbs. Often considered the great masterpiece of animal anatomy. Full reproduction of 1766 edition, plus prospectus; original text and modernized text. 36 plates. Introduction by Eleanor Garvey. 121pp. 11 x 14¾. 23402-9 Pa. $6.00

BRIDGMAN'S LIFE DRAWING, George B. Bridgman. More than 500 illustrative drawings and text teach you to abstract the body into its major masses, use light and shade, proportion; as well as specific areas of anatomy, of which Bridgman is master. 192pp. 6½ x 9¼. (Available in U.S. only) 22710-3 Pa. $3.50

ART NOUVEAU DESIGNS IN COLOR, Alphonse Mucha, Maurice Verneuil, Georges Auriol. Full-color reproduction of *Combinaisons ornementales* (c. 1900) by Art Nouveau masters. Floral, animal, geometric, interlacings, swashes—borders, frames, spots—all incredibly beautiful. 60 plates, hundreds of designs. 9⅜ x 8-1/16. 22885-1 Pa. $4.00

FULL-COLOR FLORAL DESIGNS IN THE ART NOUVEAU STYLE, E. A. Seguy. 166 motifs, on 40 plates, from *Les fleurs et leurs applications decoratives* (1902): borders, circular designs, repeats, allovers, "spots." All in authentic Art Nouveau colors. 48pp. 9⅜ x 12¼. 23439-8 Pa. $5.00

A DIDEROT PICTORIAL ENCYCLOPEDIA OF TRADES AND INDUSTRY, edited by Charles C. Gillispie. 485 most interesting plates from the great French Encyclopedia of the 18th century show hundreds of working figures, artifacts, process, land and cityscapes; glassmaking, papermaking, metal extraction, construction, weaving, making furniture, clothing, wigs, dozens of other activities. Plates fully explained. 920pp. 9 x 12. 22284-5, 22285-3 Clothbd., Two-vol. set $40.00

HANDBOOK OF EARLY ADVERTISING ART, Clarence P. Hornung. Largest collection of copyright-free early and antique advertising art ever compiled. Over 6,000 illustrations, from Franklin's time to the 1890's for special effects, novelty. Valuable source, almost inexhaustible.
Pictorial Volume. Agriculture, the zodiac, animals, autos, birds, Christmas, fire engines, flowers, trees, musical instruments, ships, games and sports, much more. Arranged by subject matter and use. 237 plates. 288pp. 9 x 12. 20122-8 Clothbd. $14.50

Typographical Volume. Roman and Gothic faces ranging from 10 point to 300 point, "Barnum," German and Old English faces, script, logotypes, scrolls and flourishes, 1115 ornamental initials, 67 complete alphabets, more. 310 plates. 320pp. 9 x 12. 20123-6 Clothbd. $15.00

CALLIGRAPHY (CALLIGRAPHIA LATINA), J. G. Schwandner. High point of 18th-century ornamental calligraphy. Very ornate initials, scrolls, borders, cherubs, birds, lettered examples. 172pp. 9 x 13. 20475-8 Pa. $7.00

ART FORMS IN NATURE, Ernst Haeckel. Multitude of strangely beautiful natural forms: Radiolaria, Foraminifera, jellyfishes, fungi, turtles, bats, etc. All 100 plates of the 19th-century evolutionist's *Kunstformen der Natur* (1904). 100pp. 9⅜ x 12¼. 22987-4 Pa. $5.00

CHILDREN: A PICTORIAL ARCHIVE FROM NINETEENTH-CENTURY SOURCES, edited by Carol Belanger Grafton. 242 rare, copyright-free wood engravings for artists and designers. Widest such selection available. All illustrations in line. 119pp. 8⅜ x 11¼. 23694-3 Pa. $4.00

WOMEN: A PICTORIAL ARCHIVE FROM NINETEENTH-CENTURY SOURCES, edited by Jim Harter. 391 copyright-free wood engravings for artists and designers selected from rare periodicals. Most extensive such collection available. All illustrations in line. 128pp. 9 x 12. 23703-6 Pa. $4.50

ARABIC ART IN COLOR, Prisse d'Avennes. From the greatest ornamentalists of all time—50 plates in color, rarely seen outside the Near East, rich in suggestion and stimulus. Includes 4 plates on covers. 46pp. 9⅜ x 12¼. 23658-7 Pa. $6.00

AUTHENTIC ALGERIAN CARPET DESIGNS AND MOTIFS, edited by June Beveridge. Algerian carpets are world famous. Dozens of geometrical motifs are charted on grids, color-coded, for weavers, needleworkers, craftsmen, designers. 53 illustrations plus 4 in color. 48pp. 8¼ x 11. (Available in U.S. only) 23650-1 Pa. $1.75

DICTIONARY OF AMERICAN PORTRAITS, edited by Hayward and Blanche Cirker. 4000 important Americans, earliest times to 1905, mostly in clear line. Politicians, writers, soldiers, scientists, inventors, industrialists, Indians, Blacks, women, outlaws, etc. Identificatory information. 756pp. 9¼ x 12¾. 21823-6 Clothbd. $40.00

HOW THE OTHER HALF LIVES, Jacob A. Riis. Journalistic record of filth, degradation, upward drive in New York immigrant slums, shops, around 1900. New edition includes 100 original Riis photos, monuments of early photography. 233pp. 10 x 7⅞. 22012-5 Pa. $7.00

NEW YORK IN THE THIRTIES, Berenice Abbott. Noted photographer's fascinating study of city shows new buildings that have become famous and old sights that have disappeared forever. Insightful commentary. 97 photographs. 97pp. 11⅜ x 10. 22967-X Pa. $5.00

MEN AT WORK, Lewis W. Hine. Famous photographic studies of construction workers, railroad men, factory workers and coal miners. New supplement of 18 photos on Empire State building construction. New introduction by Jonathan L. Doherty. Total of 69 photos. 63pp. 8 x 10¾. 23475-4 Pa. $3.00